IET ELECTROMAGNETIC WAVES 534

Emerging Evolutionary Algorithms for Antennas and Wireless Communications

Related titles on electromagnetic waves:

Dielectric Resonators, 2nd Edition Kajfez and Guillon
Electronic Applications of the Smith Chart Smith
Fiber Optic Technology Jha
Filtering in the Time and Frequency Domains Blinchikoff and Zverev
HF Filter Design and Computer Simulation Rhea
HF Radio Systems and Circuits Sabin
Microwave Field-Effect Transistors: Theory, Design and Application, 3rd Edition Pengelly
Microwave Semiconductor Engineering White
Microwave Transmission Line Impedance Data Gunston
Optical Fibers and RF: A Natural Combination Romeiser
Oscillator Design and Computer Simulation Rhea
Radio-Electronic Transmission Fundamentals, 2nd Edition Griffith,Jr
RF and Microwave Modeling and Measurement Techniques for Field Effect Transistors Gao
RF Power Amplifiers Albulet
Small Signal Microwave Amplifier Design Grosch
Small Signal Microwave Amplifier Design: Solutions Grosch
2008+ Solved Problems in Electromagnetics Nasar
Antennas: Fundamentals, Design, Measurement, 3rd Edition Blake and Long
Designing Electronic Systems for EMC Duff
Electromagnetic Measurements in the Near Field, 2nd Edition Bienkowski and Trzaska
Fundamentals of Electromagnetics with MATLAB®, 2nd Edition Lonngren *et al.*
Fundamentals of Wave Phenomena, 2nd Edition Hirose and Lonngren
Integral Equation Methods for Electromagnetics Volakis and Sertel
Introduction to Adaptive Arrays, 2nd Edition Monzingo et al
Microstrip and Printed Antenna Design, 2nd Edition Bancroft
Numerical Methods for Engineering: An Introduction using MATLAB® and Computational Electromagnetics Warnick
Return of the Ether Deutsch
The Finite Difference Time Domain Method for Electromagnetics: With MATLAB® Simulations Elsherbeni and Demir
Theory of Edge Diffraction in Electromagnetics Ufimtsev
Scattering of Wedges and Cones with Impedance Boundary Conditions Lyalinov and Zhu
Circuit Modeling for Electromagnetic Compatibility Darney
The Wiener–Hopf Method in Electromagnetics Daniele and Zich
Microwave and RF Design: A Systems Approach, 2nd Edition Steer
Spectrum and Network Measurements, 2nd Edition Witte
EMI Troubleshooting Cookbook for Product Designers Andre and Wyatt
Transmission Line Transformers Mack and Sevick
Electromagnetic Field Standards and Exposure Systems Grudzinski and Trzaska
Practical Communication Theory, 2nd Edition Adamy
Complex Space Source Theory of Spatially Localized Electromagnetic Waves Seshadri
Electromagnetic Compatibility Pocket Guide: Key EMC Facts, Equations and Data Wyatt and Jost
Antenna Analysis and Design Using FEKO Electromagnetic Simulation Software Elsherbeni, Nayeri and Reddy
Scattering of Electromagnetic Waves by Obstacles Kristensson
Adjoint Sensitivity Analysis of High Frequency Structures with MATLAB® Bakr, Elsherbeni and Demir
Developments in Antenna Analysis and Synthesis Vol. 1 and Vol. 2 Mittra
Advances in Planar Filters Design Hong
Post-processing Techniques in Antenna Measurement Castañer and Foged
Nano-Electromagnetic Communication at Terahertz and Optical Frequencies, Principles and Applications Alomainy, Yang, Imran, Yao and Abbasi
Nanoantennas and Plasmonics: Modelling, Design and Fabrication Werner, Campbell and Kang
Electromagnetic Reverberation Chambers: Recent Advances and Innovative Applications Andrieu
Radio Wave Propagation in Vehicular Environments Azpilicueta, Vargas-Rosales, Falcone and Alejos
Advances in Mathematical Methods for Electromagnetics Kobayashi and Smith

Emerging Evolutionary Algorithms for Antennas and Wireless Communications

Sotirios K. Goudos

The Institution of Engineering and Technology

Published by SciTech Publishing, an imprint of The Institution of Engineering and Technology, London, United Kingdom

The Institution of Engineering and Technology is registered as a Charity in England & Wales (no. 211014) and Scotland (no. SC038698).

First published 2021

The Institution of Engineering and Technology
Michael Faraday House
Six Hills Way, Stevenage
Herts, SG1 2AY, United Kingdom

www.theiet.org

British Library Cataloguing in Publication Data
A catalogue record for this product is available from the British Library

ISBN 978-1-78561-552-8 (hardback)
ISBN 978-1-78561-553-5 (PDF)

Typeset in India by MPS Limited

For my wife Athena that inspires me, and my daughters Mary and Mandy
For my mother and for the memory of my father

Contents

About the author

Dr Sotirios K. Goudos is an Associate Professor at the Department of Physics of Aristotle University of Thessaloniki, Thessaloniki, Greece. He received his B.Sc. degree in Physics in 1991 and his M.Sc. of Postgraduate Studies in Electronics in 1994 both from the Aristotle University of Thessaloniki. In 2001, he received his Ph.D. degree in Physics from the Aristotle University of Thessaloniki and in 2005 the Master in Information Systems from the University of Macedonia, Greece. In 2011, he obtained a Diploma degree in Electrical and Computer Engineering from the Aristotle University of Thessaloniki. His research interests include antenna and microwave structures design, evolutionary algorithms, wireless communications, and semantic web technologies. He is the director of the ELEDIA@AUTH lab member of the ELEDIA Research Center Network. He is the founding Editor-in-Chief of the Telecom open access journal (MDPI publishing). He is currently serving as an Associate Editor for IEEE ACCESS and IEEE open journal of the communication society. He is also a member of the Editorial Board of the *International Journal of Antennas and Propagation* (IJAP), the *EURASIP Journal on Wireless Communications and Networking*, and the *International Journal on Advances on Intelligent Systems*. He is also member of the topic board of the Electronics open access journal. He is currently serving as a Chapter/AG coordinator for IEEE Greece Section. He was the Lead Guest Editor in the 2016 and 2017 Special Issues of the IJAP with topic "Evolutionary Algorithms Applied to Antennas and Propagation: Emerging Trends and Applications". He was also the Lead Guest Editor in the 2018 Special Issue of the *EURASIP Journal on Wireless Communications and Networking* with topic "Optimization methods for Key Enabling Technologies: 5G, IoT and Big Data". He was a Guest Editor in the *IEEE Antennas and Wireless Propagation Letters*, Special Cluster in 2019 with topic "Machine Learning Applications in Electromagnetics, Antennas, and Propagation". He has served as the Technical Program Chair in the International Conference on Modern Circuits and Systems Technologies (MOCAST). He was a sub-committee chair in the *Asian-Pacific Microwave Conference (APMC 2017)* in the track of Smart and reconfigurable antennas. He has also served as a member of the Technical Program Committees in several IEEE and non-IEEE conferences.

Preface

Several evolutionary algorithms (EAs) have emerged in the past decades that mimic biological entities' behavior and evolution. EAs are widely used for the solution of single- and multi-objective optimization engineering problems. The EAs have also been applied to a variety of microwave component, antenna design, radar design, and wireless communications problems.

These techniques, among others, include genetic algorithms, evolution strategies, particle swarm optimization (PSO), differential evolution (DE), and ant colony optimization (ACO). In addition, new innovative algorithms that are not only biology based but also physics based or music based have emerged. The use of the above algorithms has an increasing impact on antenna design and wireless communications problems. EAs combined in several occasions with numerical methods in electromagnetics have obtained significant and successful results. Additionally, hybrid combinations of EAs are also emerging.

This book aims to present some of the emerging EAs and their variants. We present design cases using different EAs applied to popular design problems in antennas and wireless communications. The book contains both cases of single- and multi-objective optimization. Additionally, this book also aims to supply the reader with the performance metrics that can be used for algorithms performance evaluation. Thus, this book aims to provide a complete framework for antenna and wireless communications design using EAs.

The target audience of this book includes the students in electrical engineering disciplines, the multidisciplinary students, and particularly those pursuing postgraduate studies in advanced topics and the PhD students. Additionally, this book will be useful to the antenna and the wireless communications engineers, and to those faculty members with research interest in antennas and wireless communications.

I hope that the algorithms and the design cases presented in this book will be able to motivate the readers to further develop new efficient algorithms and to further explore the application of these algorithms to new optimization problems in the antennas and the wireless communication domain.

The book contains seven chapters altogether. The first chapter is the "Introduction," the second presents the details of the EAs, and the rest of the chapters contain the details of the design cases.

More accurately, the chapter organization is as follows:

Chapter 1 introduces the optimization methods in general and the EAs. Furthermore, it gives some of the features of EAs for both single- and multi-objective optimization. Also, performance metrics and indices are presented.

Chapter 2 briefly presents some of the most popular EAs such as PSO, DE, and ACO as well as some of the emerging ones like grey wolf optimizer, salp swarm algorithm, and monarch butterfly optimization. Multi-objective algorithms are also discussed in this chapter.

Chapter 3 focuses on antenna array synthesis, which constitutes a wide range of antenna design problems. Several different array design cases are presented and optimized using different EAs, which include linear, planar, conformal, thinned, and shaped-beam arrays.

Chapter 4 gives an overview of patch antenna design using EAs. Moreover, antenna design cases for various applications in wireless communications like 5G, Wi-Fi are discussed here.

Chapter 5 presents design cases from different microwave structure cases. These include microwave broadband absorbers, dielectric filters, and microstrip filters that are optimized using single- and multi-objective algorithms.

Chapter 6 discusses on various representative design problems in wireless communications. The peak to average power ratio reduction in orthogonal frequency division multiplexing systems, the antenna selection in MIMO systems, the cognitive radio engine design, the spectrum allocation in cognitive radio networks, and the optimization of wireless sensor networks are the cases provided in this chapter.

Furthermore, Chapter 7 deals with design cases for 5G and beyond. The multi-objective optimization in 5G massive MIMO wireless networks, and the joint power allocation and user association in non-orthogonal multiple access networks examples are included in this chapter.

One may read first Chapters 1 and 2 and then read any of the other chapters with any order.

Finally, I would like to dedicate this book to my family—my wife Athena, and our daughters Mary and Mandy, for their unconditional love and support for making this book a reality. Also, to my mother and to the memory of my father.

Sotirios K. Goudos
Thessaloniki, Greece
October 2020

Chapter 1
Introduction

This chapter briefly introduces basic optimization concepts. Optimization problems exist in our everyday life. We often wonder about the route with less traffic to work, the less money to spend in various activities, or the best work schedule. Optimization can be thought as the art of making good decisions. In more precise manner, we may define optimization as the process of changing a system's state (or values of some unknown variables) toward a state of a minimum or maximum property. Several optimization problems exist in the whole engineering domain. A common problem in antennas design is to find the best geometry for a given antenna, while in wireless communications the association between a base station and user in an optimal manner is a very popular optimization problem.

The general form of a constrained optimization problem can be defined mathematically as

Find $\mathbf{x}^* = \{x_1^*, x_2^*, \ldots, x_t^*\} \in D^t = D_1 \cap D_2 \cap \cdots \cap D_t$ where

$$F_{\min}(\mathbf{x}^*) \leq F_{\min}(\mathbf{x}) \; \forall \mathbf{x} \in D^t \quad \text{or} \quad F_{\max}(\mathbf{x}^*) \geq F_{\max}(\mathbf{x}) \; \forall \mathbf{x} \in D^t$$

$$\text{subject to}$$

$$\begin{aligned} g_j(\mathbf{x}^*) &< 0 \; j = 1, 2, \ldots, m \\ l_j(\mathbf{x}^*) &> 0 \; j = 1, 2, \ldots, p \\ h_j(\mathbf{x}^*) &= 0 \; j = 1, 2, \ldots, n \end{aligned} \tag{1.1}$$

where $\mathbf{x}^*$ is a t-dimensional vector (design vector) in the search space D^t that represents the optimal solution, t is the problem dimension or the number of optimization parameters (or decision variables or unknowns). Moreover, D_k is the search space domain in the k-dimension, $F_{\min}(\mathbf{x}^*)$ in the case of a minimization problem or $F_{\max}(\mathbf{x}^*)$ in the case of a maximization problem is called the objective function, $g_j(\mathbf{x}^*)$ is the ith negative constraint function, $l_j(\mathbf{x}^*)$ is the ith positive constraint function, and $h_j(\mathbf{x}^*)$ is the ith equality constraint function. If the decision variables are bounded (which is the most probable in engineering domain) then have also side constraints $x_{i,up} \geq x_i \geq x_{i,low}$ $i = 1, 2, \ldots, t$, where $x_{i,up}$, $x_{i,low}$ are the upper and lower limits in the ith dimension, respectively.

1.1 Optimization algorithms

1.1.1 Deterministic algorithms

There are several ways to classify the optimization algorithms or methods. The simplest way is to divide them according to the algorithm nature, which can be deterministic or stochastic. Deterministic algorithms always obtain the same solution when starting from the same point. The major drawback of deterministic algorithms is that their performance depends on the form of the objective function. The development of most of these algorithms is based on the assumption that the objective function is bowl shaped. These algorithms fall into local optimums when the objective function form is different. They are considered local optimizers, because they can obtain a local optimum. In most of cases, these algorithms do not perform well in constrained optimization problems. The advantage of such algorithms is the fact that they are not usually computationally intensive. Some of these algorithms require calculation of objective function derivatives, which is not always possible.

Representative examples of deterministic algorithms are the downhill Simplex method [1], the Grid Search algorithm [2], Newton's method [3], and the conjugate gradient algorithm [3]. Additionally, popular deterministic algorithms include sequential quadratic programming (SQP) methods that perform very well in cases of constrained nonlinear optimization. The basic requirement for SQP methods to work is that the objective function and the constraints are twice continuously differentiable. A survey of SQP algorithms can be found in [4].

Overall, local optimization methods have several disadvantages. These methods require an initial starting point for the optimization variables. This starting point is very important because its selection can greatly affect the objective function value of the local solution obtained. Moreover, local optimizers are often sensitive to algorithm parameter values, which may need to be fine-tuned for a specific problem.

1.1.2 Stochastic algorithms

The main characteristic of a stochastic optimizer is randomness. The solutions obtained will be different in each run. Additional features of stochastic optimizers are as follows [5]:

- *Simplicity*. Stochastic algorithms are in general simpler than deterministic counterparts. Their implementation in any programming language is generally easier than the deterministic algorithms.
- *Efficiency*. They are more computationally expensive than deterministic algorithms. They require much more objective function evaluations than the deterministic algorithms.
- *Robustness*. Although the finding of the global optimum solution is in general uncertain, they are in most cases able to obtain a quasi-optimal solution. This type of solution may be adequate for most optimization problems.

- *Versatility*. The stochastic optimization problems can be applied to optimization problems that have solutions not only in the real domain, but they also can easily be applied to discrete domain problems or even mixed integer optimization problems.
- *Wide applicability*. The stochastic algorithms are not limited to any objective function type. They can be applied to any type of optimization problem.

Stochastic optimizers are considered to be global optimizers in a way that they have the ability to obtain the global optimum. This is due to the fact of randomness that allows larger exploration of the search space than determinist algorithms. However, there is no mathematical proof that a stochastic algorithm will find the global optimum.

A very simple form of a stochastic optimizer is the Monte Carlo algorithm [6]. In this case, a random solution is generated in each iteration and the objective function value is calculated. The best value in each run is selected. Moreover, simulated annealing is another example of a popular stochastic algorithm [7]. This algorithm mimics the annealing in metallurgy, a technique involving heating and controlled cooling of a material to increase the size of its crystals and reduce their defects.

Evolutionary algorithms (EAs) belong to the family of stochastic optimizers. The basic characteristic of EAs is that they mimic the behavior and evolution of biological entities inspired mainly by Darwin's theory of evolution and its natural selection mechanism. Additional advantages of EAs include the fact that EAs can be very often hybridized with deterministic methods, and that EAs due to its evolving population have been easily parallelized in multiple cores or different CPUs.

1.2 Evolutionary algorithms

The research community started to study the EAs in the 1960s. Several researchers independently developed four mainstream EA paradigms: the genetic algorithms (GAs) [8,9], the genetic programming (GP) [10], evolutionary programming (EP) [11], and evolutionary strategies (ES) [12] (Figure 1.1). EAs are widely used for the solution of single- and multi-objective optimization problems (MOOPs), and Figure 1.1 depicts some of the main algorithmic families.

GAs, the most popular EAs, are inspired by Darwin's natural selection. GAs can be real or binary coded. In a binary-coded GA, each chromosome encodes a binary string [13,14]. The most commonly used operators are crossover, mutation, and selection. The selection operator chooses two parent chromosomes from the current population according to a selection strategy. Most popular selection strategies include roulette wheel and tournament selection. The crossover operator combines the two parent chromosomes in order to produce one new child chromosome. The mutation operator is applied with a predefined mutation probability to a new child chromosome. Popular books about the GAs in design problems in electromagnetics can be found in [13,14].

GP is family of algorithms that uses a variable-sized tree for the representation of functions and values. GP can be used to discover a functional relationship between

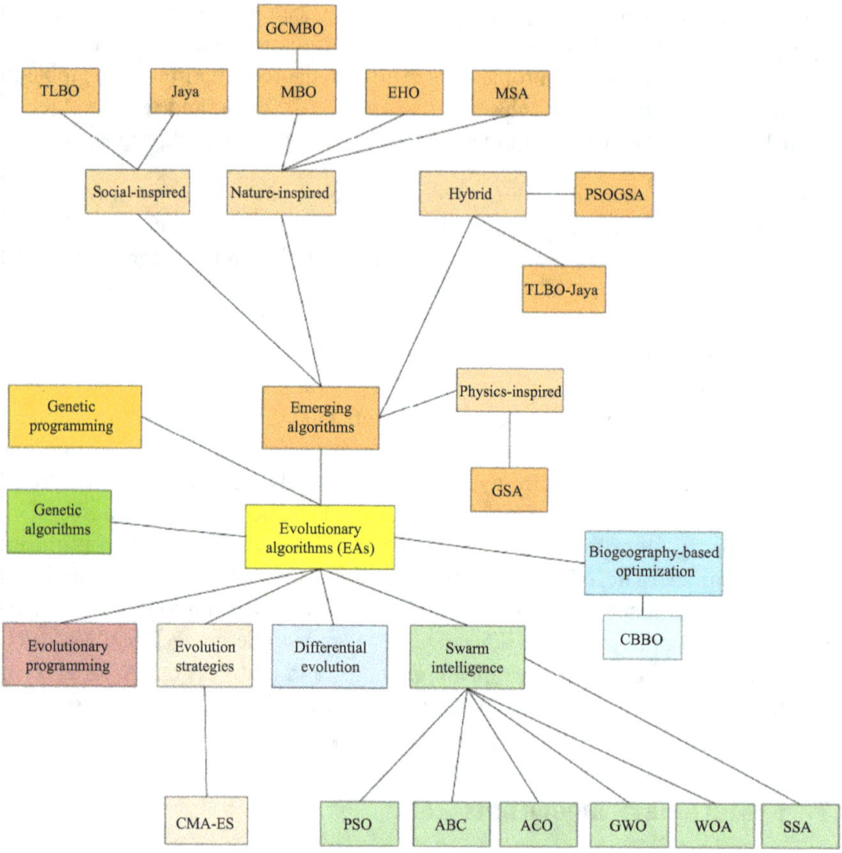

Figure 1.1 A diagram depicting main families of evolutionary algorithms

features in data (symbolic regression), to group data into categories (classification), and to assist in the design of electrical circuits, antennas, and quantum algorithms. GP as the GAs uses mutation and crossover operators to evolve. There are different types of GP; the first one initially introduced by Koza [10] is tree-based GP. In tree-based GP, each leaf in the tree represents a label from a fixed set of value labels. Each internal node in the tree represents a label from a set of function labels.

ES use fixed length real-valued vectors for representation. ES use as main reproduction operator a Gaussian mutation operator. This operator selects a random value from a Gaussian distribution and adds it to every member of a vector to generate a new child vector. Moreover, an intermediate recombination operator is also applied to two vectors. In this case, the two parent vectors are averaged to generate a new child vector. Depending on the specific ES type, the recombination or the mutation operators may exist or not.

EP is similar to GP, but the structure of the program to be optimized is fixed, while its numerical parameters are allowed to evolve. The basic idea in EP is to represent the members of the population as finite state machines capable of responding to environmental stimuli and developing operators for effecting structural and behavioral changes over time. EP has been used to a wide range of problems, including prediction problems, optimization, and machine learning. If the EP uses a fixed-length real-valued vector, then the algorithm is similar to ES (without recombination operator) with a major difference. That is, in EP no crossover operator exists so that no exchange of information between members of the population is made. Therefore, EP uses only mutation operators.

Swarm intelligence (SI) algorithms are also a special type of EAs. The SI can be defined as the collective behavior of decentralized and self-organized swarms. SI algorithms include particle swarm optimization (PSO) [15], ant colony optimization (ACO) [16–18], and artificial bee colony (ABC) [19]. PSO mimics the swarm behavior of birds flocking and fish schooling [15]. The most common PSO algorithms include the classical inertia weight PSO and the constriction factor PSO [20]. The PSO algorithm is easy to implement and is computational efficient; it is typically used only for real-valued problems. The ABC algorithm models and simulates the behaviors of honey bees foraging for food. The ACO algorithm mimics the behavior of real ants. Moreover, there are several emerging algorithms that belong to the SI category. These include the whale optimization algorithm [21], the salp swarm algorithm [22], and the grey wolf optimizer [23]. Additional emerging nature-inspired algorithms include the Monarch butterfly optimization [24], the moth search algorithm [25], the firefly algorithm [26], and the elephant herding optimization [27].

Differential evolution (DE) [28] is a population-based stochastic global optimization algorithm that has been used in several real-world engineering problems utilizing several variants of the DE algorithm. DE is not inspired by a real natural phenomenon, but steams from a mathematical model of evolution. A very interesting book about the application of DE in electromagnetics is [5].

Additionally, another category of EAs is the physics-based algorithms. These are based on concepts from physics, and they have an evolving population. Some of these include the wind-driven optimization [29] and the gravitational search algorithm [30]. Chemistry-based algorithms also exist in the literature that are based on chemical phenomena like chemical reaction optimization [31,32].

Other evolutionary techniques applied to antenna problems include the biogeography-based optimization (BBO), invasive weed optimization [33–37], teaching–learning-based optimization [38], Jaya [39], EP [40,41], and the covariance matrix adaptation evolution strategy [42,43].

It must be pointed that most of the above-described algorithms are developed initially for real-valued problems. However, versions of them made for discrete-valued problems also exist in the literature. For example, one of the most popular discrete-valued algorithms is binary PSO introduced in [44].

The application of EAs to antennas and wireless communications domain can be easily verified by performing a simple search in the Scopus database. We limit the search time from 2015 to 2020 and look for papers that address a problem in the

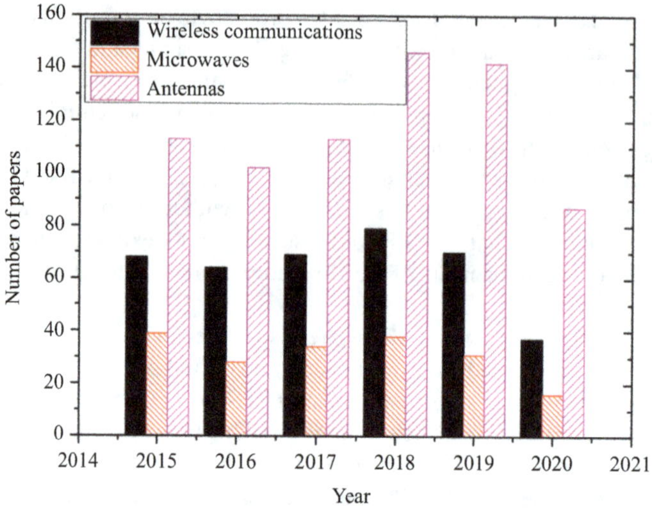

Figure 1.2 Scopus database results for papers that use EAs in different domains from 2015 to 2020

antenna, microwave, and wireless communications domain using EAs. The results show that there are in these years totally 703 antenna-related papers, 186 microwave-related papers, and 387 wireless communication papers that apply EAs. This result is depicted in Figure 1.2. We notice that there is a growing interest of the research community for antenna papers using EAs over the last years.

1.2.1 Encoding

The encoding of the problem variables is important for the solution. The suitable encoding also depends on the EA used. For example, it is preferable to use binary encoding if the problem variables take only two possible values. Binary encoding refers to have a binary string for each individual of the population. This is very useful in problems like thinned antenna arrays, where the array elements can be turned on off, or in user association problems, where a user is connected or not to a base station. An extension of binary encoding is integer encoding where all unknown variables are integers. For example, let us consider an optimization problem, where a base station transmission power is an integer between 0 dBm and 43 dBm. Moreover, we can think of an antenna array design problem where the array excitations should be integers from 1 to 10. The index of the antenna to select in a MIMO antenna selection problem is another case for integer encoding.

Real-valued encoding refers to have vectors of real numbers as members of the populations. Real-valued encoding is also very common in antenna and wireless communications domain. The common problem of optimizing an antenna geometry

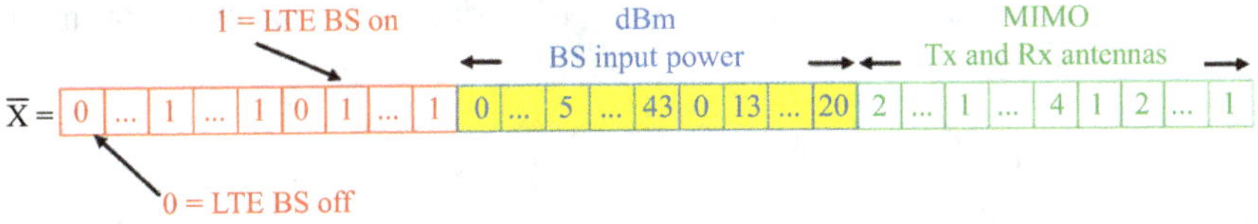

Figure 1.3 Example vector encoding with three different unknown variable domains [45]

requires the representation of real variables. Additionally, in a wireless sensor network the suitable gains of the sensor nodes in order for total network power minimization are real valued.

Furthermore, in an optimization problem some variables could be real and some others could be integer. These types of problems are called mixed-integer problems and require the application of a special algorithm. This means, this algorithm should handle the real numbers using real operators and the integers using integer or binary operators. These algorithms are often called mixed integer. An example of a mixed representation taken from [45] is depicted in Figure 1.3. We notice that the individual of the population requires the use of three different domains for unknown variables. The first is a binary domain, while the other two are different integer domains.

1.2.2 Boundary conditions constraint handling methods

Since all of the optimization problems in antennas and wireless communications use boundary constraints for each variable of the problem, it is important how an EA handles the solutions that are found outside the problem boundaries. There are several different boundary conditions handling methods that can be applied to any EA. The most common boundary constraint handling methods include [46,47] the following:

1. *Reflection method*

$$x_j^v = \begin{cases} 2x_{L,j} - x_j & \text{if} x_j < x_{L,j} \\ 2x_{U,j} - x_j & \text{if} x_j > x_{U,j} \end{cases} \tag{1.2}$$

   where x_j^v is a valid value, x_j is the value that violates the bound constraint, $x_{L,j}$, and $x_{U,j}$ are the lower and upper bounds for the jth variable, respectively.

2. *Projection method*

$$x_j^v = \begin{cases} x_{L,j} & \text{if } x_j < x_{L,j} \\ x_{U,j} & \text{if } x_j > x_{U,j} \end{cases} \tag{1.3}$$

   In this case, the variables that violate the bound constraints are trimmed to the lower and upper bounds, respectively.

3. *Reinitialization by position.* In this case, each variable that violates the constraints is randomly reinitialized with

$$x_j^v = rand_{j[0,1]}\left(x_{U,j} - x_{L,j}\right) + x_{L,j} \tag{1.4}$$

   where $rand_{j[0,1]}$ is a uniformly distributed random number between 0 and 1.
4. *Reinitialize all.* In this case, if at least one of the solution variables violates the boundaries, a complete new vector is generated within the allowed boundaries using (1.4).
5. *Conservatism.* This technique was proposed to work particularly with DE. In this case, the infeasible solution is rejected and it is replaced by the original feasible vector.

1.2.3 The no free lunch theorem

The performance of optimization algorithms is relevant to the so-called no free lunch (NFL) theorem. This theorem involves the average behavior of optimization algorithms over given spaces of optimization problems. The authors in [48] have proven that when averaged over all possible optimization problems defined over some search space X, no algorithm has a performance advantage over any other.

Moreover, the authors show in [49] that it is theoretically impossible to have a best general-purpose universal optimization strategy. Thus, a best algorithm for all cases does not exist. The only way for one algorithm to perform better than the others is when it centers on a specific problem class.

Correspondingly, in [50] the authors define the NFL theorem for multi-objective optimization. The main conclusion from the NFL theorem for Pareto fronts is that we ultimately learn nothing from the result that algorithm A outperforms algorithm B on a problem with Pareto front P_1. This is due to the fact that the NFL result guarantees that this will be compensated by the algorithm B that will outperform A on some other problem or set of problems that share the same Pareto front P_2.

Having the previous in mind, the question that might someone ask could be if it is useful to introduce new EAs. Moreover, this question can be expanded to the usage of new EAs to antennas and wireless communication problems.

A wide variety of optimization problems exist in the antenna domain, the selection of the most suitable algorithm for each problem to solve is usually a difficult task. Thus, it is worthwhile to explore new optimization algorithms if we find that they can work well for the problem at hand. Optimization problems arising in the design and synthesis of antennas can benefit considerably from an application of the EAs. The vast majority of antenna design problems can be grouped into two large types. One type is to find the best geometry for a specific antenna element. The other is to search for the appropriate positions and excitations for an antenna array. Other optimization cases may be a combination of the abovementioned problems. The application domain includes all types of wireless communications, like mobile communications, cognitive radio applications, wireless networks, wireless broadcast networks, and radio frequency identifications (RFIDs).

1.3 Objective functions

Any optimization problem is modeled mathematically using an appropriate objective function. Thus, the objective function $F(\mathbf{x})$ is what we need to optimize. The optimization problem can be a minimization or a maximization problem. A maximization problem is transformed to a minimization one by selecting as objective function the negative of the original function $-F(\mathbf{x})$. Each objective function has at least one global optimum that corresponds to the optimal solution vector $\mathbf{x}^*$ and may have several local optima. In some cases, an objective may have more than one optimal solution points but the objective function value at all these points should be the same. For example, we could consider the $F(x) = \cos(x), x \in [-2\pi, 2\pi]$. The objective functions have certain properties.

1. *Continuity*. The objective function could be continuous or not.
2. *Differentiability*. A differentiable function of one real variable is a function, derivative of which exists at each point in its domain. If the objective function is continuous, then it could be differentiable or non-differentiable.
3. *Modality*. An objective function is unimodal, if it has only one global optimum and it has not local optima. This fact can be defined mathematically for a function of one variable. If for the function $F(x)$ there is a value x_{opt} such that the function value is monotonically decreasing for $x \leq x_{opt}$ and monotonically increasing for $x \geq x_{opt}$. In this case, the value x_{opt} is the global minimum of $F(x)$ and there are no other local minima. Figure 1.4 depicts such an example unimodal function of one variable. Moreover, Figure 1.5 shows the same function example with two variables. If an objective function is not unimodal, then it is multimodal. This means that the function has at least one global optimum and multiple local optima. For example, Figure 1.6 shows a multimodal function of one variable, while the same function with two dimensions is illustrated in Figure 1.7. Problems with unimodal objective function are generally easier to solve with any optimization method. However, multimodal objective functions are difficult to solve, while

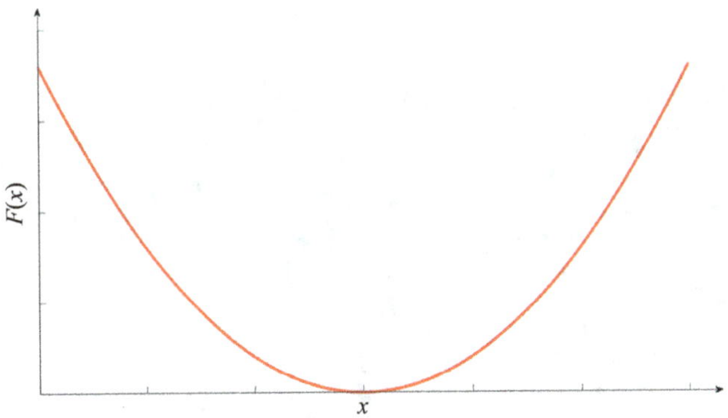

Figure 1.4 Example unimodal function of one variable

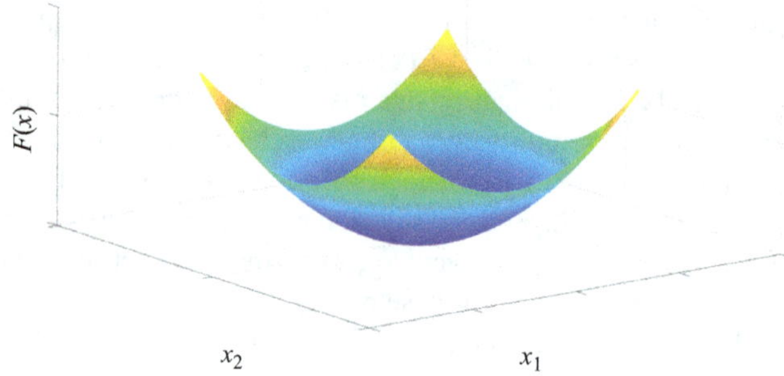

Figure 1.5 Example unimodal function of two variables

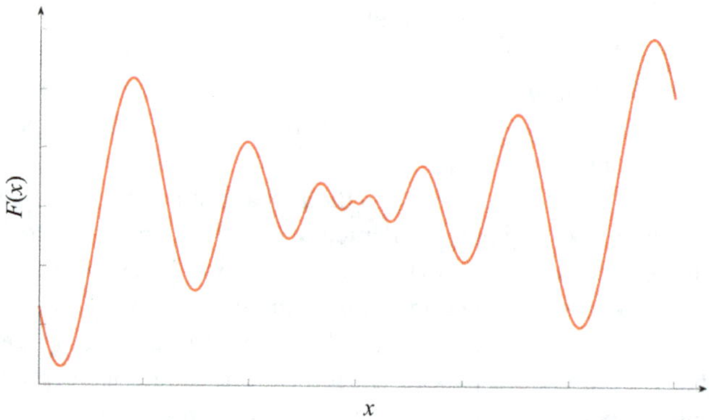

Figure 1.6 Example multimodal function of one variable

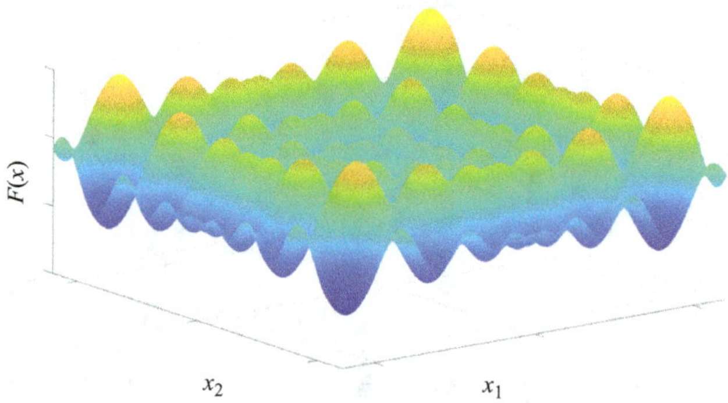

Figure 1.7 Example multimodal function of two variables

it is possible for an EA to get trapped in an local optimum. In this case, it is preferable to have an EA with large population diversity, so that it can explore more portion of the search space.

4. *Separability*. An objective function $F(\mathbf{x})$ is separable if [51] can be expressed as the sum of m single-variable functions, $f_1(x_1), f_2(x_2), \ldots, f_m(x_m)$. Otherwise it is non-separable. If the objective function is separable, then the optimization problem can be decomposed to m optimization problems of an objective function with one variable. Thus, the problem is in general easier to solve.

1.3.1 Common benchmark functions

In order to evaluate an EA and compare its performance with others, we use benchmark functions (or test functions). We usually define a test suite of common benchmark functions that contain both unimodal and multimodal functions. It is important also to test the EAs in one or two cases with high dimensions, e.g., 50, 100, or more. However, the most accurate way to evaluate an algorithm's performance is to use the IEEE Congress on Evolutionary Computation (CEC) test suites. The conference every year gives a test suite of about 30 functions that are difficult to solve. An interesting review paper about the CEC benchmark functions can be found in [52]. Some common benchmark functions are described briefly next.

1.3.1.1 Unimodal functions

1. *Sphere function*

$$f_1(x) = \sum_{j=0}^{D-1} x_j^2, \quad |x_j| \leq 100 \text{ and } f_1\,(0, 0, \ldots, 0) = 0$$

2. *Rosenbrock's function*

$$\begin{aligned} f_2(x) &= \sum_{j=0}^{D-2} \left[100\left(x_{j+1} - x_j^2\right)^2 + \left(x_j - 1\right)^2 \right] \\ |x_j| &\leq 2.048 \text{ and } f_2\,(1, 1, \ldots, 1) = 0 \end{aligned}$$

1.3.1.2 Multimodal functions

3. *Ackley's function*

$$\begin{aligned} f_3(x) &= -20 \exp\left(-\tfrac{1}{5} \sqrt{\tfrac{1}{D} \sum_{j=0}^{D-1} x_j^2} \right) \\ &\quad - \exp\left(\tfrac{1}{D} \sum_{j=0}^{D-1} \cos\left(2\pi x_j\right) \right) + 20 + e, \\ |x_j| &\leq 32 \text{ and } f_3\,(0, 0, \ldots, 0) = 0 \end{aligned}$$

4. *Griewanks's function*

$$\begin{aligned} f_4(x) &= \sum_{j=0}^{D-1} \tfrac{x_j^2}{4{,}000} - \prod_{j=0}^{D-1} \cos\left(\tfrac{x_j}{\sqrt{j}}\right), \\ |x_j| &\leq 600 \text{ and } f_4\,(0, 0, \ldots, 0) = 0 \end{aligned}$$

5. *Weierstrass function*

$$
\begin{aligned}
f_5(x) &= \sum_{j=0}^{D-1}\left(\sum_{k=0}^{k_{\max}}\left[a^k\cos\left(2\pi b^k\left(x_j+0.5\right)\right)\right]\right)\\
&\quad -D\left(\sum_{k=0}^{k_{\max}}\left[a^k\cos\left(\pi b^k\right)\right]\right),\\
a &= 0.5, b=3, k_{\max}=20\ \left|x_j\right| \le 0.5 \text{ and } f_5\left(0,0,\ldots,0\right)=0
\end{aligned}
$$

6. *Rastrigin's function*

$$
\begin{aligned}
f_6(x) &= \sum_{j=0}^{D-1}\left[x_j^2-10\cos\left(2\pi x_j\right)+10\right],\\
\left|x_j\right| &\le 5.12 \text{ and } f_6\left(0,0,\ldots,0\right)=0
\end{aligned}
$$

7. *Noncontinuous Rastrigin's function*

$$
\begin{aligned}
f_7(x) &= \sum_{j=0}^{D-1}\left[y_j^2-10\cos\left(2\pi y_j\right)+10\right],\\
y_j &= \begin{cases} x_j & \left|x_j\right| < 0.5\\ round(2x_j)/2 & \left|x_j\right| \ge 0.5\end{cases}\\
\text{for } j &= 0,1,\ldots,D-1 \text{ and } f_7\left(0,0,\ldots,0\right)=0
\end{aligned}
$$

8. *Schwefel's function*

$$
\begin{aligned}
f_8(x) &= 418.9829\times D-\sum_{j=0}^{D-1}\left[x_j\sin\left(\left|x_j\right|^{1/2}\right)\right],\\
\left|x_j\right| &\le 500 \text{ and } f_8\left(420.96,420.96,\ldots,420.96\right)=0
\end{aligned}
$$

9. *Generalized penalized function 1*

$$
\begin{aligned}
f_9(x) &= \frac{\pi}{n}\times\left\{10\sin^2\left(\pi y_1\right)\right.\\
&\quad\left.+\sum_{i=1}^{n-1}\left(y_i-1\right)^2\left[1+10\sin^2\left(\pi y_{i+1}\right)\right]+\left(y_n-1\right)^2\right\}\\
&\quad+\sum_{i=1}^{n}u\left(x_i,a,k,m\right)
\end{aligned}
$$

where

$$
\begin{aligned}
y_i &= 1+\frac{1}{4}\left(x_i+1\right),\\
u\left(x_i,a,k,m\right) &= \begin{cases} k\left(x_i-a\right)^m & \text{if} x_i > a\\ 0 & \text{if } -a \le x_i \le a\\ k\left(-x_i-a\right)^m & \text{if} x_i < -a\end{cases}\\
a &= 10, k=100, m=4\\
&\quad -50 \le x_i \le 50,\quad i=1,2,\ldots,n\\
&\quad f_9\left(-1,-1,\ldots,-1\right)=0
\end{aligned}
$$

10. *Generalized penalized function 2*

$$\begin{aligned} f_{10}(x) &= 0.1 \times \{\sin^2(3\pi x_1) \\ &\quad + \sum_{i=1}^{n-1} (x_i - 1)^2 \left[1 + \sin^2(3\pi x_{i+1})\right] \\ &\quad + (x_n - 1)^2 \left[1 + \sin^2(2\pi x_n)\right]\} + \sum_{i=1}^{n} u(x_i, a, k, m) \end{aligned}$$

$$u(x_i, a, k, m) = \begin{cases} k(x_i - a)^m & \text{if} x_i > a \\ 0 & \text{if} - a \le x_i \le a \\ k(-x_i - a)^m & \text{if} x_i < -a \end{cases}$$

$$a = 5, k = 100, m = 4$$

$$-50 \le x_i \le 50, i = 1, 2, \ldots, n$$

$$f_{10}(1, 1, \ldots, 1) = 0$$

1.4 Comparison metrics

We compare algorithms in several benchmark functions in terms of the average objective function values obtained after at least 50 independent trials. For example, if we consider three EAs A, B, C, then Table 1.1 lists their comparative results in some test functions. We notice that at first the algorithm C looks to perform better, since it outperforms the others in seven out of ten test functions. So, the result regarding the best algorithm is quite clear in this case. However, there could be cases where the result is not clear at first sight and we cannot decide which algorithm performs best. In this case, the nonparametric statistical tests provide us with the solution [53,54].

Table 1.1 Example of a comparison between three algorithms

Benchmark function	Algorithm *A*	Algorithm *B*	Algorithm *C*
Sphere	**1.51E−55**	4.65E−08	2.89E−165
Rosenbrock	5.64E+01	**5.30E+01**	**5.30E+01**
Ackley	2.54E−14	2.45E+00	**3.84E−15**
Generalized Griewank	1.10E−03	**2.72E−05**	1.20E−03
Weierstrass	1.71E−15	2.52E+01	**0.00E+00**
Generalized Rastrigin	4.18E−01	6.43E+01	**0.00E+00**
Noncontinuous Rastrigin	1.48E+00	9.94E+01	**0.00E+00**
Schwefel	1.45E+04	1.03E+04	**8.90E+02**
Generalized penalized 1	**1.34E−08**	4.98E+00	3.60E−04
Generalized penalized 2	**1.40E−02**	2.04E+01	**1.40E−02**

The bold font indicates the smaller values.

1.4.1 Nonparametric tests

Nonparametric statistical tests provide an accurate mean for evaluating an algorithm's performance. More details can be found in [53,54]. The Friedman ranking test is the first to apply. The algorithms according to Friedman ranking test obtain a rank according to their performance in each test function. For example, the algorithm with the best performance is assigned the rank 1, and the algorithm with the second best performance is assigned the rank 2. If two or more algorithms perform equally in a test function, then we add the corresponding ranks and divide by the number of algorithms, e.g., if two algorithm perform equally best as the algorithm *B* and the algorithm *B* in the Rosenbrock function of Table 1.1, then both the algorithms are in tie so then they both get the rank $1+2=3, 3/2=1.5$. Finally, we sum all ranks of each algorithm and divide by the number of test functions to obtain the average rank. This average rank clearly shows the ranking of the algorithms. The rankings according to Friedman test are reported in Table 1.2 for the results of the Table 1.1.

Additionally, in order if an algorithm is significantly better than another we can perform the Wilcoxon signed-rank test with significance level of 0.05 or less. The Wilcoxon signed-rank test is a nonparametric statistical hypothesis test used to compare two related samples, matched samples, or repeated measurements on a single sample to assess whether their population mean ranks differ (i.e., it is a paired difference test).

The H_0 hypothesis is that both population mean is the same, which in the case of algorithm comparison means that algorithm *A* is not significantly better than algorithm *B*. The null hypothesis is true when the test result is higher than the significance level. Otherwise, if the *p*-value result from the test is lower than the significance level then this means that the H_1 hypothesis is true and the algorithm at hand is significantly better than the other.

An example is shown in Table 1.3. We notice that algorithm *C* is significantly better than algorithm *B*, the *p*-value result is $0.035 < 0.05$. However, algorithm *C* is not significantly better than algorithm *B* because the obtained *p*-value result is $0.117 > 0.05$.

Table 1.2 Average rankings achieved by Friedman test

Method	Average rank	Normalized values	Rank
Algorithm *A*	2.05	1.46	2
Algorithm *B*	2.55	1.82	3
Algorithm *C*	**1.40**	**1.00**	**1**

The bold font indicates the smaller values.

Table 1.3 Wilcoxon signed-rank test between algorithm C and the other algorithms

Algorithm *C* vs	***p*-Values**
Algorithm *A*	0.117
Algorithm *B*	**0.035**

The bold font indicates value below significant level.

1.4.2 Signature of an algorithm

Clerc in [55] introduced the idea of obtaining the intrinsic bias or the signature of an algorithm. This is achieved by setting as objective function a problem impossible to solve. For example, if we consider the objective function $F(x_1, x_2) = 10, x_1, x_2 \in [-10, 10]$. This type of problem will produce random solutions. In order to obtain as much as possible random points, we may set the population size to 100, the maximum number of iterations to 1,000, and the maximum number of runs is set to 100. This configuration obtains 10,000 random points as it is suggested in [55]. That way we get the signature of an algorithm and we reveal certain intrinsic characteristics. Clerc suggests to obtain the algorithm's signature prior to running the test functions. We may distinguish among the following bias cases [55].

1. *Central bias*. This means that the solutions are gathered around the center of the search space.
2. *Edge bias*. This means that it is more probable for the algorithm to search for solutions near the edges of the search space.
3. *Axial bias*. This means that there is an increased possibility to search solutions along a coordinate axis and variation of this bias also increased possibility to search solution along a diagonal of the search space.
4. *Exploitation bias*. This means that there could be a gathering of solution points near a position that has no special characteristics.

Clearly, an unbiased optimizer is preferable and such an optimizer should obtain the same signature as a random search method. Figures 1.8–1.10 show example signatures of two different algorithms. We notice that from Figure 1.8, it seems that the algorithm is unbiased, while Figure 1.10 shows a slight center bias. This means that this algorithm should perform better in objective functions where the optima are located in the close vicinity of the center of the search space. Moreover, Figure 1.10 depicts a slight exploitation bias at certain areas of the search space.

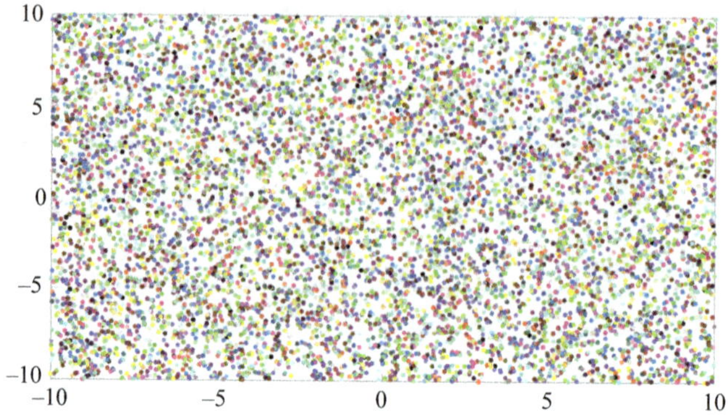

Figure 1.8 Example signature of an algorithm. Unbiased case

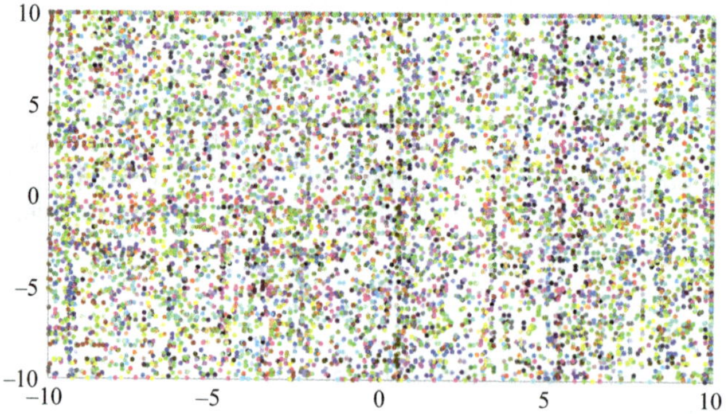

Figure 1.9 Example signature of an algorithm. Exploitation bias case

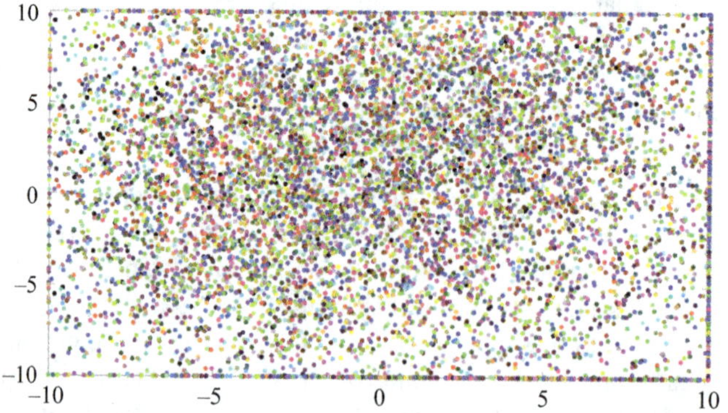

Figure 1.10 Example signature of an algorithm. Center bias case

1.5 Multi-objective algorithms

The general constrained MOOP definition is [56]:
minimize

$$F(\bar{x}) = [F_1(\bar{x}), F_2(\bar{x}), \ldots, F_n(\bar{x})]$$

subject to

$$g_i(\bar{x}) \leq 0 \; i = 1, 2, \ldots, k$$

where $F(\bar{x})$ is the vector of the objective functions, g_i are the constraint functions, n is the number of objective functions, and k is the number of constraint functions. If $\mathscr{X}$ is the search space and $\mathscr{Z}$ is the objective space, then the vector of the objective functions $\bar{f} : \mathscr{X} \rightarrow \mathscr{Z}$ assigns to each vector $\bar{x} \in \mathscr{X}$ a corresponding objective vector $\bar{z} = \bar{f}(\bar{x}) \in \mathscr{Z}$. An ideal point (also called utopia point) $\bar{z}^*$ is called a vector of the objective space composed of the best objective function values:

$$z_j^* = \min \left\{ z_j | \bar{z} \in \mathscr{Z} \right\} \;\; j = 1, \ldots, n$$

Generally, multi-objective optimization is quite different than single-objective optimization. The basic concept in single-objective optimization is that a single global optimum solution exists and we try to find that solution. However, when there are more than one objective function, which have conflicting objectives, no single best solution exists that is the optimum with respect to all objectives.

Thus, in the multi-objective optimization we obtain not only one single solution but a set of solutions. This set of solutions contains points that are non-dominated and belong to the set that is called Pareto front. Thus, we consider that the predominant concept regarding defining an optimal point is that of Pareto optimality. We define the Pareto-optimal solutions as those solutions (from the set of feasible solutions) that cannot be improved in any objective without causing degradation in at least one other objective.

Thus, the MOOP can be solved in two ways. The first one is to transform it to a single-objective optimization problem. This will be fulfilled by utilizing weight values for each objective function and penalty terms for each constraint function. Therefore, the problem is formulated as a weighted sum. Solving this problem using an EA will obtain a single solution. The second way is to apply Pareto optimization. This means that all the objective functions are optimized at the same time, and each is having equal importance. This type of optimization will finally obtain a set of solutions. If none of the objective function values can be further improved without impairing the value of at least one objective for a given solution, then this solution is Pareto optimal and belongs to the set of non-dominated solutions which is called Pareto front.

Thus, the fundamental objective of a multi-objective algorithm is to derive a set of solutions that belong to the Pareto front. An example of a Pareto front for the case of two conflicting minimization objective functions is illustrated in Figure 1.11. It also

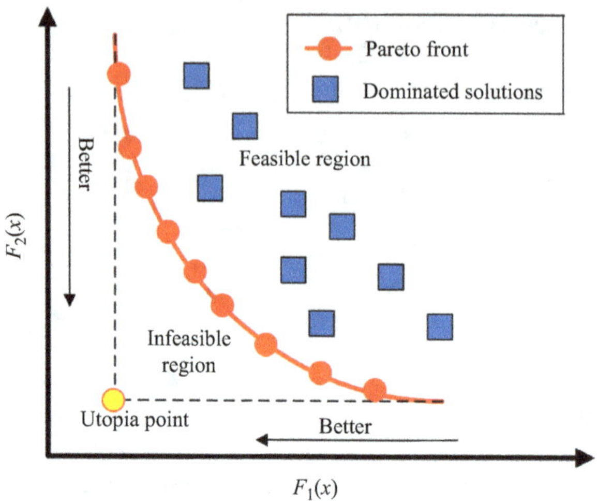

Figure 1.11 An example of solution of MO problem with two conflicting minimization objective functions

depicts the utopia point defined as *utopia point*, which is defined for minimization objectives as

$$\bar{u}_{\text{utopia}} = [u_1 \cdots u_M]^T = \begin{bmatrix} \min_{\bar{x}\in\mathscr{X}} F_1(\bar{x}) \\ \vdots \\ \min_{\bar{x}\in\mathscr{X}} F_M(\bar{x}) \end{bmatrix} \tag{1.5}$$

It is well known that the EAs model the possible solutions using a vector representation. For the purpose of distinguishing the members of the non-dominated set from the population members, we refer to the first as solutions and to the second ones as vectors. Later, we provide the definitions of dominance relations between two vectors (or individuals of the population). The weak dominance $\preceq$ relation between two vectors $\bar{x}_1, \bar{x}_2$ in the search space is defined as [57]

$\bar{x}_1$ weakly dominates $\bar{x}_2$

$$\bar{x}_1 \preceq \bar{x}_2 \text{ iff } \forall i : F_i(\bar{x}_1) \leq F_i(\bar{x}_2) \tag{1.6}$$

while the dominance $\prec$ relation is defined as

$\bar{x}_1$ dominates $\bar{x}_2$

$$\bar{x}_1 \prec \bar{x}_2 \text{ iff } \bar{x}_1 \preceq \bar{x}_2 \wedge \exists i : F_i(\bar{x}_1) < F_i(\bar{x}_2) \tag{1.7}$$

The previous relations can be extended to include constraint dominance $\prec_c$[57]:

$\bar{x}_1$ constraint-dominates $\bar{x}_2$

$$\bar{x}_1 \prec_c \bar{x}_2 \tag{1.8}$$

when any of the following conditions are true:

1. $\bar{x}_1$ belongs to the feasible design space and is infeasible.
2. $\bar{x}_1, \bar{x}_2$ are both infeasible but dominate in constraint function space.
3. $\bar{x}_1, \bar{x}_2$ both belong to the feasible design space but $\bar{x}_1$ dominates $\bar{x}_2$ in objective function space.

1.5.1 Fuzzy decision maker

In order to obtain the best compromise solution from the Pareto front, we can use a methodology from the fuzzy set theory [58]. The basic idea is to map each nth objective function value to a satisfaction degree that is modeled by a linear fuzzy membership function. We can define this satisfaction degree for maximization problems as [58]

$$sf_n = \begin{cases} 1, & \text{if } z_n \geq z_n^{\max}, \\ 1 - \dfrac{z_n^{\max} - z_n}{z_n^{\max} - z_n^{\min}}, & \text{if } z_n^{\min} < z_n < z_n^{\max} \\ 0, & \text{if } \ z_n \leq z_n^{\min} \end{cases} \tag{1.9}$$

where z_n is the value of nth objective function, $z_n^{\min}$, $z_n^{\max}$ are the minimum and maximum values of the nth objective function, respectively. For each non-dominated solution j, a normalized membership function or degree of satisfaction can be derived with the following expression:

$$s_j = \frac{\sum_{n=1}^{N_{\text{obj}}} sf_n}{\sum_{m=1}^{M_{\text{par}}} \sum_{n=1}^{N_{\text{obj}}} sf_n} \tag{1.10}$$

where M_{par} is the number of non-dominated solutions in the Pareto front. Therefore, we compute using (1.10) for each non-dominated vector of the Pareto front. The best compromise solution is obtained by finding the solution vector with the maximum s_j value. In the case of minimization problems, (1.10) becomes

$$sf_n = \begin{cases} 1, & \text{if } z_n \leq z_n^{\min}, \\ \dfrac{z_n^{\max} - z_n}{z_n^{\max} - z_n^{\min}}, & \text{if } z_n^{\min} < z_n < z_n^{\max} \\ 0, & \text{if } \ z_n \geq z_n^{\max} \end{cases} \tag{1.11}$$

1.5.2 Performance indicators for MOEAs

A specific methodology with suitable performance indicators for comparing different multi-objective EAs (MOEAs) is presented in [59]. A function that maps to each non-dominated solution set (or Pareto set approximation) a real number is called (unary) quality indicator. Therefore, the mapping from the set of all Pareto sets approximation to the set of real numbers $QI : \Omega \rightarrow \mathbb{R}$ defines a unary quality indicator QI. In this book, we will apply to the results the following performance indicators [59]:

1. *The hypervolume difference to a reference set.* The hypervolume indicator QI_H measures the hypervolume of the portion of objective space that is weakly

dominated by a Pareto set approximation Y. The hypervolume difference to a reference set $QI_{\bar{H}}(Y)$ for a given Pareto set approximation Y is given mathematically as

$$QI_{\bar{H}}(Y) = QI_H(R) - QI_H(Y) \tag{1.12}$$

where we define R as the reference set obtained by the union of the Pareto set approximations found by all algorithms that we evaluate.

2. *The unary ε-indicator* $QI_{\varepsilon 1}$ that comes from the definition of the binary ε-indicator. The definition of the binary ε-indicator $QI_\varepsilon(Y_1, Y_2)$ is that it provides the maximum factor by which each point in Y_2 can be multiplied in order that the resulting Pareto set approximation is weakly dominated by Y_1, i.e.,

$$QI_\varepsilon(Y_1, Y_2) = \inf_{\varepsilon \in R} \{\forall z_2 \in Y_2 \;\; \exists z_1 \in Y_1 \; : z_1 \preceq_\varepsilon z_2\} \tag{1.13}$$

where the ε-dominance relation $\preceq_\varepsilon$ for the vectors z_1, z_2 of the objective space is defined as

$$z_1 \preceq_\varepsilon z_2 \Leftrightarrow z_{i,1} \leq \varepsilon z_{i,2}, \; \forall i \in 1, \ldots, n \tag{1.14}$$

The unary ε-indicator $QI_{\varepsilon 1}(Y)$ for a given Pareto set approximation Y is defined by $QI_{\varepsilon 1}(Y) = QI_\varepsilon(Y, R)$.

3. *The R_2 indicator* that is applied for the comparison of approximation sets on the basis of a set of utility functions. A utility function u is a mapping from the set Z of objective space to the real number set $\mathbb{R}$, i.e., $u: \; Z \to \mathbb{R}$. The R_2 indicator is defined by

$$QI_{R_2}(Y) = \frac{\sum_{\lambda \in \Lambda} u^*(\bar{\lambda}, Y) - u^*(\bar{\lambda}, R)}{|\Lambda|} \tag{1.15}$$

where u^* is the maximum value reached by the utility function u with weight vector $\bar{\lambda}$ on a Pareto set approximation Y, i.e., $u^*(\bar{\lambda}, Y) = \max_{z \in Y} u(z)$, where Λ is a set of normalized weight vectors derived by the following rule:

$$\Lambda = \left\{ \bar{\lambda} \in \mathbb{R}^N | \lambda_j \geq 0 \;, j = 1, \ldots, N \;\wedge \sum_{j=1}^{N} \lambda_j = 1 \right\} \tag{1.16}$$

where N denotes the number of objective functions. For all indicators, the smaller the value, the better the Pareto set approximation.

1.6 Discussion-open issues

In order to select the most suitable algorithm for every optimization problem, we need to consider the problem-specific characteristics. After the algorithm selection, a critical issue is setting of the algorithms' control parameters. This is not an easy task, since in most cases the best control parameters can also be problem dependent.

A practical initial approach is to use the control parameters for these algorithms that commonly perform well regardless of the characteristics of the problem to be solved. The use of parameter-free algorithms offers an additional solution that helps one toward the direction of the solution of the optimization problem.

Moreover, the choice of the stopping criterion is another issue that is applicable to all EAs. More often, we use the maximum iteration number or the maximum number of objective function evaluations. Furthermore, there could be a criterion to avoid stagnation and save computation time. This way the algorithm could be set to stop if after a specific number of iterations there is no further improvement of the objective function value.

Additionally, there is a rapid growth in the research domain of EAs. The hybridization of EAs with local optimizers constitutes a current and growing research trend. These hybrid algorithms are named memetic algorithms (MAs) [60], and they are mainly inspired by Dawkins' notion of meme. Such an approach has an advantage. It can establish that specific regions of the search space can be explored using fewer evaluations and good-quality solutions can be generated early during the search.

Thus, the global optimizer may generate good initial solutions for the local one. MAs have the potential to be highly efficient because of this combination of global exploration and local exploitation.

As we mentioned earlier according to the NFL theorem, there exists no optimization algorithm that is able to solve every problem effectively and efficiently. There are different optimizers that have the capabilities for obtaining a solution in different types of optimization problems. Thus, it is very difficult to find the best-suited optimization algorithm for every possible optimization problem. An interesting idea is the ensemble of different optimization algorithms. The basic concept in an ensemble of algorithms is the population is divided into different subpopulations to enhance population diversity. Each subpopulation evolves using a different algorithm. Therefore, the ensemble of different optimization algorithms could be more efficient than applying one single algorithm for solving complex problems. In this context, the use of an ensemble strategy for EAs was proposed to benefit from both the availability of diverse approaches and the need to tune the associated parameters.

The authors in [61] propose an ensemble of different PSO algorithms called the ensemble PSO to solve real-parameter optimization problems. Moreover, the authors in [62] apply an ensemble of mutation strategies and control parameters with the DE. Ensemble of algorithms can also be expanded to multi-objective problems like in [63]. The general applicability of the ensemble strategy in solving diverse problems can be found in the literature. Additionally, the usage of different populated optimization algorithms is also considered.

Moreover, a paper that conceptually studies the equivalences of various popular EAs like GAs, PSO, DE, and BBO is presented in [64]. The authors conclude that the conceptual equivalence of the algorithms is supported by the fact that modifications in algorithms result in very different performance levels.

References

[1] Nelder JA and Mead R. A simplex method for function minimization. Computer Journal. 1965;7:308–313.

[2] Deb K, Pratap A, Agarwal S, *et al.* A fast and elitist multiobjective genetic algorithm: NSGA-II. IEEE Transactions on Evolutionary Computation. 2002;6(2):182–197.

[3] Ravindran A, Ragsdell KM, and Reklaitis GV. Engineering Optimization: Functions of Several Variables. John Wiley & Sons, Ltd; 2007. Doi: 10.1002/9780470117811.ch3.

[4] Boggs PT and Tolle JW. Sequential quadratic programming. Acta Numerica. 1995;4:1–51.

[5] Qing A. Differential Evolution: Fundamentals and Applications in Electrical Engineering. Singapore; Hoboken, NJ; [Piscataway, NJ]: J. Wiley & Sons Asia; IEEE Press; 2009.

[6] Rubinstein RY and Kroese DP. Simulation and the Monte Carlo Method. 3rd ed. NJ: Wiley Publishing; 2016.

[7] Kirkpatrick S, Gelatt CD, and Vecchi MP. Optimization by simulated annealing. 1983;220(4598):671–680.

[8] Goldberg DE. Genetic Algorithms in Search, Optimization and Machine Learning. New York, NY: Addison Wesley; 1989.

[9] Holland JH. Adaptation in Natural and Artificial Systems. Ann Arbor, MI: The University of Michigan Press; 1975.

[10] Koza JR. Genetic Programming: On the Programming of Computers by Means of Natural Selection. Cambridge, MA: MIT Press; 1992.

[11] Fogel DB. Evolutionary Computation: Toward a New Philosophy of Machine Intelligence. Piscataway, NJ: IEEE Press; 1995.

[12] Beyer HG and Schwefel HP. Evolution strategies: A comprehensive introduction. Natural Computing. 2002;1(1):3–52.

[13] Haupt RL and Haupt SE. Practical Genetic Algorithms. Practical Genetic Algorithms. New York, NY: Wiley and Sons; 1998.

[14] Haupt RL and Werner DH. Genetic Algorithms in Electromagnetics. NJ: Wiley-Interscience; 2007.

[15] Kennedy J and Eberhart R. Particle swarm optimization. In: IEEE International Conference on Neural Networks. 1995;4:1942–1948.

[16] Dorigo M and Gambardella LM. Ant colonies for the travelling salesman problem. BioSystems. 1997;43(2):73–81.

[17] Dorigo M, Maniezzo V, and Colorni A. Ant system: Optimization by a colony of cooperating agents. IEEE Transactions on Systems, Man, and Cybernetics, Part B: Cybernetics. 1996;26(1):29–41.

[18] Dorigo M and Stutzle T. Ant Colony Optimization. Cambridge, MA: The MIT Press; 2004.

[19] Karaboga D and Basturk B. A powerful and efficient algorithm for numerical function optimization: Artificial bee colony (ABC) algorithm. Journal of Global Optimization. 2007;39(3):459–471.

[20] Clerc M. The swarm and the queen: Towards a deterministic and adaptive particle swarm optimization. In: Proceedings of the 1999 Congress on Evolutionary Computation, 1999. CEC 99. vol. 3; 1999. p. 1951–1957.

[21] Mirjalili S and Lewis A. The whale optimization algorithm. Advances in Engineering Software. 2016;95:51–67.

[22] Mirjalili S, Gandomi AH, Mirjalili SZ, *et al.* Salp Swarm Algorithm: A bio-inspired optimizer for engineering design problems. Advances in Engineering Software. 2017;114:163–191.

[23] Mirjalili S, Mirjalili SM, and Lewis A. Grey wolf optimizer. Advances in Engineering Software. 2014;69:46–61.

[24] Wang GG, Deb S, and Cui Z. Monarch butterfly optimization. Neural Computing and Applications. 2015;31:1995–2014.

[25] Wang GG. Moth search algorithm: A bio-inspired metaheuristic algorithm for global optimization problems. Memetic Computing. 2018;10(2):151–164.

[26] Yang XS. In: Watanabe O, Zeugmann T, editors. Firefly Algorithms for Multimodal Optimization. Berlin, Heidelberg: Springer Berlin Heidelberg; 2009. p. 169–178.

[27] Wang G, Deb S, and dos S Coelho L. Elephant herding optimization. In: 2015 3rd International Symposium on Computational and Business Intelligence (ISCBI); 2015. p. 1–5.

[28] Storn R and Price K. Differential evolution – A simple and efficient heuristic for global optimization over continuous spaces. Journal of Global Optimization. 1997;11(4):341–359.

[29] Bayraktar Z, Komurcu M, Bossard JA, *et al.* The wind driven optimization technique and its application in electromagnetics. IEEE Transactions on Antennas and Propagation. 2013;61(5):2745–2757.

[30] Rashedi E, Nezamabadi-pour H, and Saryazdi S. GSA: A gravitational search algorithm. Information Sciences. 2009;179(13):2232–2248.

[31] Lam AYS and Li VOK. Chemical-reaction-inspired metaheuristic for optimization. IEEE Transactions on Evolutionary Computation. 2010;14(3):381–399.

[32] Lam AYS, Li VOK, and Yu JJQ. Real-coded chemical reaction optimization. IEEE Transactions on Evolutionary Computation. 2012;16(3):339–353.

[33] Karimkashi S and Kishk AA. Invasive weed optimization and its features in electromagnetics. IEEE Transactions on Antennas and Propagation. 2010;58(4):1269–1278.

[34] Roy GG, Das S, Chakraborty P, *et al.* Design of non-uniform circular antenna arrays using a modified invasive weed optimization algorithm. IEEE Transactions on Antennas and Propagation. 2011;59(1):110–118.

[35] Yan-Ying B, Shaoqiu X, Changrong L, *et al.* A hybrid IWO/PSO algorithm for pattern synthesis of conformal phased arrays. IEEE Transactions on Antennas and Propagation. 2013;61(4):2328–2332.

[36] Zaharis ZD, Lazaridis PI, Cosmas J, *et al.* Synthesis of a near-optimal high-gain antenna array with main lobe tilting and null filling using Taguchi initialized invasive weed optimization. IEEE Transactions on Broadcasting. 2014;60(1):120–127.

[37] Zaharis ZD, Skeberis C, Xenos TD, *et al.* Design of a novel antenna array beamformer using neural networks trained by modified adaptive dispersion invasive weed optimization based data. IEEE Transactions on Broadcasting. 2013;59(3):455–460.

[38] Rao RV, Savsani VJ, and Vakharia DP. Teaching-learning-based optimization: A novel method for constrained mechanical design optimization problems. CAD Computer Aided Design. 2011;43(3):303–315.

[39] Rao RV. Jaya: A simple and new optimization algorithm for solving constrained and unconstrained optimization problems. International Journal of Industrial Engineering Computations. 2016;7(1):19–34.

[40] Hoorfar A. Evolutionary programming in electromagnetic optimization: A review. IEEE Transactions on Antennas and Propagation. 2007;55(3):523–537.

[41] Hoorfar A, Zhu J, and Nelatury S. Electromagnetic optimization using a mixed-parameter self-adaptive evolutionary algorithm. Microwave and Optical Technology Letters. 2003;39(4):267–271.

[42] Boudaher E and Hoorfar A. Electromagnetic design optimization using mixed-parameter and multiobjective CMA-ES. In: 2013 IEEE Antennas and Propagation Society International Symposium, APSURSI 2013; 2013. p. 406–407.

[43] BouDaher E and Hoorfar A. Electromagnetic optimization using mixed-parameter and multiobjective covariance matrix adaptation evolution strategy. IEEE Transactions on Antennas and Propagation. 2015;63(4):1712–1724.

[44] Kennedy J and Eberhart RC. Discrete binary version of the particle swarm algorithm. In: Proceedings of the IEEE International Conference on Systems, Man and Cybernetics. 1997;5:4104–4108.

[45] Goudos SK, Deruyck M, Plets D, *et al.* Optimization of power consumption in 4G LTE networks using a novel barebones self-adaptive differential evolution algorithm. Telecommunication Systems. 2017;66(1):109–120.

[46] Arabas J, Szczepankiewicz A, and Wroniak T. Experimental comparison of methods to handle boundary constraints in differential evolution. In: Schaefer R, Cotta C, Kołodziej J, *et al.*, editors. Parallel Problem Solving from Nature, PPSN XI. Berlin, Heidelberg: Springer Berlin Heidelberg; 2010. p. 411–420.

[47] Juárez-Castillo E, Pérez-Castro N, and Mezura-Montes E. A novel boundary constraint-handling technique for constrained numerical optimization problems. In: 2015 IEEE Congress on Evolutionary Computation (CEC); 2015. p. 2034–2041.

[48] Wolpert DH and Macready WG. No free lunch theorems for optimization. IEEE Transactions on Evolutionary Computation. 1997;1(1):67–82.

[49] Ho YC and Pepyne DL. Simple explanation of the no-free-lunch theorem and its implications. Journal of Optimization Theory and Applications. 2002;115(3):549–570.

[50] Corne DW and Knowles JD. No free lunch and free leftovers theorems for multiobjective optimisation problems. In: Fonseca CM, Fleming PJ, Zitzler E, *et al.*, editors. Evolutionary Multi-Criterion Optimization. Berlin, Heidelberg: Springer Berlin Heidelberg; 2003. p. 327–341.

[51] Singiresu SR. Engineering Optimization: Theory and Practice. John Wiley & Sons, Ltd; 2019. Doi: 10.1002/9781119454816.ch1.

[52] Hellwig M and Beyer HG. Benchmarking evolutionary algorithms for single objective real-valued constrained optimization – A critical review. Swarm and Evolutionary Computation. 2019;44:927–944.

[53] García S, Molina D, Lozano M, *et al.* A study on the use of non-parametric tests for analyzing the evolutionary algorithms' behaviour: A case study on the CEC'2005 Special Session on Real Parameter Optimization. Journal of Heuristics. 2009;15(6):617–644.

[54] García S, Fernández A, Luengo J, *et al.* Advanced nonparametric tests for multiple comparisons in the design of experiments in computational intelligence and data mining: Experimental analysis of power. Information Sciences. 2010;180(10):2044–2064.

[55] Clerc M. Guided Randomness in Optimization. John Wiley & Sons, Ltd; 2015.

[56] Marler RT and Arora JS. Survey of multi-objective optimization methods for engineering. Structural and Multidisciplinary Optimization. 2004;26(6): 369–395.

[57] Kukkonen S and Lampinen J. GDE3: The third evolution step of generalized differential evolution. In: Proceedings of the 2005 IEEE Congress on Evolutionary Computation (CEC 2005). vol. 1; 2005. p. 443–450.

[58] Carlsson C and Fullér R. Fuzzy multiple criteria decision making: Recent developments. Fuzzy Sets and Systems. 1996;78(2):139–153.

[59] Knowles J, Thiele L, and Zitzler E. A Tutorial on the Performance Assessment of Stochastic Multiobjective Optimizers. Zurich: Computer Engineering and Networks Laboratory (TIK), Swiss Federal Institute of Technology (ETH); 2005.

[60] Neri F and Cotta C. Memetic algorithms and memetic computing optimization: A literature review. Swarm and Evolutionary Computation. 2012;2:1–14.

[61] Lynn N and Suganthan PN. Ensemble particle swarm optimizer. Applied Soft Computing. 2017;55:533–548.

[62] Mallipeddi R, Suganthan PN, Pan QK, *et al.* Differential evolution algorithm with ensemble of parameters and mutation strategies. Applied Soft Computing. 2011;11(2):1679–1696.

[63] Zhao S, Suganthan PN, and Zhang Q. Decomposition-based multiobjective evolutionary algorithm with an ensemble of neighborhood sizes. IEEE Transactions on Evolutionary Computation. 2012;16(3):442–446.

[64] Ma H, Simon D, Fei M, *et al.* On the equivalences and differences of evolutionary algorithms. Engineering Applications of Artificial Intelligence. 2013;26(10):2397–2407.

Chapter 2
Evolutionary algorithms

This chapter describes some of the evolutionary algorithms (EAs) that will be used in the book.

2.1 Swarm intelligence algorithms

Swarm intelligence (SI) algorithms are special category of EAs. The essential concept of SI is the collective behavior of decentralized and self-organized swarms. There are several popular SI algorithm families that among others include particle swarm optimization (PSO) [1], ant colony optimization (ACO) [2], and artificial bee colony (ABC) [3].

The popularity of SI algorithms is due to the fact that they in general can handle efficiently arbitrary optimization problems. Additionally, SI algorithms as it can be found from the literature have been widely utilized to solve several problems in antennas [4] and in wireless communications [5].

The swarm behavior of bird flocking and fish schooling is modeled mathematically by the PSO algorithm [1]. One may find several PSO variants in the literature. The most frequently applied PSO variants include the inertia weight PSO (IWPSO) and the constriction factor PSO (CFPSO) [6]. Moreover, comprehensive learning particle swarm optimizer (CLPSO) [7,8] is a PSO algorithm that has been applied to antenna design problems. The PSO algorithm is intrinsically suitable for application to real-valued problems. Thus, binary PSO (BPSO) versions should be used for solving discrete-valued problems. BPSO [9] is one of the most popular discrete PSO algorithms. BPSO maps real values to the discrete set [0, 1] by using a sigmoid transfer function. Additionally, several new transfer functions that perform better than the original algorithm have been introduced by the authors in [10]. Furthermore, Boolean PSO [11] is another BPSO version with main characteristic the usage of binary operators for velocity and position update. Several authors have applied Boolean PSO to antenna design problems [12–15].

ABC is also a popular SI algorithm. The basic concept in ABC algorithm is the fact that it models mathematically and simulates the honey bee behavior in food foraging [3]. The ABC algorithm models each potential solution by a food source, where the food source denotes the quality (objective function fitness value) of the corresponding solution. There are several papers in the literature that solve problems in antennas and wireless communications with the ABC algorithm [16–18]. Additionally, several new ABC extensions, variants, and hybrids exist in the literature [19,20].

ACO was introduced by Dorigo in [2]. The fundamental concept of ACO is the modeling of the behavior of real ants. ACO like PSO and ABC is also a population-based metaheuristic. The ACO algorithm utilizes the ant colonies behavior when they are looking for nourishment. Precisely, ACO relies on the ants' behavior when they are looking for the shortest path between their nest and a food source. The ants deposit and react to pheromones while they are exploring their environment. ACO is inherently best suited for solving combinatorial optimization problems. There are several combinatorial optimization problems in antenna and the wireless communication domain.

In this section, we briefly present some of the most popular SI algorithms.

2.1.1 Initialization

Initialization is the common starting point for all EAs. In terms of SI algorithms, we may define a population (or swarm) of PS vectors (or particles) $\bar{x}_{G,k}, k = 1, 2, \ldots, PS$, where G denotes the iteration number. The population is initialized using random numbers sampled from a uniform distribution. We assume that each M-dimensional vector is a possible solution. Thus, we can express the kth member of the population or vector as $\bar{x}_{G,k} = (x_{G,1k}, x_{G,2k}, \ldots, x_{G,jk}, \ldots, x_{G,Mk})$. We initialize the population by

$$x_{0,mk} = rand_{m[0,1)}\left(x_{m,Upper} - x_{m,Lower}\right) + x_{m,Lower} \; m = 1, 2, \ldots, M \tag{2.1}$$

where $x_{m,Upper}$ and $x_{m,Lower}$ are the M-dimensional vectors of the upper and lower bounds in the mth dimension, respectively, and $rand_{m[0,1)}$ is a uniformly distributed random number within [0, 1).

2.1.2 Inertia weight particle swarm optimization

IWPSO [1] is the original and most popular PSO algorithm. The fundamental concept of IWPSO is the movement of a swarm of particles in the search space. PSO uses the information taken from two optimum values to modify each particle position in every iteration. The first optimum is the best solution (objective function value) that has been achieved so far by each particle in the swarm. This optimum is denoted as *pbest* and is called personal best. The second optimum is the global best value obtained by any particle in the swarm. This optimum value is denoted as *gbest*. Thus, PSO for every iteration finds these values the *pbest* and *gbest*. Additionally, a velocity update rule that models the swarm movement is used. This rule plays a very important role for obtaining the global best. We can formulate this velocity update rule of the kth particle in the mth dimension as

$$\begin{aligned} u_{G+1,mk} = w u_{G,mk} &+ c_1 rand_{1,mk}(pbest_{G+1,mk} - x_{G,mk}) \\ &+ c_2 rand_{2,mk}(gbest_{G+1,m} - x_{G,mk}) \end{aligned} \tag{2.2}$$

where $u_{G+1,mk}$ denotes the kth particle velocity in the mth dimension, $G+1$, G represent the current and the previous iterations, respectively, $x_{G,mk}$ is the kth particle position in the mth dimension, in current iteration, $rand_{1,mk}, rand_{2,mk}$ are uniformly distributed random numbers in $(0, 1)$, w is the inertia weight, c_1 and c_2 are called the learning factors. The new particle position in the $G + 1$ iteration of the kth particle in the mth dimension is $x_{G+1,nk}$ and is derived by

$$x_{G+1,mk} = x_{G,mk} + u_{G+1,mk} \tag{2.3}$$

Thus, the IWPSO algorithm requires the setting of three control parameters. These are the inertia weight w and the positive constants c_1, c_2. Commonly, it is for the inertia weight $w \in [0, 1]$. The inertia weight represents the particle's fly without any external influence. Therefore, the higher value of w, the more the particle stays more unaffected from *pbest* and *gbest*. Inertia weight controls exploration and exploitation of the search space. Thus, large inertia weight values favor exploration, and small values favor exploitation.

The parameter c_1 is the cognitive learning factor. This parameter models the influence of the particle memory on its personal best position (*pbest*). Additionally, the other parameter c_2 is the social learning factor. This parameter symbolizes the influence of the swarm global best position (*gbest*).

Overall, the IWPSO algorithm needs the selection of the following control parameters: the swarm size (or population size, usually set to 80 or less), the cognitive learning factor c_1, the social learning factor c_2 (usually both are set to equal to 2.0), the inertia weight w, and the maximum number of iterations. The inertia weight usually does not obtain a constant value, but it usually decreases as the number of iterations increases. This is expressed as

$$w_G = (w_{\max} - w_{\min}) \times frac(G_{\max} - G)G_{\max} + w_{\min} \tag{2.4}$$

where $w_{\min}$ and $w_{\max}$ are the minimum and the maximum allowable inertia weight values, respectively, G is the current iteration, $G_{\max}$ is the maximum number of iterations. Usually, the initial value for $w_{\max}$ is 0.9 or 0.95 and the $w_{\min}$ is set to 0.4. Thus, the algorithms' behavior in the initial iterations favors exploration, while afterward as the number of iterations increases the inertia weight value decreases. Thus, it will favor exploitation in this case.

2.1.3 Constriction factor particle swarm optimization

Different velocity update rules have been proposed in various occasions in the literature. The most popular is the one in [6] proposed by Clerc. The basic difference is that in this case, the velocity update rule introduced a new parameter the constriction factor, which is denoted as *KF*. This velocity uses the constriction factor and not the inertia weight. The constriction factor advantage is that it favors convergence when the swarm stops moving. The velocity update rule for this case is formulated as

$$\begin{aligned} u_{G+1,mk} = KF \big[& u_{G,mk} + k_1 rand_{1,mk}(pbest_{G+1,mk} - x_{G,mk}) \\ & + k_2 rand_{2,mk}(gbest_{G+1,m} - x_{G,mk}) \big] \end{aligned} \tag{2.5}$$

where KF is the constriction factor. The definition is given by

$$KF = \frac{2}{\left|2 - \vartheta - \sqrt{\vartheta^2 - 4\vartheta}\right|} \tag{2.6}$$

where

$$\vartheta = \begin{cases} k_1 + k_2 & \text{if } k_1 + k_2 > 4 \\ 1 & \text{if } k_1 + k_2 \leq 4 \end{cases}$$

Common values for k_1, k_2 are to set both to 2.05. This PSO algorithm is called CFPSO because of the constriction factor parameter.

2.1.4 *Comprehensive learning particle swarm optimizer*

The authors in [21] introduced a new PSO variant the CLPSO that improves the convergence speed of the classical PSO algorithms. The strategy involved in the CLPSO algorithm ensures that the diversity of the swarm is preserved. This helps one to avoid premature convergence. To accomplish this, the algorithm uses a velocity update rule that remembers all particles' previous experiences, i.e., best solutions, as a potential during the calculation of the particle's new velocity. Therefore, a particle may learn from a different exemplar in each dimension. The CLPSO velocity update rule is expressed as [21]

$$u_{G+1,mk} = w \times u_{G,mk} + c \times rand_{mk}(pbest_{f_k(i)n} - x_{G,mk}) \tag{2.7}$$

where $f_k = [f_k(1), f_k(2), \ldots, f_k(i), \ldots, f_k(M)]$ defines which particle's *pbest* the particle k should follow, $pbest_{G,f_k(i)m}$ is the corresponding dimension of any particle's *pbest*, including its own *pbest*. It is very easy to implement the CLPSO algorithm. However, two disadvantages when with the original PSO algorithm are the increased complexity and the higher computational load.

2.1.5 *PSO for discrete-valued problems*

The PSO is an algorithm that is inherently defined to solve real-valued problems. However, there are several real-world engineering problems that require discrete-valued variables. Several researchers have proposed PSO variants that work for discrete-valued problems. Two very popular methods are the BPSO and the Boolean PSO that are described next briefly.

2.1.5.1 Binary PSO variants

The BPSO algorithm is introduced by the authors of the original PSO (Kennedy and Eberhart) in [9]. The simple modification of the original PSO in order to support binary values is to map real values to the interval [0, 1] with a transfer function. Therefore, BPSO uses real-valued velocities remain real-valued, but the particle positions are binary (bit strings of length M). The mth coordinate of each particle's position is a bit, state of which is expressed by

$$x_{G+1,nk} = \begin{cases} 1 & \text{if} \quad rnd < T(u_{G+1,mk}) \\ 0 & \text{otherwise} \end{cases} \tag{2.8}$$

where rnd denotes a uniformly distributed random number in [0, 1], and $T(y)$ is a sigmoid limiting transformation that maps real numbers to the interval [0, 1]. This transfer function is expressed mathematically by the sigmoid function

$$T(y) = \frac{1}{1+e^{-y}} \tag{2.9}$$

which defines an S-shaped transfer function. This version is denoted as **S2** transfer function according to [10]. The authors in [10] have extended the original BPSO definition by proposing additional S-shaped transfer functions. These are

$$\textbf{S1} \quad T(y) = \frac{1}{1+e^{-2y}} \tag{2.10}$$

$$\textbf{S3} \quad T(y) = \frac{1}{1+e^{-(y/2)}} \tag{2.11}$$

$$\textbf{S4} \quad T(y) = \frac{1}{1+e^{-(y/3)}} \tag{2.12}$$

Additionally, in [10] the authors also propose new BPSO variants that use V-shaped transfer functions. These V-shaped transfer functions are given by [10]

$$\textbf{V1} \quad T(y) = \left| erf\left(\frac{\sqrt{\pi}}{2}y\right) \right| \tag{2.13}$$

$$\textbf{V2} \quad T(y) = |\tanh(y)| \tag{2.14}$$

$$\textbf{V3} \quad T(y) = \left| \frac{y}{\sqrt{1+y^2}} \right| \tag{2.15}$$

$$\textbf{V4} \quad T(y) = \left| \frac{2}{\pi} \arctan\left(\frac{\pi}{2}y\right) \right| \tag{2.16}$$

In accordance with [10], the BPSO variant that performs best is that with the **V4** transfer function. Figures 2.1 and 2.2 illustrate the plots of these transfer functions.

2.1.5.2 Boolean PSO

Boolean PSO was introduced in [11,15]. It is a discrete-valued PSO variant that is based on Boolean algebra concepts. Boolean PSO is inherently very different from BPSO. Boolean PSO uses an entirely different velocity update formula from that in the BPSO that is based exclusively on Boolean algebra. The basic idea in Boolean PSO is that both positions and particle velocities are binary coded. Moreover, Boolean PSO uses a fundamental mechanism found in an artificial immune system called "negative selection." The basic idea is to correct the velocity binary values in terms

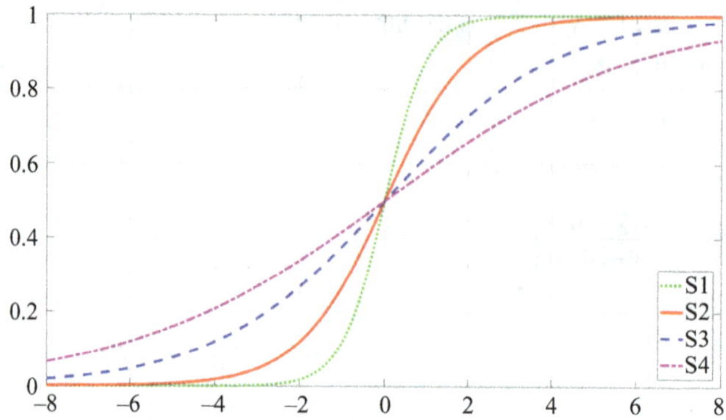

Figure 2.1 S-shaped transfer functions for binary PSO

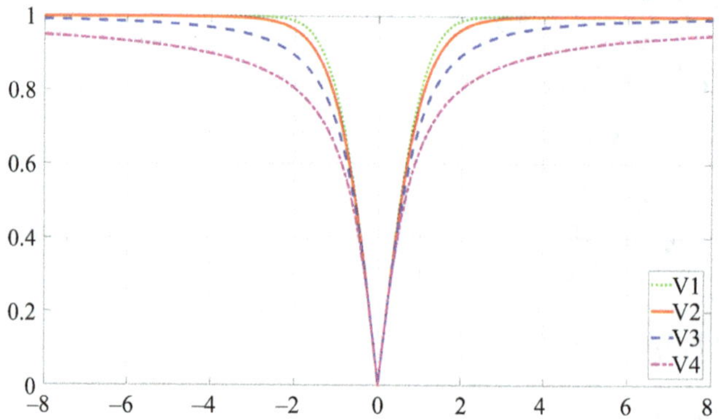

Figure 2.2 V-shaped transfer functions for binary PSO

of a maximum allowed velocity value. The velocity and particle update rules are given by

$$u_{G+1,mk} = w \times u_{G,mk} + c_1 \times (pbest_{G,mk} \oplus x_{G,mk}) + c_2 \times (gbest_{G,mk} \oplus x_{G,mk}) \tag{2.17}$$

$$x_{G+1,mk} = x_{G,mk} \oplus u_{G+1,mk} \tag{2.18}$$

where earlier relations use the Boolean operators "AND" ($\bullet$), "XOR" ($\oplus$), and "OR" ($+$). In [11,15] w, c_1, and c_2 are bits randomly chosen and their probabilities of being "1" are determined by the respective parameters Ω, C_1, and C_2. Boolean PSO is applied to antenna problems in several papers in the literature [11–14,22].

2.1.6 Artificial bee colony algorithm

The authors in [3] introduced the ABC algorithm. The algorithm is based on modeling the honey bee behavior in food foraging. A possible solution vector in ABC is the position of a food source in ABC terminology. Additionally, in ABC we call the objective function fitness as the nectar amount of a food source. In the ABC algorithm, there are three different classes or types of bees: employed, onlooker, and scout. The different types of bee behave differently. The employed bees look for food sources. Correspondingly, the onlooker bees share the information provided by the employed bees and then decide to choose the food sources. The scout bees search for new food sources by randomly initialize to new positions. Thus, they favor exploration of the search space.

The ABC algorithm maps each employed bee to one food source. This means that the number of the employed bees is equal to the number of solutions. The employed bees look for new neighbor food source near their hive. We generate each new position $\bar{x}_p$ with the following expression:

$$u_{p,n} = x_{p,n} + \varphi_{p,n}\left(x_{p,n} - x_{m,n}\right) \tag{2.19}$$

where m is a randomly chosen index from the population $\{1, 2, \ldots, SN\}$ different than p, n is a randomly chosen index from $\{1, 2, \ldots, NP\}$, SNP is the number of food sources, and $\varphi_{p,n}$ is a uniformly distributed random number within $[-1, 1]$. The ABC algorithm uses then a greedy selection operator to generate a new position vector, this is given by

$$\bar{x}'_p = \begin{cases} \bar{u}_p, & \text{if } f(\bar{u}_p) < f(\bar{x}_p) \\ \bar{x}_p, & \text{otherwise} \end{cases} \tag{2.20}$$

where $\bar{x}'_p$ represents the new position of the food source. The selection of a food source by an onlooker bee depends on a probability value assigned to that food source, S_p, expressed as

$$S_p = \frac{fit_p}{\sum_{i=1}^{SN} fit_i} \tag{2.21}$$

where fit_p is the fitness value of the pth solution. This is proportional to the nectar amount of the food source in the pth position. The ABC has an additional control parameter, which is called *limit*. This parameter gives the predetermined number of trials after which a solution vector cannot be further improved. Thus, this parameter is used to avoid stagnation and to increase the population diversity. The *limit* number increases by one, if the algorithm does improve the solution. If this variable has reached the *limit* number, then the ABC algorithm generates new random solutions. These new solutions are obtained using

$$x_{p,n} = rnd_n\left(x_{n,Upper} - x_{n,Lower}\right) + x_{n,Lower} \qquad n = 1, 2, \ldots, NP \tag{2.22}$$

where $x_{n,Lower}$ and $x_{n,Upper}$ are the lower and upper bounds of the nth dimension, respectively, and rnd_n is a uniformly distributed random number within (0, 1).

2.1.6.1 Gbest-guided ABC

The Gbest-guided ABC (GABC) [20] is an ABC variant that uses a different formula to generate new solutions. In this case, each new position of the pth solution is created by

$$u_{p,n} = x_{p,n} + \varphi_{p,n}\left(x_{p,n} - x_{m,n}\right) + \psi_{p,n}\left(y_n - x_{p,n}\right) \tag{2.23}$$

where $m \in \{1, 2, \ldots, SN\}$, $m \neq k$, $n \in \{1, 2, \ldots, NP\}$ are randomly selected indices, SN denotes the number of food sources, $\varphi_{p,n}$ denotes a uniformly distributed random number within $[-1, 1]$, y_n represents the value of the nth dimension of the global best solution, $\psi_{p,n}$ represents a uniformly distributed random number in $[0, C]$, and C is a nonnegative constant. The reader may refer to [20] for more details about GABC.

2.1.7 Ant colony optimization

Another popular SI algorithm is the ACO algorithm [2,23,24]. ACO models the ants' foraging behavior. ACO uses the fact that ants communicate with each using pheromone trails. This type of communication enables them to find the shortest paths between their nest and food sources. The ants have the ability to sense the chemical trails of the pheromone. The ants that are based on strong pheromone intensities select to go through paths toward the nest or the food source. Therefore, the shorter paths have the potential to gather more pheromone than longer ones. The ACO algorithms model this exact characteristic of ants' behavior. ACO is utilized for solving combinatorial optimization problems that are considered to be *NP*-hard.

2.1.8 Emerging nature-inspired swarm algorithms

These include three different emerging nature-inspired algorithms with low complexity. Namely, these are the whale optimization algorithm (WOA) [25], the salp swarm algorithm (SSA) [26], and the grey wolf optimizer (GWO) [27].

2.1.8.1 Grey wolf optimizer

The GWO algorithm is based on mathematical modeling of the basic mechanisms (hierarchy and hunting) of grey wolves in nature. Its main characteristic is the preservation information about the search space over the iteration process. In GWO, it is not required to set any control parameters. The GWO algorithm divides the population into four categories. The first three best solutions/grey wolves are deemed the alpha (α), beta (β), and delta (δ) categories. All the unclassified solutions are grouped into omega (ω) category. Thus, the group hunting (optimization process), as a social behavior in a pack of wolves, is oriented by the aforementioned categories (α, β, δ) of the population. The mathematical description of prey encirclement in the hunting process is given by the following expressions:

$$\vec{V} = |\vec{C}^2 \times \vec{P}_G - \vec{W}_G| \tag{2.24}$$

$$\vec{W}_{G+1} = \vec{P}_G - \vec{C}^1 \times \vec{V} \tag{2.25}$$

where $\vec{C}^1$ and $\vec{C}^2$ are coefficient vectors, $\vec{P}$ is the position vector of the prey, $\vec{W}$ corresponds to the position vector of the grey wolf, and G indicates the current generation. The vectors $\vec{C}^1$ and $\vec{C}^2$ are given by (2.26) and (2.27):

$$\vec{C}^1 = 2\vec{u} \times \vec{v}_1 - \vec{u} \tag{2.26}$$

$$\vec{C}^2 = 2 \times \vec{v}_2 \tag{2.27}$$

where $\vec{u} \in [2, 0]$ and $\vec{v}_1, \vec{v}_2 \in [0, 1]$ (random vectors). The hunting process of the GWO algorithm, as social behavior of a grey wolve's pack, can be described by the following set of equations:

$$\begin{aligned} \vec{V}_\alpha &= |\vec{C}_1^2 \times \vec{W}_\alpha - \vec{W}| \\ \vec{V}_\beta &= |\vec{C}_2^2 \times \vec{W}_\beta - \vec{W}| \\ \vec{V}_\delta &= |\vec{C}_3^2 \times \vec{W}_\delta - \vec{W}| \end{aligned} \tag{2.28}$$

$$\begin{aligned} \vec{W}_1 &= \vec{W}_\alpha - \vec{C}_1^1 \times (\vec{V}_\alpha) \\ \vec{W}_2 &= \vec{W}_\beta - \vec{C}_2^1 \times (\vec{V}_\beta) \\ \vec{W}_3 &= \vec{W}_\delta - \vec{C}_3^1 \times (\vec{V}_\delta) \end{aligned} \tag{2.29}$$

$$\vec{W}_{G+1} = \frac{\vec{W}_1 + \vec{W}_2 + \vec{W}_3}{3} \tag{2.30}$$

The pseudocode of GWO algorithm is outlined in Algorithm 1. The GWO algorithm has been applied to the user association problem in THz networks in [28]. In [29], the authors use GWO for linear array design. Moreover, the authors in [30] use among other algorithms the GWO for leaf-shaped patch antenna design.

Algorithm 1: GWO algorithm

```
Initialize the population of grey wolves W_i (i = 1, 2, ..., n)
Initialize u, C^1, and C^2
Calculate the category position vectors: W_α, W_β, and W_δ
while (t < max no. of iterations) do
    for each member of the pack do
        Update the position vector of the current member by (2.39)
    end for
    Update u, C^1, and C^2
    Calculate the position vectors of all members
    Update W_α, W_β, and W_δ
    Set: t = t + 1
end while
```

2.1.8.2 Binary GWO versions

For the binary GWO2 (second approach) presented in [31], a sigmoid function is utilized to map the real numbers ∈ [0, 1] in (2.39). The updated expression for the ith dimension is then given by

$$W_{G+1,i} = \begin{cases} 1 & \text{if } S\left(\dfrac{W_{1,i} + W_{2,i} + W_{3,i}}{3}\right) \geq rnd \\ 0 & \text{otherwise} \end{cases} \tag{2.31}$$

where rnd is random and uniform distributed number ∈ [0, 1]. In (2.31), the sigmoid function $S()$ is expressed by

$$S(x) = \frac{1}{1 + e^{-10(x-0.5)}} \tag{2.32}$$

The GWO was applied in [32] for dual-band E-shaped patch antenna design for RF energy harvesting applications.

2.1.8.3 Whale optimization algorithm

The WOA is a nature-inspired SI algorithm that is proposed in [25]. WOA basic concept is the social and the hunting behavior of humpback whales. In their natural environment, the whales after they identify the prey location, they perform specific movements to encircle them. In WOA terminology, the prey denotes the best solution vector obtained in each iteration. The population members (whales) are trying to come close to that best solution and they are updating their positions accordingly. This type of whale behavior (prey encirclement) is in WOA expressed mathematically by

$$D_k = |C_k \times x_{k,G}^{best} - x_{k,G}^{m}| \tag{2.33}$$

$$x_{k,G+1}^{i} = x_{k,G}^{best} - A_k D_k \tag{2.34}$$

where G denotes the current iteration, $x_{k,G}^{m}$ is the mth population member in the kth dimension, $x_{k,G}^{best}$ is the best solution found in the kth dimension, D_k denotes the distance vector of the current whale to the prey, and C_k, A_k are the kth dimension coefficient vectors. These latter coefficient vectors are derived as

$$A_k = 2a_k rnd_k - a_k \tag{2.35}$$

$$C_k = 2 \times rnd_k \tag{2.36}$$

where a_k denotes a number ∈ [2, 0] that is linearly decreased during the iteration process, and rnd_k is a uniformly distributed random number ∈ [0, 1].

Moreover, the exploitation phase of the optimization algorithm corresponds to the bubble-net behavior of humpback whales. This phase is modeled mathematically by WOA. WOA accomplishes this modeling using the combination of two different movement mechanisms: the encircling mechanism with a shrinking radius and the updating position mechanism with a spiral trajectory. The latter is expressed mathematically as

$$x^{m}_{k,G+1} = B_k e^{ps} \cos(2\pi s) + x^{best}_{k,G} \tag{2.37}$$

where B_k is the kth coordinate of the distance vector of the mth solution to the best solution, p is a constant number that defines the shape of the logarithmic spiral, and s is a uniformly distributed random number $\in [-1, 1]$.

Simultaneously, the whales make two types of movements; they swim toward the prey in a circle with a shrinking radius and along a spiral-shaped trajectory. This fact is modeled by authors in [25] using a 50% probability that can be written as

$$x^{m}_{k,G+1} = \begin{cases} x^{best}_{k,G} - A_k D_k, & \text{if } b < 0.5 \\ B_k e^{ps} \cos(2\pi s) + x^{best}_{k,G}, & \text{otherwise} \end{cases} \tag{2.38}$$

where b is a uniform random number $\in [0, 1]$. Additionally, the exploration phase of WOA corresponds to the humpback whales random search for prey. This can be modeled mathematically as

$$D_k = |C_k \times x^{r}_{k,G} - x^{i}_{k,G}| \tag{2.39}$$

$$x^{i}_{k,G+1} = x^{r}_{k,G} - A_k D_k \tag{2.40}$$

where r, with $r \neq i$, is a randomly selected member of the population that the ith member will follow.

The pseudocode of WOA is given in Algorithm 2. WOA was utilized for wearable Wi-Fi antenna design in [33].

2.1.8.4 Salp swarm algorithm

The SSA models mathematically swarming behavior of salps when navigating and foraging in oceans [26]. In SSA, each salp is a D-dimensional solution vector. Additionally, the salp (solution vector) that obtains the best objective function value found in every iteration represents the food position. In SSA, the salps behavior is modeled by partitioning the population into two groups. These are the leaders and the followers. The salps in front of the salp chain are called leaders. Moreover, the rest of the salps are called followers. This means that SSA names leaders of the salps (solution vectors) with the best objective function values, while the algorithm considers as followers are the solution vectors with the worse objective function values.

SSA divides the population into two groups of equal size $NP/2$, where NP denotes the population size.

The leaders' position follows the update rule given next:

$$x_{G+1,m,k} = \begin{cases} Food_k + C_1 \times \left[rand_{2k}\left(x_{k,U} - x_{k,L}\right) + x_{k,L}\right] & rand_{3k} < 0.5 \\ Food_k - C_1 \times \left[rand_{2k}\left(x_{k,U} - x_{k,L}\right) + x_{k,L}\right] & rand_{3k} \geq 0.5 \end{cases} \tag{2.41}$$

where $x_{G+1,m,k}$ denotes the position of the mth leader salp for $m < NP/2$ in the kth dimension for generation $G+1$, $Food_k$ represents the food source position (or the

Algorithm 2: WOA algorithm

```
Initialize randomly a population of whales (m = 1, 2, ..., NP)
Compute the objective function values for each population member x^m
Find the best solution vector x^best
while (i < G_max) do
    for (k = 1 : D) do
        Calculate a_k, rand_k, A_k, C_k
        for (m = 1 : NP) do
            if (b < 0.5) then
                if (|A_k| ≥ 1) then
                    Calculate distance vector D_k using (2.39)
                    Calculate position vector x^m_{k,i+1} using (2.40)
                else
                    Calculate distance vector D_k using (2.33)
                    Calculate position vector x^m_{k,i+1} using (2.34)
                end if
            else
                Calculate position vector x^m_{k,i+1} using (2.37)
            end if
        end for
    end for
    Set: i = i + 1
end while
```

best vector in the population) in the kth dimension, $x_{k,U}$ and $x_{k,L}$ are the upper and lower bounds in the kth dimension, respectively. The parameter C_1 is an important one for SSA defined by

$$C_1 = 2 \times e^{-(4G/G_{\max})^2} \tag{2.42}$$

where G is current iteration, and $G_{\max}$ is the maximum number of iterations. According to [26], the importance of the C_1 parameter lies on the fact that it balances exploration and exploitation. Additionally, $rand_{2k}, rand_{3k}$ are uniform, distributed random numbers within the interval [0, 1]. These random numbers help the algorithm to decide if the next position in kth dimension will be toward positive infinity or negative infinity as well as the step size. SSA defines a different-position update rule for the rest $NP/2$ vectors which is expressed as

$$x_{G+1,m,k} = \frac{\left(x_{G,m,k} + x_{G,m-1,k}\right)}{2} \tag{2.43}$$

Algorithm 3 describes briefly the SSA. SSA has been applied for solving the joint power allocation and user association problem in non-orthogonal multiple access networks [34]. Additionally, SSA was used for MIMO antenna design for 5G communication systems in [35].

Algorithm 3: SSA algorithm

```
Select the population size NP, and the maximum number of iterations G_max
Initialize a random population of NP salps
Compute the objective function value for all salps
Select the position of the salp with the best objective function value as the initial
  food position Food
i = 1
while i < G_max do
   Update C_1 parameter using (2.42)
   for m=1 to NP do
      if m <= NP/2 then
         Use (2.41) to update leaders positions
      else
         Use (2.43) to update followers positions
      end if
   end for
   Sort salps from best to worst
   Update food vector if better found
   i = i + 1
end while
```

2.2 Differential evolution

Differential evolution (DE) [36,37] is a very popular EA with very wide application. DE does not model any nature phenomenon but it is a mathematical construct. DE algorithms apply three operators to evolve the population in each generation. These are mutation, crossover, and selection. Several different DE variants can be found in the literature [37,38]. The distinction between these variants is in the mathematical formulation of the DE operators. One may select the most suitable DE strategy or variant to use based on the characteristics of a specific optimization problem; it is problem dependent [39]. The notation of the form *DE/x/y/z* means that for mutation we select vector(s) using x (possible values include "rand" for random vector, and "best" for the best vector), y is the number of vectors of type x involved in mutation, and z is the crossover type (usually "bin" for binary, or "exp," for exponential). Some of the most popular DE strategies include the *DE/rand/1/bin* and *DE/best/1/bin* strategies. Then, we can formulate the mth mutant vector $\bar{V}_{t,m}$ for each target vector $\bar{X}_{t,m}$ in the t iteration as

$$\begin{aligned}
&\textit{DE/rand/1/bin} \\
&\bar{V}_{t,m} = \bar{X}_{t,r_1} + F \times (\bar{X}_{t,r_2} - \bar{X}_{t,r_3}), r_1 \neq r_2 \neq r_3 \\
&\textit{DE/best/1/bin} \\
&\bar{V}_{t,m} = \bar{X}_{t,best} + F \times (\bar{X}_{t,r_1} - \bar{X}_{t,r_2}), r_1 \neq r_2
\end{aligned} \tag{2.44}$$

where r_1, r_2, and r_3 are randomly selected population members, and F is the mutation control parameter or mutation scale factor. Then, we use the crossover operator to generate a trial vector $\bar{U}_{t,m}$ with components $U_{t,km}$ written as

$$U_{t,km} = \begin{cases} V_{t,km}, & \text{if } rand_{k[0,1)} \leq CR \text{ or } k = rn(m) \\ X_{t,km}, & \text{if } rand_{k[0,1)} > CR \text{ and } k \neq rn(m) \end{cases} \tag{2.45}$$

where $k = 1, 2, \ldots, D$, $rand_{k[0,1)}$ is a uniformly distributed random number in the interval [0, 1), $rn(m)$ is a randomly chosen index from $(1, 2, \ldots, D)$, and CR is another DE control parameter, the crossover constant from the interval [0, 1]. Moreover, the DE algorithms use a greedy selection operator, which for minimization problems is expressed as

$$\bar{X}_{t+1,m} = \begin{cases} \bar{U}_{t,m}, & \text{if } f(\bar{U}_{t,m}) < f(\bar{X}_{t,m}) \\ \bar{X}_{t,m}, & \text{otherwise} \end{cases} \tag{2.46}$$

where $f(\bar{U}_{t,m}), f(\bar{X}_{t,m})$ are the objective function values of the trial and the old vector, respectively. Therefore, the new trial vector $\bar{U}_{t,m}$ replaces the old vector $\bar{X}_{t,m}$ if it is better than the previous one. Otherwise, the old vector remains in the population.

2.2.1 Self-adaptive DE algorithms

DE requires the setting of the mutation and the crossover control parameters F, and CR. Thus, it is a DE advantage that only two control parameters are required. However, the fine-tuning of these control parameters may require a time-consuming trial-and-error procedure. Taking that into consideration, there are several adaptive and self-adaptive DE variants proposed in the literature.

A DE strategy (named jDE by the authors) that self-adapts the F, CR parameters is proposed in [40]. The jDE algorithm has been utilized in the electromagnetics domain solving problems like the microwave absorber design problem [41] and the linear array synthesis [42,43].

Another DE algorithm that self-adapts both control parameters and strategy is self-adaptive differential evolution (SaDE) [44–46]. SaDE was utilized for microwave filter design [47], and for linear arrays synthesis [48].

Moreover, the composite DE (CoDE) [49] is another adaptive DE variant. CoDE uses mainly three different trial vector generation strategies. Each one of these strategies selects the control parameters F, and CR using a specified control parameter pool. CoDE presents advantages, such as it is based on a simple structure and can be easily expanded with additional strategies. CoDE is in general easy to be implemented in any programming language. CoDE obtained a good performance for optimal neural network design in [50]. Additionally, a CoDE variant, CoDE-EIG was proposed in [51] and applied to shaped beam array synthesis.

Moreover, in [52], the authors introduce the barebones DE. This variant uses only one control parameter and it is a hybrid of DE and the barebones particle swarm optimizer [53]. Additionally, Wang *et al.* [54] proposed two new barebones DE variants, the Gaussian barebones DE (GBDE) and the modified GBDE (MGBDE).

In [55], the authors apply different self-adaptive DE algorithms for waveform design for wireless power transfer.

2.2.1.1 jDE algorithm

It is suggested in [37] to set the DE control parameters F and CR from the intervals $[0.5, 1]$ and $[0.8, 1]$, respectively. The authors in [40] propose a self-adaptive strategy, which they call jDE. The jDE variant uses the same mutation operator as the *DE/rand/1/bin* scheme. The jDE concept is based on the idea to have two additional unknowns in each vector. These are its own F and CR values. jDE evolves these control parameters. The jDE strategy obtains new vectors with the improved values of the control parameters. These vectors are considered to be more likely to survive and produce offspring in the next generation. Therefore, the better new vectors keep the improved values of the control parameters to the next generation. The control parameters are self-adjusted in every generation for each vector of the population with the formulas:

$$
\begin{aligned}
F_{t+1,m} &= \begin{cases} F_l + rnd_{1[0,1]} \times F_u & \text{if } rnd_{2[0,1]} < prob_1 \\ F_{t,m}, & \text{otherwise} \end{cases} \\
CR_{t+1,m} &= \begin{cases} rnd_{3[0,1]} & \text{if } rnd_{4[0,1]} < prob_2 \\ CR_{t,m}, & \text{otherwise} \end{cases}
\end{aligned}
\tag{2.47}
$$

where $rnd_{k[0,1]}, k = 1, 2, 3, 4$ represent uniform random numbers $\in [0, 1]$, F_l and F_u are the lower and the upper limits of F set to 0.1 and 0.9, respectively, and $prob_1$ and $prob_2$ the probabilities of adjusting the control parameters. In the jDE original paper, the authors set both these probabilities to 0.1 after several trials. Additionally, the conclusion from the jDE strategy performance is that it is better or at least comparable to the classical DE *DE/rand/1/bin* strategy [40]. Furthermore, the use of jDE does not increase the time complexity. The interested reader may look for more details about the jDE strategy in [40].

2.2.1.2 Barebones DE

Omran *et al.* in [52] introduced a DE variant named barebones DE (BBDE). BBDE is based on the idea that the trial vector generation is formulated as

$$
U_{t,km} = \begin{cases} P_{t,km} + rand_{2k[0,1]} \times \left(X_{t,kr_1} - X_{t,kr_2}\right), \text{if } rand_{k(0,1)} > CR \\ X_{t,kr_3}, \text{ otherwise} \end{cases} \tag{2.48}
$$

where r_1, r_2, r_3 , $r_1 \neq r_2 \neq r_3$ are randomly chosen indices from the population, and $rand_{k[0,1]}$ are uniform random numbers $\in [0, 1]$, for the kth dimension, and $P_{t,km}$ is given by

$$
P_{t,km} = X_{t,km} \times rand_{1k[0,1]} + X_{t,kbest}(1 - rand_{1k[0,1]}) \tag{2.49}
$$

where $rand_{1k[0,1]}$ is a random value within [0, 1] for the kth dimension. One may observe that BBDE requires only one control parameter the CR. Furthermore, Wang *et al.* [54] proposed two new BBDE variants the GBDE and a modification of it named MGBDE. The basic concept of GBDE and MGBDE is to use Gaussian sampling for mutant vector generation and for dynamically updating the *CR* parameter.

2.2.1.3 Composite DE

CoDE uses three different strategies to generate the trial vector [49]. These are *DE/rand/1/bin*, *DE/rand-to-best/2/bin*, and *DE/current-to-rand/1/bin*. Thus, a mutant vector $\bar{v}_{t+1,m}$ for each target vector $\bar{x}_{t,m}$ is derived by

$$
\begin{aligned}
&DE/rand/1/bin\\
&\bar{v}_{t+1,m} = \bar{x}_{t,r_1} + F(\bar{x}_{t,r_2} - \bar{x}_{t,r_3}),\ r_1 \neq r_2 \neq r_3\\
&DE/rand/2/bin\\
&\bar{v}_{t+1,m} = \bar{x}_{t,r_1} + F(\bar{x}_{t,r_2} - \bar{x}_{t,r_3}) + F(\bar{x}_{t,r_4} - \bar{x}_{G,r_5}),\\
&\quad r_1 \neq r_2 \neq r_3 \neq r_4 \neq r_5\\
&DE/current - to - rand/1\\
&\bar{v}_{t+1,m} = \bar{x}_{t,i} + rand_{[0,1)}(\bar{x}_{t,r_1} - \bar{x}_{t,i}) + F(\bar{x}_{t,r_2} - \bar{x}_{t,r_3}),\ r_1 \neq r_2 \neq r_3
\end{aligned}
\tag{2.50}
$$

where r_1, r_2, r_3, r_4, r_5 are randomly chosen indices from the population, different than index m, and $rand_{[0,1)}$ is a randomly generated number from a uniform distribution within the interval [0, 1). CoDE generates a trial vector for each of the abovementioned strategies. The control parameters for each strategy are randomly chosen from a parameter candidate pool with fixed values. This parameter candidate pool is given by the settings:

$$
\begin{aligned}
&(F = 1.0,\ CR = 0.1)\\
&(F = 1.0,\ CR = 0.9)\\
&(F = 0.8,\ CR = 0.2)
\end{aligned}
\tag{2.51}
$$

The strategies used by CoDE are those that have been widely used and therefore their properties have been extensively studied. Moreover, in [49] the authors have modified the *DE/rand/2/bin* strategy by using a random number from the interval [0, 1). CoDE selects the best trial vector out of the three generated. This trial vector is then compared with the old vector and if its objective function value is better, it replaces it.

After mutation in each generation for every vector of the population, CoDE applies the crossover operator to generate a trial vector using (2.45). One additional feature in CoDE is the fact that the crossover operator is not applied for the *DE/current-to-rand/1/bin* strategy. As the original DE CoDE uses a greedy selection operator using (2.46).

2.2.1.4 CoDE with eigenvector-based crossover operator (CoDE-EIG)

The DE-EIG variant is proposed in [56], which uses an eigenvector-based crossover operator, which utilizes eigenvectors of covariance matrix of individual solutions. Therefore, this crossover is rotationally invariant. In order to keep the diversity of the population high, the new trial vector can be stochastically born from the parents with either the standard coordinate system or the rotated coordinate system.

Furthermore, in [56] a new parameter to control the probability of selecting one of the coordinate systems is introduced. This scheme has the potential to increase the population diversity and prevent premature convergence. Moreover, this eigenvector-based crossover operator presents a significant advantage, such as it can be applied to any crossover strategy with minimal changes. Thus, this operator may enhance any existing DE variant.

The author in [51] based on that idea proposes a new DE variant of the combination of the CoDE with eigenvector-based crossover operator called CoDE-EIG. The fundamental idea of CoDE-EIG is to exchange the information between the target and the mutant vectors in the eigenvector basis instead of the natural basis. The covariance between nth and mth dimension of the population in the tth iteration is expressed as [56].

$$\mathrm{cov}\,(n,m) = \frac{\sum_{k=1}^{NP} \left(x_{t,nk} - \mu_{t,n}\right)\left(x_{t,mk} - \mu_{t,m}\right)}{NP - 1} \tag{2.52}$$

where $\mu_{t,n}$,$\mu_{t,m}$ are the variables' mean values in the nth and mth dimension, respectively. In order to calculate the eigenvector basis, one has to factorize the covariance matrix $\mathbf{C}_t = (c_{nm}, c_{nm} = \mathrm{cov}\,(n,m))$ into a canonical form

$$\mathbf{C}_t = \mathbf{Q}_t \Lambda_t \mathbf{Q}_t^{-1} \tag{2.53}$$

where $\mathbf{Q}_t$ is the square matrix $(D \times D)$ whose nth column is the eigenvector $\bar{q}_{t,n}$ of $\mathbf{C}_t$, and Λ_t is the diagonal matrix, diagonal elements of which are the corresponding eigenvalues.

We call eigen decomposition the factorization of a matrix into a canonical. In [56], they utilize Jacobi's method [57] for eigen decomposition. If the eigenvector basis is found, the nth target vectors $\bar{x}_{t,n}$ can be expressed by $\left(\mathbf{Q}_t \times \bar{x}_{t,n}\right)$; the nth mutant vectors $\bar{v}_{t+1,n}$ can be expressed by $\left(\mathbf{Q}_t \times \bar{v}_{t+1,n}\right)$. Then, a binomial crossover operator exchanges some of the elements of the mutant vector with some of the elements of its target vector. Thus, a trial vector is created. The trial vector is formulated mathematically as [56]

$$\bar{u}_{t+1,n} = \begin{cases} \mathbf{Q}_t^* \, xover\left(\mathbf{Q}_t \bar{x}_{t,n}, \mathbf{Q}_t \bar{v}_{t+1,n}\right) & \text{if } rand_{[0,1]} \leq P \\ xover\left(\bar{x}_{t,n}, \bar{v}_{t+1,n}\right) & \text{otherwise} \end{cases} \tag{2.54}$$

where $\mathbf{Q}_t^*$ is the conjugate transpose of the eigenvector basis $\mathbf{Q}_t$, and $xover\,(\bar{x}, \bar{y})$ is a crossover operator on two vectors $\bar{x}$ and $\bar{y}$. Furthermore, P is a control parameter (eigenvector ratio) introduced in [56]. The value of P varies between 0 and 1 and determines the ratio of the eigenvector-based crossover operator and the other crossover operator. The basic characteristic of this approach is that in spite of everything the crossover operator exchanges the elements in the eigenvector basis. Thus, the crossover behavior will become rotationally invariant in the natural basis.

Thus, we generate the CoDE-EIG variant using (2.50) for mutant vector generation and (2.54) for trial vector creation. Moreover, CoDE-EIG requires the setting of the eigenvector ratio P. One may observe that when $P = 1$ then only the eigenvector crossover operator is used, while when $P = 0$ then only the binary crossover operator is used. The CoDE-EIG algorithm is briefly presented in Algorithm 4.

Algorithm 4: CoDE-EIG algorithm

Select the population size NP, the maximum number of iterations $G_{\max}$, and the eigenvector ratio P.
Initialize a uniformly distributed random population of NP vectors.
Compute the objective function value of all population members.
$i = 1$
while $i < G_{\max}$ **do**
  for n=1 to NP **do**
    for s=1 to 3 **do**
      Select randomly a control parameter setting from the parameter candidate pool.
      Create the s mutant vector $\bar{v}_{i+1,n\_s}$ using one of the three strategies given in (2.50)
      Generate the s trial vector $\bar{u}_{i+1,n\_s}$ using the eigenvector-based crossover operator given in (2.54)
      Compute the objective function value for the trial vector s
    end for
    Find the best out of the three trial vectors.
    Apply the greedy selection operator according to (2.46).
    Update best vector if better vector found.
  end for
  $i = i + 1$
end while

2.2.1.5 The SaDE algorithm

The SaDE [58] algorithm self-adapts not only the control parameters but also the trial vector generation strategies. The self-adapting scheme is based on previous learning experiences. In SaDE, there is the option to use one of four commonly used strategies. SaDE assigns to every strategy is a probability of selection. The initial values of these probabilities are 0.25 and they are gradually adjusted from generation to generation. SaDE defines the probability of applying the nth strategy as $p_n, n = 1, 2, \ldots, N$, where N is the total number of strategies.

SaDE stores the number of successful trial vectors generated by the nth strategy in each generation and updates a variable which is called $ks_{n,t}$. Furthermore, SaDE also keeps the number of trial vectors that fail to replace the old vectors in the next generation, $kf_{n,t}$. Moreover, SaDE uses another control parameter the learning period (LP). The LP refers to the number of the previous generations that store the success and fail numbers. Thus, every LP iterations, the algorithm updates the strategy selection probabilities with the following formula:

$$p_{n,t} = \frac{S_{n,t}}{\sum_{n=1}^{N} S_{n,t}} \tag{2.55}$$

where $S_{n,t}$ is the success rate of the trial vectors generated by the *n*th strategy within the previous *LP* generation and is expressed as

$$S_{n,t} = \frac{\sum_{g=t-LP}^{t-1} ks_{n,g}}{\sum_{g=t-LP}^{t-1} ks_{n,g} + \sum_{t=t-LP}^{t-1} kf_{n,g}} + \varepsilon \tag{2.56}$$

where ε is a constant set equal to 0.01 to avoid possible null success rates. Therefore, higher probability is assigned to the strategies with high success rates. Additional information about the SaDE algorithm is found in [58].

2.2.1.6 The JADE algorithm

JADE [59] uses the *DE/current-to-pbest/1/bin* strategy. This strategy is combined with an optional external archive. The role of the optional archive is to utilize the previous data so that the population diversity is maintained. The previous strategy diversifies the population and improves convergence. The JADE uses a normal and a Cauchy distribution to create the new crossover constant and mutation control parameter for each target vector. JADE uses the recent successful control parameters to generate the new ones. The JADE strategy uses not only the best vector of the current population but also additional good vectors. Furthermore, the mutation operation may use also vectors from the external archive in order to generate new trial vectors.

2.2.2 Novel binary differential evolution

Novel binary DE (NBDE) is a binary DE variant that can be used for discrete-valued problems [60]. NBDE was applied to discrete-valued antenna design problems in [61]. NBDE has the same operators as those in the original DE, i.e., the mutation operator, the selection operator, and the crossover operator. NBDE uses binary coding for each vector of the population. The mutation operator in (2.45) generates real vectors so the NBDE introduces a new probability estimator operator into which the mutant operator is integrated.

This probability estimator operator was inspired by the concept of population-based incremental learning (PBIL) used in [62]. However, there is difference between NBDE and PBIL. This is the fact that the NBDE constructs the multiple probability models at each iteration according to the information extracted by the parents using the mutant operator. The probability estimator operator is mathematically formulated as

$$\begin{aligned} P(x_{t+1,km}) &= \frac{1}{1+e^{(-2b(MO-0.5))/(1+2F)}} \\ MO &= x_{t,k,r_1} + F(x_{t,k,r_2} - x_{t,k,r_3}), r_1 \neq r_2 \neq r_3 \end{aligned} \tag{2.57}$$

where b is a positive constant named bandwidth (BW) factor, and x_{t,k,r_1}, x_{t,k,r_2} are the kth bits of randomly chosen indices from the population.

NBDE's main feature is that it maintains the mutant operator of the original DE algorithm, and this is denoted as *MO*. *MO* is embedded into the probability estimation operator. This probability estimation operator uses the differential information of three random parent vectors to construct the probability distribution model of the mutant vector for the kth bit to be "1." Additionally, NBDE uses an additional control parameter the BW factor b. The role of the BW factor is to tune the range and shape of the probability distribution model. Thus, if we select an appropriate value for b, then the algorithm can simultaneously maintain population diversity and improve the search efficiency. Thus, NBDE generates the kth bit of the mutant vector using the probability estimation operator given by

$$v_{t+1,km} = \begin{cases} 1, & \text{if } rand_{j[0,1)} \leq P(x_{t+1,km}) \\ 0, & \text{otherwise} \end{cases} \tag{2.58}$$

Figure 2.3 depicts the mutant vector creation. NBDE uses the same crossover operator as the original DE defined previously in (2.45) for trial vector generation. One may

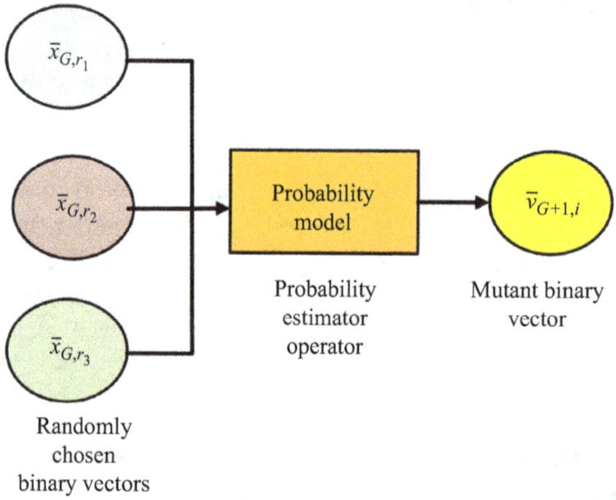

Figure 2.3 Mutant vector creation in NBDE

Algorithm 5: NBDE algorithm

1: Set the values for the NBDE control parameters. These are the mutation control parameter MO, the crossover constant CR, and the BW factor b.
2: Set the population size NP, the maximum number of iterations $MAXG$, and the eigenvector ratio P
3: Initialize a random population of NP vectors
4: Calculate the objective function value of all population vectors
5: $G = 1$
6: **while** $G < MAXG$ **do**
7:     Apply the binary mutation operator according to (2.57) and (2.58) to generate a mutant vector $\bar{v}_{t+1,i}$.
8:     Generate a trial vector $\bar{u}_{t+1,i}$ using the binary crossover operator using (2.45).
9:     Evaluate objective function value for the trial vector.
10:     Apply the greedy selection operator according to (2.46).
11:     Update best vector if better vector found.
12:     $G = G + 1$
13: **end while**

---

observe from (2.45) that at least one bit of the trial vector is inherited from the mutant vector. NBDE uses also the same selection operator as the original DE defined in (2.46). Thus, NBDE requires the setting of an additional control parameter, the BW factor b, compared with the original real-valued DE algorithm. The NBDE algorithm is briefly presented in Algorithm 5. More details about the NBDE strategy can be found in [60].

2.3 Biogeography-based optimization

Biogeography-based optimization (BBO) [63] is based on the mathematical models of biogeography for predicting animal migration patterns. These patterns help one to make an estimation of the number of species in a habitat. BBO defines a specific terminology, a geographically isolated area is called a habitat. In BBO, a habitat suitability index (HSI) value is assigned to each habitat. Additionally, in BBO models the habitat features are called suitability index variables (*SIV*). Thus, the decision variables in a optimization problem are the *SIV*s, while *HSI* represents the objective function value.

Thus, in BBO we represent every possible solution to a D-dimensional problem as a vector of SIV variables $[SIV_1, SIV_2, \ldots, SIV_D]$. This vector models the habitat or island. The *HSI* value of a habitat or vector denotes the value of the objective function that corresponds to that solution and it is derived by

$$HSI = F\ (habitat) = F(SIV_1, SIV_2, \ldots, SIV_D) \tag{2.59}$$

As a consequence, vectors that have high *HSI* values are the good solutions of the objective function. The emigration rate ψ of these solutions is high because they have a large number of species, while the immigration rate ξ is low. On the other hand, the poor solution vectors have low *HSI* value (low objective function values), i.e., they have low emigration ψ and high immigration ξ rates, respectively.

The immigration rate ξ and emigration rate ψ, for the original BBO linear model, are formulated as [63]

$$\psi_k = E\left(\frac{k}{Q_{\max}}\right) \tag{2.60}$$

$$\xi_k = I\left(1 - \frac{k}{Q_{\max}}\right) \tag{2.61}$$

where E is the maximum possible emigration rate, k is the species number of the kth habitat, $Q_{\max}$ is the maximum number of species, and I is the maximum possible immigration rate. One may observe from (2.60) and (2.61) that the immigration rate ξ and emigration rate ψ are linear functions of the number of species k in the habitat. This is illustrated in Figure 2.4 graphically. Hence, if the number of species in a habitat increases then the emigration rate linearly increases and the immigration rate linearly decreases, i.e., the more crowed the habitat the fewer species can be successful

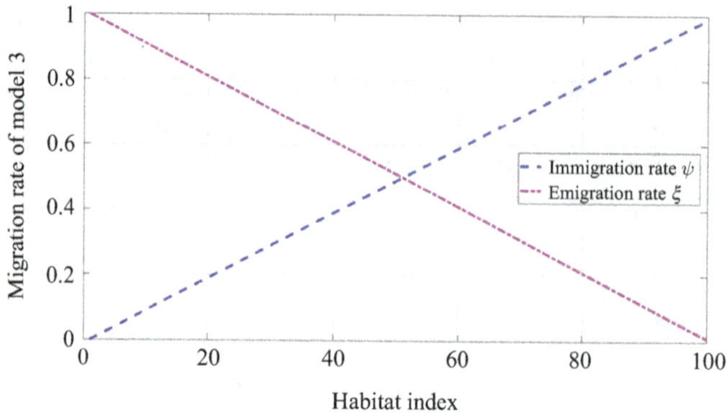

Figure 2.4 Migration model 3: habitat index versus migration rates for population size NP = 100

immigrants to that habitat. At the same time, more species are more probable to leave the habitat to other habitats. However, the linear models represent a simple approximation of complicated phenomena in nature-like migration [64].

This fact requires the creation of nonlinear migration models in order to model more accurately the migration phenomenon. In [65], two new nonlinear migration models are proposed. Furthermore, the authors in [65] present a theoretical analysis of BBO and they define the quantity

$$C = \frac{\sum_{n=1}^{KP} \sum_{m=1}^{n} \psi_m}{NP \sum_{m=1}^{KP} \psi_m}. \tag{2.62}$$

where KP is the population size. In accordance with [65], the larger the C the better the BBO variant performance. Thus, new nonlinear models are created on the basis of the previous principle.

One may notice that immigration rate ξ is the same in both models and is a sinusoidal functions of the number of species k, thus resulting in a bell-like shape. The natural phenomena in a habitat are described more accurately by this type of nonlinear model. The emigration rate ψ is different in both models, and it is a convex function of the number of species k (Figure 2.5).

Model 6:

$$\psi_k = \frac{E}{2}\left(-\cos\left(\frac{k\pi}{Q_{\max}}\right) + 1\right), \ \xi_k = \frac{I}{2}\left(\cos\left(\frac{k\pi}{Q_{\max}}\right) + 1\right) \tag{2.63}$$

The nonlinear models, called model 7 (BBO7) and model 8 (BBO8), the emigration rate ψ, and the immigration rate ξ are defined by [65]

Model 7

$$\psi_k = E\left(\frac{k}{Q_{\max}}\right)^4, \ \xi_k = \frac{I}{2}\left(\cos\left(\frac{k\pi}{Q_{\max}}\right) + 1\right) \tag{2.64}$$

Model 8

$$\psi_k = E\left(\frac{k}{Q_{\max}}\right)^{16}, \ \xi_k = \frac{I}{2}\left(\cos\left(\frac{k\pi}{Q_{\max}}\right) + 1\right) \tag{2.65}$$

Figures 2.6 and 2.7 show the correlation between habitat index and migration rates for models 7 and 8, respectively. The migration for K habitats is presented in Algorithm 6.

In Algorithm 6, X_n and X_m denote the habitats n and m, respectively. Furthermore, other than the migration operator, BBO uses also a mutation operator. A mutation rate

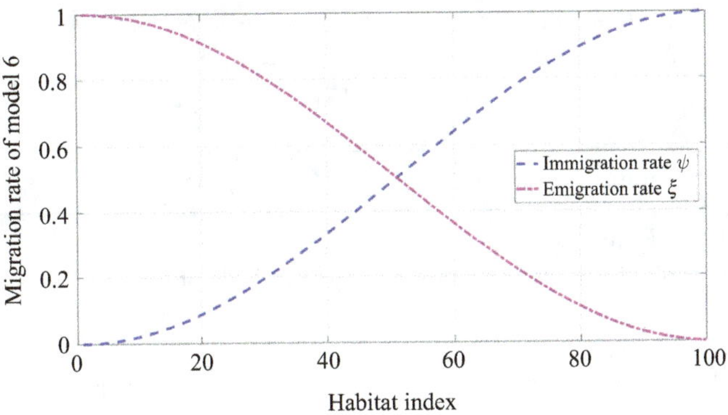

Figure 2.5 Migration model 6: habitat index versus migration rates for population size NP = 100

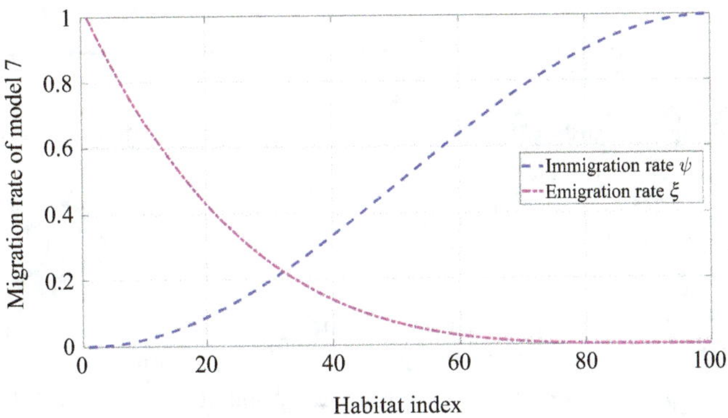

Figure 2.6 Migration model 7: habitat index versus migration rates for population size NP = 100

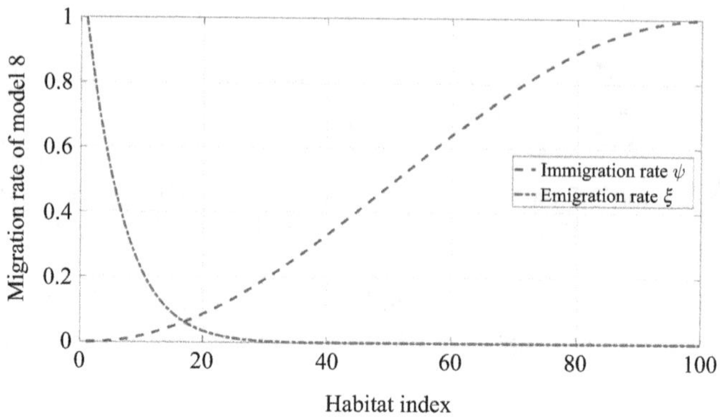

Figure 2.7 Migration model 8: habitat index versus migration rates for population size NP = 100

Algorithm 6: Habitat migration

---

for k=1 to KP **do**
  Using probability that depends on ξ_n select X_k habitat
  if $rand\,(0,1) < \xi_n$ **then**
    for m=1 to KP **do**
      Using probability that depends on ψ_m select X_m Habitat
      if $rand\,(0,1) < \psi_m$ **then**
        Pick a SIV σ randomly from X_m
        Pick a SIV in X_k Habitat randomly to replace it with σ
      end if
    end for
  end if
end for

---

m is defined for a possible solution Q. This mutation rate is a function of the solution probability, which is modeled mathematically as [63]

$$m\,(Q) = m_{\max}\left(\frac{1-P_q}{P_{\max}}\right) \tag{2.66}$$

where P_q is the probability that habitat contains Q species, $P_{\max}$ is the maximum P_q value over all $q \in [1, Q_{\max}]$, and $m_{\max}$ is a user-defined parameter.

The mutation operator goal is to increase the population diversity. If the mutation operator was not integrated into BBO, then the algorithm would favor the most probable solutions. Thus, these solutions would become predominant among the other population habitats, and the population diversity would have been reduced. The basic idea in this mutation operator is that it favors the low HSI (low fitness) solutions to

Algorithm 7: Habitat mutation

for m=1 to KP **do**
  Calculate the probability P_m
  Choose the SIV X_p^m using probability that depends on P_m
  if $random(0, 1) < p_m$ **then**
    Generate randomly a SIV and then replace X_p^m
  end if
end for

mutate, thus they could be improved. Furthermore, the good high HSI solutions are also probable to mutate and to further improve. The mutation procedure is presented in Algorithm 7 where X_p^m is the pth SIV of the mth habitat. The concept of elitism is also used in BBO. This requires the setting of an additional control parameter that represents the number of elites to survive in each generation.

2.3.1 Chaotic BBO

Chaotic BBO (CBBO) [66] is a BBO variant that utilizes random variables generated by chaotic maps instead of uniform distribution. As it is reported in [66], the authors have evaluated different CBBO algorithms with ten different chaotic maps. In this case, the main conclusion was that CBBO with Gauss/mouse chaotic map performed better than the other CBBO variants. This could be due to the fact that random values returned by this chaotic map are mostly less than 0.5.

Thus, this derives to a higher emigration probability than one in original BBO and the other CBBO variants. Hence, CBBO with Gauss/mouse chaotic map favors exploration of the search space, without preventing exploitation. For this reason, in next chapter we use the CBBO Gauss/mouse chaotic map. This chaotic map is expressed by

$$x_{i+1} = \begin{cases} 1, & \text{when } x_i = 0 \\ \dfrac{1}{\bmod (x_1, 1)}, & \text{otherwise.} \end{cases} \tag{2.67}$$

The final value from the chaotic map is within the interval [0, 1]. A sample chaotic map for 400 iterations is depicted in Figure 2.8.
The CBBO habitat migration is presented in Algorithm 8, where $c(t)$ is the value obtained by the chaotic map in the tth iteration.

We can define the following types of CBBO algorithms based on the use of chaotic operators or not.

- CBBO0 which is BBO algorithm with chaotic selection operator.
- CBBO1 which is BBO algorithm with chaotic migration operator.
- CBBO2 which is BBO algorithm with chaotic mutation operator.
- CBBO3 which is BBO algorithm with selection/migration operators combined.
- CBBO4 which is BBO algorithm selection/migration/mutation operators combined.

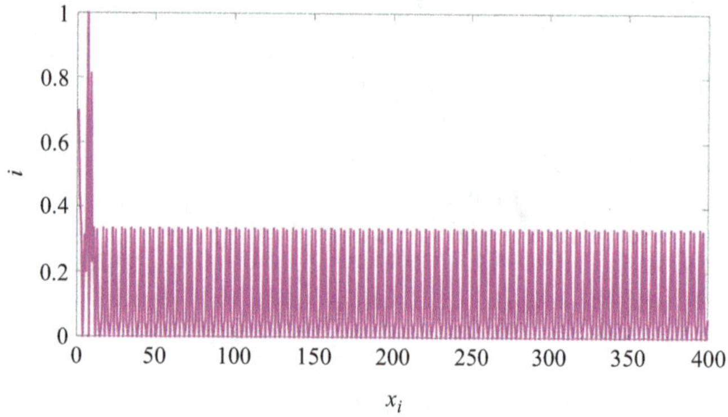

Figure 2.8 Gauss/mouse chaotic map

Algorithm 8: CBBO habitat migration

---

for s=1 to KP **do**
  Pick X_s Habitat using probability that depends on ξ_s
  if $c\,(t) < \xi_s$ **then**
    for k=1 to KP **do**
      Pick X_k Habitat using probability that depends on ψ_k
      if $random(0, 1) < \psi_k$ **then**
        Pick a SIV σ from X_k randomly
        Pick a SIV in X_p Habitat randomly to replace it with σ
      end if
    end for
  end if
end for

2.4 Emerging evolutionary algorithms

In this section, we present some of the emerging nature-inspired algorithms. These can be grouped into categories depending on the source of inspiration they model. Biology-based emerging algorithms include the firefly algorithm (FA) [67], the monarch butterfly optimization (MBO) [68], the moth search algorithm (MSA) [69], the elephant herding optimization (EHO) [70], and the shuffled frog-leaping algorithm (SFLA) [71].

Physics-based algorithms are also emerging like the gravitational search algorithm (GSA) [72], the hybrid PSOGSA [73], and the wind-driven optimization (WDO) [74]. Additionally, there are algorithms that model human social behavior like the teaching–learning-based optimization (TLBO) [75] and the hybrid TLBO–Jaya [76]. Furthermore, the Jaya algorithm [77] is simple mathematical model that does not model a natural phenomenon. Moreover, the harmony search (HS) [78] is music-based algorithm.

2.4.1 Biology-based algorithms

2.4.1.1 Firefly algorithm

The FA models the fireflies' behavior [67]. FA assigns a value of brightness to each firefly (solution vector) of the population, depending on its objective function value and the distance between two fireflies. Each firefly is attracted toward the fireflies that are more bright, i.e., the attraction is proportional to their brightness.

FA was originally created for real-valued problems. However, binary FA algorithms have been proposed in the literature. In [79,80], a tanh function is used to map real numbers within the interval [0, 1]. This is used to map the firefly's position x_p to a binary value as

$$F(x_p) = \tanh(x_p) = \frac{e^{2x_p} - 1}{e^{2x_p} + 1} \tag{2.68}$$

Then a binary value of "0" or "1" is obtained using

$$x_p = \begin{cases} 1 & \text{if } rnd < F(x_p) \\ 0 & \text{otherwise} \end{cases} \tag{2.69}$$

where *rnd* denotes a uniformly distributed random number in [0, 1].

2.4.1.2 Monarch butterfly optimization

MBO models the migration of monarch butterflies that live in North America [68]. In [81], the authors use MBO for patch antenna design for C2C communication systems, in [82] MBO is applied to the cognitive engine design problem. Moreover, MBO is also used in several occasions for different problems in wireless communications like localization in wireless sensor networks [83].

Additionally, another MBO version is introduced in [84]. The main characteristic of this MBO version is that it uses a greedy selection strategy and has an additional operator the self-adaptive crossover operator.

The migration behavior of the monarch butterflies was the source of inspiration of MBO. MBO splits the whole population in two subpopulations, which are called lands in MBO terminology. The problem unknowns are represented by the positions of the monarch butterflies. MBO uses different operators for position update. These operators update the vector positions successively.

At first the migration operator is applied. This operator generates child vectors that can be adjusted by the migration ratio. Then, the butterfly adjusting operator is applied. This operator tunes the butterflies' positions. Hence, these operators guide the exploration of the search space. Moreover, MBO can be easily implemented in parallel according to the authors claim in [68]. In this case, these two operators can also be applied at the same time.

The migration behavior of monarch butterflies follows the following rules:

1. The two subpopulations are located in Land 1 and Land 2.
2. The new offspring (butterflies) are generated by applying the migration operator in either Land 1 or Land 2.
3. MBO keeps the best individuals to the next generation.

The ratio r_{pop} of monarch butterflies in subpopulation 1 (Land 1) is an MBO control parameter of MBO. Hence, if the total population size consists of N vectors then the population size of Land 1 is $N_{pop1} = \text{ceil}(r_{pop} \times N)$ where the ceil() operator denotes the rounding to nearest integer greater or equal.

Correspondingly, the population size of subpopulation 2 (Land 2) is $N_{pop2} = N - N_{pop1}$. For a D-dimensional optimization problem, we define the vector of butterfly positions as $\bar{b} = (b_1, b_2, \ldots, b_D)$.

During the migration process, MBO generates for each individual a random number using

$$R = rnd_{(0,1)} \times Per \tag{2.70}$$

where $rnd_{(0,1)}$ is a uniformly distributed number within the interval (0,1), and Per is the migration period. The authors in [68] suggest that the migration period should be set to 1.2. The migration process is then expressed as [68]

$$\begin{aligned} b_{i,j}^{t+1} &= b_{r1,j}^{t} \text{ if } R \le r_{pop} \\ b_{i,j}^{t+1} &= b_{r2,j}^{t} \text{ otherwise} \end{aligned} \tag{2.71}$$

where $b_{i,j}^{t+1}$ is the value of the ith butterfly position at the jth dimension at the next generation, $b_{r1,j}^{t}$ is the value of newly generated position at the jth dimension of a randomly selected butterfly $r1$ from subpopulation 1, and $b_{r2,j}^{t}$ is the value of newly generated position at the jth dimension of a randomly selected butterfly $r2$ from subpopulation 2, respectively.

Additionally, the butterfly adjusting operator of MBO is given by [68]

$$\begin{aligned} b_{i,j}^{t+1} &= b_{best,j}^{t} \quad \text{if } rand_{(0,1)} \le r_{pop} \\ b_{i,j}^{t+1} &= b_{r3,j}^{t} \quad \text{otherwise} \end{aligned} \tag{2.72}$$

where $b_{best,j}^{t}$ is the value of the best so far individual at the jth dimension of the whole population, and $b_{r3,j}^{t}$ is the value of the position at the jth dimension of a randomly selected butterfly $r3$ from subpopulation 2.

Additionally, we use the formulation given next to modify the butterflies position [68]

$$b_{i,j}^{t+1} = b_{i,j}^{t} + a\left(db_j - 0.5\right) \quad \text{if } R_{(0,1)} > BAR \tag{2.73}$$

where BAR is the butterfly adjusting rate, db_j is the walk step of the individual i that can be calculated using Lévy flight, and a is a weighing factor calculated as

$$a = \frac{W_{\max}}{t^2} \tag{2.74}$$

where $W_{\max}$ denotes the maximum walk step of a monarch butterfly in a single move.

2.4.1.3 Greedy strategy and self-adaptive crossover MBO (GCMBO)

GCMBO, which is an MBO algorithm enhanced with greedy strategy and self-adaptive crossover operator, was introduced in [84]. GCMBO changes the original MBO operators like the migration and the butterfly adjusting operator. This is accomplished as follows. A greedy strategy is applied during the migration process, which for minimization problems can be defined as [84]

$$\bar{b}_{t+1,i} = \begin{cases} \bar{b}_{t+1,new}, \text{if} F(\bar{b}_{t+1,new}) < F(\bar{b}_{t,i}) \\ \bar{b}_{t,i}, \text{otherwise} \end{cases} \tag{2.75}$$

where $F(\bar{b}_{t+1,new})$ and $F(\bar{b}_{t,i})$ are the objective function values of a new child vector and the old one for the ith butterfly, respectively.

Moreover, GCMBO has a crossover operator in the butterfly adjusting phase. According to [84], the crossover operator modifies the ith butterfly by

$$\overline{b^{CR}}_{t+1,i} = \bar{b}_{t+1,i} \times (1 - CR) + \bar{b}_{t,i} \times CR \tag{2.76}$$

where $\overline{b^{CR}}_{t+1,i}$ is the new generated butterfly with crossover operator and CR is a self-adapting crossover parameter defined by

$$CR = 0.8 + 0.2 \times \frac{F(\bar{b}_{t,i}) - F\left(\bar{b}_{t,best}\right)}{F\left(\bar{b}_{t,worst}\right) - F\left(\bar{b}_{t,best}\right)} \tag{2.77}$$

where $F\left(\bar{b}_{t,best}\right)$ and $F\left(\bar{b}_{t,worst}\right)$ represent the objective function values of the best and worst individual, respectively.

Moreover, after the crossover generated offspring GCMBO uses a greedy selection strategy expressed as

$$\bar{b}_{t+1,i} = \begin{cases} \overline{b^{CR}}_{t+1,i}, & \text{if } F(\overline{b^{CR}}_{t+1,i}) < F(\bar{b}_{t,+1i}) \\ \bar{b}_{t+1,i}, & \text{otherwise} \end{cases} \tag{2.78}$$

where $F(\overline{b^{CR}}_{t+1,i})$ denote the objective function value of the new crossover-generated butterfly.

2.4.1.4 Moth search algorithm

The moths are a family of insects associated with butterflies belonging to the order Lepidoptera. In biology, the phenomenon of phototaxis is defined as the movement of an organism toward or away from a source of light. It has been shown that moths in nature have the tendency to follow Lévy flights and have also the feature of phototaxis. Having the previous in mind, the authors in [69] proposed a new metaheuristic algorithm, called MSA. MSA was used for patch antenna design for RF energy harvesting applications in [85]. Algorithm 9 describes the MSA.

Lévy flights are random walks where the numbers are taken from a Lévy distribution of the form

$$L(x) \sim |x|^{-\beta} \tag{2.79}$$

where β is an index taken from $1 < \beta \leq 3$. In the case of moth flight, the authors in [69] set $\beta = 1.5$. The moths in MSA move using two types of moves, Lévy flights and phototaxis moves.

The moths update their position using Lévy flights using the following formula:

$$\bar{x}_{t+1,i} = \bar{x}_{t1,i} + \sigma L(d) \tag{2.80}$$

where $\bar{x}_{t+1,i}$ denotes the ith moth position in the $t+1$ iteration, σ is a scaling factor, given by $\sigma = S_{\max}/t^2$. $S_{\max}$ is the max walk step, set to different values for different problems. The Lévy distribution in (2.80) is given by

$$L(d) = \frac{(\beta - 1)\Gamma(\beta - 1)\sin\left(\frac{\pi(\beta-1)}{2}\right)}{\pi d^{\beta}} \tag{2.81}$$

where $\Gamma()$ denotes the gamma function.

A moth may also flight toward the light source. This is modeled in MSA as

$$\bar{x}_{t+1,i} = \lambda \times (\bar{x}_{t,i} + \phi \times (\bar{x}_{t,best} - \bar{x}_{t,i})) \tag{2.82}$$

where $\bar{x}_{t,best}$ denotes the best moth at iteration t, ϕ is an acceleration factor, λ is scaling factor, which is set to uniformly distributed random number in [69]. If the moth flies beyond the light source, then its position update is given by

$$\bar{x}_{t+1,i} = \lambda \times \left(\bar{x}_{t,i} + \frac{1}{\phi} \times (\bar{x}_{t,best} - \bar{x}_{t,i})\right). \tag{2.83}$$

For each moth, there is a 50% probability that its position will be updated by either (2.82) or (2.83). Usual values for the MSA control parameters are as in [69] $\beta = 1.5, \phi = 0.618, S_{\max} = 1$.

2.4.1.5 Elephant herding optimization

EHO is a nature-inspired algorithm, which is presented by the authors in [70]. The source of inspiration for the EHO algorithm is the herding behavior of elephant group. The main social behaviors of elephant groups in nature are two. First, the elephants belonging to different clans live to the same group under the leadership of a matriarch, and second the male elephants will leave their family group when they grow up. EHO models mathematically the previously described behaviors using two operators; the clan updating operator and the separating operator. The clan updating operator is used to update the elephants position using the information of the current position and the matriarch position (clan leader). After the clan updating operator, the separating operator is applied in order to enhance the population diversity.

Algorithm 9: Moth search algorithm

MSA control parameters initialization.
Set maximum generations number (*MAXGEN*), population size of *NP* moths, Max walk step $S_{\max}$, the index β, the acceleration factor ϕ, and the elitism size p.
Initialize population randomly and calculate objective function value for every member of the population
Sort *NP* population members in descending order according to objective function value
Divide population into two equal subpopulations.
G=0
while $G < MAXGEN$ **do**
    for *i*=1 to *NP*/2 **do**
        Generate new solution moths $\bar{x}_{t+1,i}$ using Lévy flights according to (2.80)
    end for
    for *i*=*NP*/2+1 to *NP* **do**
        if $rand\,(0, 1) < 0.5$ **then**
            Generate new solution moths $\bar{x}_{t+1,i}$ according to (2.82)
        else
            Generate new solution moths $\bar{x}_{t+1,i}$ according to (2.83)
        end if
    end for
    Calculate objective function value for each new moth.
    Sort 2*NP* population members in descending order according to objective function value.
    Keep *NP* better moths and replace p worse moths with previous generation elites.
    $G = G + 1$
end while

2.4.1.6 Shuffled frog-leaping algorithm

SFLA [71] is a new nature-inspired MA based on frog behavior. In SFLA, the virtual population is a set of individuals (frogs). The individuals operate as hosts of memes (each host carries a single meme; each meme consists of one or more memotypes). The virtual population of frogs is partitioned into several parallel communities (memeplexes). Within each memeplex, the frogs experience a memetic evolution (hold or exchange ideas). During the evolution, the frogs may change their memes, resulting in the enhancement of the individual's performance. After a specific number of steps, a shuffling mix process is applied between memeplexes. The memetic evolution and the shuffling process are repeated until certain conditions are met. Figure 2.9 shows an example of a population distributed in four memeplexes. SFLA is applied to phased antenna array design problem in [86].

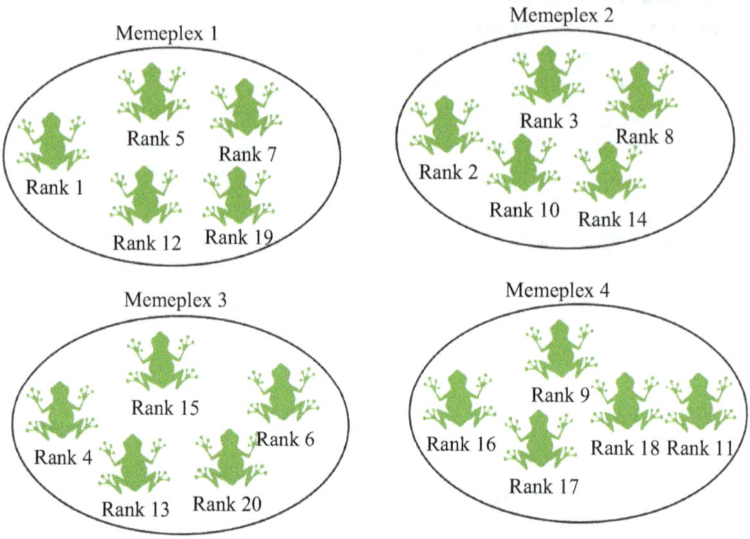

Figure 2.9 An example population of 50 frogs distributed in four memeplexes

2.4.2 Physics-based algorithms

2.4.2.1 Gravitational search algorithm

GSA is a physics-based stochastic optimization algorithm that was proposed in [72]. GSA is inspired from Newton's theory of gravity. This states that every particle in the universe attracts every other particle with a force that is directly proportional to the product of their masses and inversely proportional to the square of the distance between them [72]. GSA is a collection of agents (candidate solutions) that have masses proportional to their value of the objective function. In all iterations, all the agents masses attract each other by the gravity forces between them. The heavier the mass the larger the attraction force. Thus, the heavier masses that are probably close to the function global optimum attract the other masses proportional to their distances.

If we consider a population of NP agents, then the gravitational forces from agent k on agent n at a specific time t (or iteration) at the dth dimension can be modeled mathematically as [72]

$$F^d_{kn}(t) = G(t)\frac{M_{pk}(t) \times M_{an}(t)}{R_{kn}(t) + \varepsilon}\left(x^d_n(t) - x^d_k(t)\right) \tag{2.84}$$

where M_{an} denotes the active gravitational mass related to agent n, M_{pk} denotes the passive gravitational mass related to agent k, $G(t)$ is the gravitational constant at iteration t, ε is a very small positive constant (usually set to 2.2204×10^{-16}), and $R_{kn}(t)$ denotes the Euclidean distance between the agents k and n. We model the decision variables as the vector of the agent positions $\bar{x} = (x^1, x^2, \ldots, x^d, \ldots, X^D)$ for a problem with D dimensions.

Moreover, we can compute $G(t)$ as [72]

$$G(t) = G_0 \times \exp(-\alpha \times t/MaxIter) \tag{2.85}$$

where t is the current iteration, $MaxIter$ is the maximum number of iterations, G_0 is a positive initial gravitational constant, and α is the descending coefficient. The total force that acts on agent k can be derived as [72]

$$F_k^d(t) = \sum_{n=1, n \neq k}^{NP} rnd_n F_{kn}^d(t) \tag{2.86}$$

where rnd is a uniformly distributed random number within the interval [0, 1]. Additionally, in accordance with the law of motion, the acceleration of an agent is proportional to the result force and inverse of its mass. Thus, the acceleration of all agents can be derived as

$$ac_k^d(t) = \frac{F_k^d(t)}{M_{kk}(t)} \tag{2.87}$$

where t is the current iteration and M_{kk} is the inertial mass of the kth agent. Moreover, in GSA the agent velocity is updated using [72]

$$v_k^d(t+1) = rnd_k \times v_k^d(t) + ac_k^d(t) \tag{2.88}$$

where rnd_k is a uniformly distributed random number within the interval [0, 1]. The agent position is then updated as

$$x_k^d(t+1) = x_k^d(t) + v_k^d(t+1) \tag{2.89}$$

where $x_k^d(t+1)$ and $x_k^d(t)$ are the agents' positions in the next and current iteration, respectively. The steps of GSA are as follows:

1. Initialize all agents randomly, set $t = 0$.
2. Calculate velocities using (2.88).
3. Calculate gravitational constant using (2.85).
4. Calculate gravitational force using (2.84).
5. Calculate total forces using (2.86).
6. Calculate accelerations using (2.87).
7. Update agent positions using (2.89).
8. $t = t + 1$.
9. Repeat steps 2–8 until $t = MaxIter$.

2.4.2.2 PSOGSA

The authors in [73] have presented a hybrid combination of PSO and GSA that has become popular in the literature. The basic idea in PSOGSA is to combine the ability of social thinking (X_{best}) in PSO with the local search capability of GSA. The authors in [73] modeled mathematically this concept using

$$\begin{aligned} V_k(t+1) = {} & w \times V_k(t) + c_1' \times rnd \times ac_k(t) + c_2' \times rnd \\ & \times (X_{best} - X_k(t)) \end{aligned} \tag{2.90}$$

where $V_k(t)$ and $ac_k(t)$ are the k agent velocity and acceleration at current iteration t, respectively, c'_1 and c'_2 are weighting factors, w is a weighting function, rnd is a uniformly distributed random number within the interval [0, 1], and X_{best} is the best solution found so far. Similar to GSA, the particle position is updated as

$$X_k^d(t+1) = X_k^d(t) + V_k^d(t+1). \tag{2.91}$$

The steps of PSOGSA are as follows [73]:

1. Initialize all agents randomly, set $t = 0$.
2. Calculate gravitational constant using (2.85).
3. Calculate gravitational force using (2.84).
4. Calculate total forces using (2.86).
5. Calculate accelerations using (2.87).
6. Update best solution so far X_{best}.
7. Calculate velocities using (2.90).
8. Update agent positions using (2.91).
9. $t = t + 1$.
10. Repeat steps 2 to 8 until $t = MaxIter$.

2.4.2.3 Wind-driven optimization

WDO is recently proposed new physics-based algorithm that models the atmospheric motion. The basic idea in WDO is to model the movement of air parcels by following Newton's second law of motion. WDO uses four different control parameters that should be tuned. These are, namely, the friction coefficient, the Coriolis force contribution parameter, the universal gas constant, the air temperature, the gravitational constant, and the maximum allowable velocity. More details about tuning these parameters can be found in [74]. The authors in [74] use WDO for several design cases, which include linear arrays, artificial magnetic conducting ground planes, and E-shaped patch antenna for Wi-Fi operation at 5 GHz.

2.4.3 Human social behavior-based algorithms

2.4.3.1 Teaching–learning-based optimization

TLBO was introduced in [75]. TLBO source of inspiration is the effect of a teacher's capabilities on the growth of knowledge of students. Thus, TLBO uses a population of possible solutions that represent a class of *NP* students/learners. The best student (best solution) is assumed to be the teacher at each iteration. The basic advantage of TLBO is the fact that does require the setting of any control parameter. TLBO is applied several times for solving real-world engineering problems [87,88]. TLBO has two distinct processes or phases, the "teacher phase" and the "learner phase."

The teacher's goal in a class of students is to raise the mean grades of the class. TLBO is inspired by this process. This modeled as follows. The teacher T_k at iteration G tries to move the mean M_k of the population toward its own level. Thus, TLBO in the teacher phase updates the kth individual in the j-dimension as

$$x_{k,j}^{new} = x_{k,j}^{old} + r_k \left(x_{best,j} - T_F \times M_j\right) \tag{2.92}$$

where T_F is the teaching factor, r_k is a uniformly distributed random number within the range [0, 1], and M_j is the mean solution vector of the population. The parameter T_F belongs to the discrete set $\{1, 2\}$, and it is selected with equal probability between the two numbers.

During the learner phase, which is executed sequentially, the algorithm selects two random learners k and n that interact with each other. The learner may increase his knowledge (objective function value) if the other learner has more knowledge, which represents a solution with better objective function value. Therefore, if we consider two randomly chosen solutions $\mathbf{x_k}$ and $\mathbf{x_n}$, the modification of the existing solution $\mathbf{x_k}$ is mathematically formulated as

$$\begin{aligned} \mathbf{x_k^{new}} &= \mathbf{x_k^{old}} + r_k\left(\mathbf{x_k} - \mathbf{x_n}\right), \text{ if} F\left(\mathbf{x_k}\right) < F\left(\mathbf{x_n}\right) \\ \mathbf{x_k^{new}} &= \mathbf{x_k^{old}} + r_k\left(\mathbf{x_n} - \mathbf{x_k}\right), \text{ otherwise} \end{aligned} \tag{2.93}$$

where r_k is random uniformly distributed number within the range [0, 1]. The algorithm accepts the new solution and it replaces the old in the population only if it obtains better objective function value than the old one.

2.4.3.2 Jaya

The Jaya is a very simple algorithm with low complexity that was first introduced in [77]. Jaya uses as main concept the mathematical model that each solution vector should be updated in order to move toward the best solution and away from the worst solution obtained at each iteration. The algorithm's name origin is the Sanskrit word "Jaya" which means *victory*.

Jaya as TLBO does not require the setting of any control parameter. Jaya updates the position of each member of the population using the best and the worst solution vector at each iteration. Thus, the kth member of population $\mathbf{y_k}$ updates its position by the following rule:

$$\mathbf{y_i^{new}} = \mathbf{y_i^{old}} + rnd_1\left(\mathbf{y_{best}} - \mathbf{y_i^{old}}\right) - rnd_2\left(\mathbf{y_{worst}} - \mathbf{y_i^{old}}\right) \tag{2.94}$$

where rnd_1 and rnd_2 are uniformly distributed random numbers within the range [0, 1]. Jaya as TLBO uses a greedy selection operator. The description of the Jaya algorithm is found in Algorithm 10.

2.4.3.3 TLBO–Jaya algorithm

The authors in [76] introduce a hybrid algorithm that combines concepts from both TLBO and Jaya the TLBO–Jaya algorithm, which is applied for wireless sensor network optimization.

TLBO–Jaya sets the teaching factor, T_F, to be either one or two not in a random way, but in accordance with each vector objective function value. This teaching factor is defined as

$$T_F = \text{round}\left(1 + \frac{F_k - F_{best}}{F_{worst} - F_{best}}\right) \tag{2.95}$$

where F_k denotes the objective function value of the kth vector, and F_{best} and F_{worst} are the minimum and maximum function values obtained so far among all solutions,

Algorithm 10: Jaya algorithm

Set the population size NP, and the maximum number of objective function evaluations $FES_{\max}$
Initialize a uniformly distributed random population of NP vectors
Compute the objective function values of all population vectors
while $FES < FES_{\max}$ **do**
    Find the best and worst vectors for the current population
    for k=1 to NP **do**
        Create new solution vector using (2.94)
        Apply the greedy selection operator
        Replace old best vector if the new child vector is better
    end for
end while

respectively. An additional TLBO–Jaya characteristic is the introduction of a new phase to the original TLBO algorithm, which is inspired by the simplicity of the Jaya algorithm. This third phase is called "Jaya phase." This phase uses the second term of (2.94).

Hence, the TLBO–Jaya algorithm has three distinct phases. The first phase is the "teacher phase," where the students try to improve their average grade through the teaching process. The second phase is "learner phase," where the learners interact with each other. These phases are the same as in the original TLBO. The basic concept of the third phase is that the learners try to avoid the behavior and tactics of the worst student. The authors in [76] define the update rule for this phase for the kth solution vectors x_k as

$$\mathbf{x}_{\mathbf{k}}^{\mathbf{new}} = \mathbf{x}_{\mathbf{k}}^{\mathbf{old}} - r_k \left(\mathbf{x}_{\mathbf{worst}} - \mathbf{x}_{\mathbf{k}}^{\mathbf{old}}\right) \tag{2.96}$$

where r_k is a random number within the range [0, 1]. TLBO–Jaya uses a greedy selection operator after the third phase. Algorithm 11 describes the TLBO–Jaya algorithm.

2.4.4 Music-based algorithms

2.4.4.1 Harmony search algorithm

HS is an EA inspired by music creation [78]. A harmony in HS terminology is a possible solution vector. A harmony consists of pitches (decision variables). HS as other EAs uses control parameters. These are the harmony memory (HM) size, which is equivalent to population size in another algorithms, and the HM consideration rate (HMCR), which defines the way pitches will be selected. HMCR is a real number within [0, 1]. The pitch adjusting rate (PAR) is another control parameter that represents the probability of adjusting the original value of the selected pitches from the

Algorithm 11: TLBO–Jaya algorithm

Set the population size NP, and the maximum number of objective function evaluations $FES_{\max}$
Initialize a uniformly distributed random population of NP vectors
Compute the objective function values of all population vectors
while $FES < FES_{\max}$ **do**
  for k=1 to NP **do**
    Teacher Phase
    Compute population **Mean**
    Find T_F using (2.95)
    Create a new child solution using (2.92)
    if $F(\mathbf{x}_k^{old}) > F(\mathbf{x}_k^{new})$ **then**
      Replace vector $\mathbf{x}_k^{old}$ with $\mathbf{x}_k^{new}$
    end if
    Learner Phase
    Create a new child solution using (2.93)
    if $F(\mathbf{x}_k^{old}) > F(\mathbf{x}_k^{new})$ **then**
      Replace vector $\mathbf{x}_k^{old}$ with $\mathbf{x}_k^{new}$
    end if
    Jaya Phase
    Create a new child solution using (2.96)
    if $F(\mathbf{x}_k^{old}) > F(\mathbf{x}_k^{new})$ **then**
      Replace vector $\mathbf{x}_k^{old}$ with $\mathbf{x}_k^{new}$
    end if
  end for
end while

HM. The pitch adjustment is determined by a pitch BW and the PAR. Each new vector in the jth dimension is generated using

$$x_{new,j} = x_{old,j} + BW \times rnd \tag{2.97}$$

where rnd is a randomly generated number within $[-1, 1]$.

2.5 Opposition-based learning

Opposition-based learning (OBL) was first introduced by Tizhoosh in [89]. OBL uses as the fundamental concept not only to find the objective function value of the current solution vector but also to derive the objective function value of the opposite individual. Then the algorithm applies a greedy selection scheme between the two individuals in order to select the one with the best objective function value. Some definitions of the basic concepts of OBL are provided next [89–91].

Definition (opposite number). Let $y \in [b, c]$ be any real number. The opposite number is defined by

$$y_O = b + c - y \tag{2.98}$$

Definition (opposite point). Likewise, the earlier definition can be extended to D-dimensional space, then we consider $P(y_1, y_2, \ldots, y_D)$ a point where $y_1, y_2, \ldots, y_D \in \Re$ and $y_j \in [b_j, c_j]$ $\forall j \in \{1, 2, \ldots, D\}$. The opposite point $P_O(y_{O1}, y_{O2}, \ldots, y_{OD})$ is defined by its components

$$y_{Oj} = b_j + c_j - y_j \tag{2.99}$$

Definition (semi-opposite point) [92]. If we change the components of a point by its opposites only in some components and the other remain unchanged, then the new point is a semi-opposite point. This is expressed as

$$\begin{aligned} &P_{SO}(y_{SO1}, y_{SO2}, \ldots, y_{SOj}, \ldots, y_{SOD}) \text{ where } \forall j \in \{1, 2, \ldots, D\} \\ &y_{SOj} \in \{y_j, y_{Oj}\}. \end{aligned} \tag{2.100}$$

For example, in a two-dimensional space, where each dimension can be either one or zero. In this space, we consider the point $P1(1, 0)$. Then the points $P2(0, 0)$ and $P3(1, 1)$ are two semi-opposite points, while the opposite point is $P4(0, 1)$.

2.5.1 OBL types

Definition (quasi-opposite (QO) point). Quasi-oppositional DE (QODE) was proposed in [93]. The quasi opposition is a variant of the original OBL scheme. The authors in [93] provide the mathematical proof that the quasi-opposite point has a higher probability to be closer to the solution than the opposite point in a black-box optimization problem. The definition of the quasi opposition is given by

$$y_{QOj} = \begin{cases} rand\,(K_j, y_{Oj}), & \text{if } y_j < K_j \\ rnd\,(y_{Oj}, K_j), & \text{otherwise} \end{cases} \tag{2.101}$$

$$K_j = \frac{b_j + c_j}{2}, \tag{2.102}$$

where the notation $rnd\,(r1, r2)$ denotes a uniformly distributed random number within the interval $[r1, r2]$.

Definition (quasi-reflection opposite (QRO) point). An extension of the quasi-opposite point was proposed in [94], which is called quasi-reflection opposition. The proposed OBL scheme goal is to accelerate the exploration process. This is given by:

$$y_{QROj} = \begin{cases} rand\,(y_j, K_j), & \text{if } y_j < K_j \\ rand\,(K_j, y_j), & \text{otherwise} \end{cases}. \tag{2.103}$$

Definition (generalized opposite (GO) point). The generalized OBL (GOBL) scheme is defined in [95]. The basic idea in generalized opposition is to use a random weight in opposition definition. The definition of GOBL is given next:

$$y_{GOj} = r \times (b_j + c_j) - y_j, \tag{2.104}$$

where r is uniformly distributed random number between 0 and 1.

Definition (current optimum opposite (COO) point). Current optimum opposition (COO) is presented by the authors in [96]. The basic concept is to use the search information of the current best candidate solution in opposite point generation. This is defined as

$$y_{COOj} = 2y_{best,j} - y_j, \tag{2.105}$$

where $y_{best,j}$ is the jth dimension of the best solution in the current population.

Definition (centroid opposition (CO) point). The centroid opposition (CO) in [97] uses the centroid point to replace the current optimum in COO. This is given by

$$y_{COj} = 2CP_j - y_j, \tag{2.106}$$

$$CP_j = \frac{\sum_{i=1}^{NP} y_{i,j}}{NP}, \tag{2.107}$$

where NP is the population size.

Definition (extended opposite (EO) point). In [98], the extended opposition is defined and it is given by

$$y_{EOj} = \begin{cases} rnd\,(y_{Oj}, c_i), & \text{if } y_j < K_j \\ rand\,(b_j, y_{Oj}), & \text{if } y_j > K_j \end{cases}. \tag{2.108}$$

Definition (reflected extended opposite (REO) point). Moreover, in [98] a second OBL scheme called reflected extended opposition is defined:

$$y_{REOj} = \begin{cases} rand\,(y_j, c_i), & \text{if } y_j < K_j \\ rand\,(b_j, y_j, & \text{if } y_j > K_j \end{cases}. \tag{2.109}$$

If we define the corresponding oppositional BBO with the OBL schemes given earlier, then we would define the algorithms given next.

1. Opposition-based BBO (OBBO)
2. Quasi OBBO
3. Quasi-reflection OBBO
4. Current optimum OBBO
5. Generalized OBBO (GOBBO)
6. Centroid OBBO
7. Extended OBBO (EOBBO)
8. Reflected EOBBO

2.5.2 OBL algorithm description

The opposition-based algorithms [99–101] require the setting of an additional control parameter, the jumping rate parameter $j_r \in [0, 1]$. The jumping rate controls in each generation if the opposite population is generated or not.

The OBL algorithms add two extra parts to the original algorithm code; the opposition-based population initialization and the opposition-based generation jumping [99–101]. Algorithm 12 describes the opposition-based population initialization.

Algorithm 12: Opposition-based population initialization

Initialize a random population P of size KP solution vectors
for k=1 to KP **do**
    Create the opposite population OP of size KP solution vectors
    for n=1 to D **do**
        Replace $y_{k,n}$ with its opposite point defined by an OBL scheme
    end for
end for
Merge P and OP populations
Select the KP fittest among P and OP and set them as the current population

Algorithm 13: Opposition-based generation jumping

if $rand[0, 1] < j_r$ **then**
    for k=1 to KP **do**
        Create the opposite population OP of size KP solution vectors
        for n=1 to D **do**
            Replace $y_{k,n}$ with its opposite point defined by an OBL scheme
        end for
    end for
end if
Merge P and OP populations
Select the KP fittest among P and OP and set them as the current population

The variables low_j, $upper_j$ represent the lower and upper limits in the jth dimension, respectively. The OBL generation jumping follows a similar approach. This is described in Algorithm 13.

2.5.3 Modified generalized OBBO

In [102], the author proposes an OBBO version based on semi-opposite points. This version is called modified GOBBO (MGOBBO) or GOBBO in short. GOBBO defines an additional control parameter named opposition probability $p_o \in [0, 1]$. This parameter controls if an SIV variable in a habitat will be replaced by its opposite or not. The OBL population initialization for MGOBBO is described in Algorithm 14, where $y_{k,n}^{os}$ denotes the semi-opposite point of $y_{k,n}$, i.e., the kth vector in the nth dimension. The opposition-based generation jumping follows a similar approach. The algorithm description is given in Algorithm 15. The $\min_n$ and $\max_n$ are the minimum and maximum values of the nth dimension in the current population, respectively.

Algorithm 14: MGOBBO population initialization

Initialize a random population P of size KP solution vectors
for k=1 to KP **do**
    Create the semi-opposite population OP_s
    for n=1 to D **do**
        if $rand[0, 1] < p_o$ **then**
            $y_{k,n}^{os} = low_n + upper_n - x_{k,n}$
        else
            $x_{k,n}^{os} = x_{k,n}$
        end if
    end for
end for
Set as the initial population the KP fittest among P and OP_s

Algorithm 15: MGOBBO generation jumping

if $rnd[0, 1] < j_r$ **then**
    for k=1 to KP **do**
        Create semi-opposite population OP_s
        for n=1 to D **do**
            if $rnd[0, 1] < p_o$ **then**
                $y_{k,n}^{os} = \min_n + \max_n - x_{k,n}$
            else
                $y_{k,n}^{os} = x_{k,n}$
            end if
        end for
    end for
end if
Select fittest among current population P and OP_s

2.6 Multi-objective algorithms

2.6.1 Non-dominated sorting genetic Algorithm-II

Non-dominated sorting genetic Algorithm-II (NSGA-II) [103] is a GA that uses the concept of non-dominated ranking and sorting. NSGA-II uses crossover, mutation, and selection operators. NSGA-II uses single-point crossover and bitwise mutation for binary-coded variables. Moreover, for real-coded decision variables NSGA-II uses the simulated binary crossover operator and polynomial mutation [104] for real-coded GAs. Usual values are for crossover probability $p_c = 0.9$ and for mutation probability $p_m = 1/n$ or $p_m = 1/l$ (where n is the number of real-coded variables and l is the string length for binary-coded variables) for real-coded and binary-coded variables, respectively. Moreover, real-coded NSGA-II uses distribution indexes [104] for crossover, η_c, and mutation, η_m operators, usually set to $\eta_c = 20$, $\eta_m = 20$. NSGA-II in every generation produces a child population with the same size *NP* as the original parent population. Then NSGA-II sorts the union of both population according to their rank and selects the *NP* non-dominated solutions.

2.6.1.1 Non-dominated ranking

The non-dominated ranking is the sorting of the solution vectors regarding non-domination. This type of sorting approach is named as fast non-dominated sorting approach. The detailed description can be found in [103].

In this section, we briefly describe the basics of non-dominated ranking. The algorithm calculates at first, for each solution vector k two parameters. These are the domination count (i.e., the number of solutions which dominates the solution k), and second, the set of solution vectors that the solution vector k dominates.

This approach requires several comparisons to be performed. The algorithm sets the domination count of all solutions in the Pareto front (first non-dominated front) to zero. Next, for each solution k with $n_k = 0$, the algorithms find each member (m) of its solution set and reduce its domination count by one. If for any member m the domination count becomes zero, we include this member in a separate list M. The algorithms add these solution vectors to the second non-dominated front. Furthermore, the previous process is repeated with each solution vector of M, in order to find the third non-dominated front. The procedure ends when all fronts are found.

NSGA-II uses the crowded-comparison operator for sorting population vectors defined in [103]. We consider that every solution vector m in the population has two attributes: (a) non-domination rank (m_{rank}) and (b) crowding distance (CD) ($m_{distance}$). The definition of the crowded-comparison operator is given by

$$m \prec_n q \ \ if \ \ m_{rank} < q_{rank} \ or \ (m_{rank} = q_{rank} \ \ and \ \ m_{distance} > q_{distance}). \qquad (2.110)$$

Thus, if we have two solution vectors with different non-domination ranks, we select the solution vector with the better rank. Otherwise, if both vectors have the same rank, we select the solution vector that is located in a lesser crowded region.

2.6.1.2 Algorithm description

The NSGA-II algorithm is outlined in Algorithm 16.

2.6.2 Non-dominated sorting genetic Algorithm-III

The basic framework NSGA-III [105] is quite similar to that of the original NSGA-II algorithm [106] with significant changes in its selection operator. The main difference with the NSGA-II is the fact that NSGA-III maintains the diversity among population members by supplying and adaptively updating a number of well-spread reference points. Thus, NSGA-III uses a predefined set of reference points to ensure diversity in obtained solutions. The chosen reference points can either be predefined in a

Algorithm 16: NSGA-II algorithm

```
Initialize the NSGA-II control parameters. These are the crossover probability
  p_c, and the mutation probability p_m.
Set the population size NP, the maximum number of iterations MAXG.
Initialize a uniformly distributed random population of size NP vectors.
Compute objective function and constraint function values of all population
  vectors.
Apply non-dominated ranking to NP population vectors.
Calculate crowding distance for NP population vectors.
Sort NP population vectors in descending order using (≺_n).
G = 1
while G < MAXG do
    for k=1 to NP do
        For each population vector generate a child vector.
        Apply the crossover operator.
        Apply the mutation operator.
        Evaluate objective function and constraint function values for the child
          vector.
    end for
    Apply non-dominated ranking to 2NP population vectors.
    Calculate crowding distance for 2NP population vectors.
    Sort 2NP population vector in descending order using (≺_n)
    Keep NP non-dominated vectors.
    G = G + 1
end while
```

structured manner or supplied preferentially by the user. NSGA-III replaces the CD operator with the following approaches:

- Classification of population into non-dominated levels
- Determination of reference points on a hyperplane
- Adaptive normalization of population members
- Association operation
- Niche-preservation operation

More details about NSGA-III can be found in [105].

2.6.3 Generalized differential evolution

Multi-objective DE algorithms extend the classical DE algorithm for solving multi-objective optimization problems (MOOP). Generalized DE (GDE3) has the potential to solve problems that have n objectives and k constraint functions that was proposed in [107]. It can handle any number of objectives and any number of constraint functions, including the cases $n = 0$ (constraint satisfaction problem) and $k = 0$

(unconstraint problem). In the case of $n = 1$ and $k = 0$, the algorithm is the same as the original DE.

Thus, the original DE algorithm is a specific case of GDE3. In GDE3, one may modify the current *DE/rand/1/bin* strategy to any other exciting DE strategy or to any other method that a trial vector is compared against an old vector and the better one is kept. GDE3 has been evaluated and compared with other EAs in numerical benchmark problems [108,109]. GDE3 has been applied to Yagi antenna design in [110], to microwave filter design in [111], and to subarrayed linear antenna arrays design [112].

GDE3 transforms the selection rule of the original DE. This modified selection rule is that the trial vector replaces the old vector in the next generation if it weakly constraint dominates the old vector. The GDE3 algorithm has the following selection rules:

1. If both vectors (trial and old) are infeasible, then the trial vector is selected only if it weakly dominates the old vector in constraint violation space, otherwise the old vector is preserved.
2. If one vector is feasible and the other is unfeasible, then the feasible vector is selected.
3. If both vectors (trial and old) are feasible, then the trial is selected only if it weakly dominates the old vector in the objective function space. If the old dominates then the old vector is selected. If neither vector dominates each other in the objective function space, then both vectors are selected for the next generation.

Therefore, the population size in GDE3 may increase in the next generation. Then a sorting technique that uses the concept of CD as in NSGA-II is applied in order to set the population size back to the original size.

The vectors are sorted on the basis of non-dominance and crowdedness. The worst population members are removed and the population size is set to the original size. The basic idea is to prune a non-dominated set to have a desired number of solutions in such a way that the remaining solutions have as good diversity as possible. This means that the spread of extreme solutions is as high as possible, and the relative distance between solutions is as equal as possible. The pruning method of NSGA-II provides good diversity in the case of two objectives, but when the number of objectives is more than two, the obtained diversity declines drastically [113].

GDE3 employs a crowding estimation technique that uses the nearest neighbors of solutions in Euclidean sense, and a technique for finding these nearest neighbors quickly. The GDE3 pruning method is described with more details in [114]. Thus, GDE3 improves the selection based on CD over the original method of the NSGA-II. This way GDE3 has the potential to provide a better distributed set of vectors.

GDE3 and NSGA-II have another difference regarding the population size after a generation. NSGA-II increases the population size after a generation 2*NP*. Then non-dominated ranking is applied and *NP* non-dominated vectors are selected, while the *NP* dominated vectors are removed from the population. GDE3 uses a different approach. In this case, after a generation the population size is $NP + m$, where $m \in [0, NP]$, because the population size is increased only when the trial $\bar{u}_{i,G+1}$ and the old vector $\bar{x}_{i,G}$ are both feasible and do not dominate each other. Thus, non-dominated ranking is then applied to $NP + m$ vectors instead of $2NP$ in NSGA-II. This population size may be in general less than $2NP$. Hence, GDE3 requires less computational time than NSGA-II [107].

GDE3, and other MOEAs, can be implemented in such a way that fewer objective function evaluations are required. For example, in the case of infeasible solutions, if the first constraint function is not satisfied then no other constraint evaluation is required. However, if the solution vector is feasible then all the objective function evaluations are required. The control parameters selected for GDE3 are according to [115,116] $F = 0.5$, $CR = 0.1$. The GDE3 algorithm is outlined in Algorithm 17.

Algorithm 17: GDE3 algorithm

Initialize the GDE3 control parameters. These are the mutation control parameter, and the crossover constant.
Set the population size *NP*, the maximum number of iterations *MAXG*
Initialize a uniformly distributed random population of size *NP* vectors
Compute objective function and constraint function values of all population vectors
Apply non-dominated ranking to *NP* population vectors
Calculate Crowding Distance for *NP* population vectors.
Sort *NP* population vectors in descending order using $(\prec_n)$
$G = 1$
while $G < MAXG$ **do**
    for i=1 to *NP* **do**
        Apply the mutation operator according to *DE/rand/1/bin* strategy to generate a mutant vector $\bar{v}_{G+1,i}$.
        Create a trial vector $\bar{u}_{G+1,i}$ using the crossover operator.
        Evaluate objective function and constraint function values for the trial vector.
        Apply the selection operator according to the following criterion:

$$\bar{x}_{i,G+1} = \begin{cases} \bar{u}_{i,G+1} & \text{if } \bar{u}_{i,G+1} \underset{c}{\prec} \bar{x}_{i,G} \\ \bar{x}_{i,G}, & \text{otherwise} \end{cases}$$

        Set $m = m + 1, \bar{x}_{NP+m,G+1} = \bar{u}_{i,G+1}$ if $\forall j : g_j(\bar{u}_{i,G+1}) \leq 0 \wedge \bar{x}_{i,G+1} == \bar{x}_{i,G} \wedge \bar{x}_{i,G} \not\prec \bar{u}_{i,G+1}$
    end for
    Apply non-dominated ranking to $NP + m$ vectors.
    Calculate Crowding Distance for $NP + m$ population vectors.
    Sort $NP + m$ population vector in descending order using $(\prec_n)$
    Keep *NP* non-dominated vectors and set $m = 0$.
    $G = G + 1$
end while

2.6.4 Speed-constrained multi-objective PSO

Speed-constrained multi-objective PSO (SMPSO) is an MO Particle Swarm Optimizer introduced in [117]. SMPSO uses a polynomial mutation operator like NSGA-II and the concept of an external archive in order to store the non-dominated solutions found in each iteration. SMPSO uses the velocity update formula in (2.2). The velocity is then restricted by multiplying with the constriction factor as in (2.5). Moreover, SMPSO uses a mechanism in such a way that the accumulated velocity of kth particle

---

Algorithm 18: MOBBO algorithm

Initialize the MOBBO control parameters. Map the problem solution vectors to *SIVs* and habitats.
Set the habitat modification probability P_{mod}, the maximum immigration rate I, the maximum emigration rate E, the maximum migration rate m_{max}, and the elitism parameter p (if elitism is desired).
Set the population size *NP*, the maximum number of iterations G_{max}
Initialize a uniformly distributed random population of size *NP* habitats (solutions).
Calculate objective function and constraint function values of all habitats.
Map the *HSI* value to the number of species S.
Apply non-dominated ranking to *NP* habitats.
Calculate Crowding Distance for *NP* habitats.
Sort *NP* habitats in descending order using $(\prec_n)$
Derive the immigration rate ξ_k, the emigration rate ψ_k for each member of the population
$G = 1$
while $G < G_{\text{max}}$ **do**
    Create a new child population of *NP* habitats, which is originally the same as the parent population.
    for i=1 to *NP* **do**
        Apply the migration operator for each member of the child population based on immigration and emigration rates.
        Apply the mutation operator to each member of the child population according to (2.66).
    end for
    Apply non-dominated ranking to 2*NP* habitats (both parent and child population).
    Calculate Crowding Distance for 2*NP* habitats.
    Sort 2*NP* habitats in descending order using $(\prec_n)$
    Keep *NP* non-dominated habitats.
    $G = G + 1$
end while

---

position in the nth dimension is further bounded by defining of the following velocity constriction equation:

$$u_{G+1,nk} = \begin{cases} delta_n & \text{if } u_{G+1,nk} > delta_n \\ -delta_n & \text{if } u_{G+1,nk} \leq -delta_n \\ u_{G+1,nk} & \text{otherwise} \end{cases}$$

where $delta_n = \frac{(upper\_limit_n - lower\_limit_n)}{2}$

2.6.5 Multi-objective BBO

Multi-objective BBO (MOBBO) algorithms extend the original BBO algorithm for solving MOOP. MOBBO has been introduced and applied successfully to the indoor wireless heterogeneous networks planning problem in [119]. MOBBO is a hybrid between BBO and common MO concepts used in NSGA-II and GDE3. As NSGA-II [103] and GDE3, MOBBO includes the concept of CD.

The pruning method of NSGA-II provides good diversity in the case of two objectives, but when the number of objectives is more than two, the obtained diversity declines drastically [106]. MOBBO employs the same method used in GDE3, which is based on a crowding estimation technique using nearest neighbors of solutions in Euclidean sense, and a technique for finding these nearest neighbors quickly. More details about this pruning method can be found in [118]. Thus, the MOBBO selection using CD is improved over the original method of the NSGA-II to provide a better distributed set of vectors.

The MOBBO algorithm is outlined in Algorithm 18.

2.6.6 Computational complexity of MO algorithms

The computational complexity of one generation of NSGA-II, GDE3, MOBBO, and SMPSO algorithms is the same $\mathcal{O}(N_{\mathrm{obj}}NP^2)$, where N_{obj} is the number of objective functions and NP is the population size of each algorithm [103,117]. Moreover, the computational complexity of one generation of NSGA-III is $\mathcal{O}(N_{\mathrm{obj}}NP^2)$ or $\mathcal{O}(NP^2 log^{N_{\mathrm{obj}}-2} NP)$ [105].

References

[1] Kennedy J and Eberhart R. Particle swarm optimization. In: IEEE International Conference on Neural Networks. 1995;4:1942–1948.

[2] Dorigo M, Maniezzo V, and Colorni A. Ant system: Optimization by a colony of cooperating agents. IEEE Transactions on Systems, Man, and Cybernetics, Part B: Cybernetics. 1996;26(1):29–41.

[3] Karaboga D and Basturk B. A powerful and efficient algorithm for numerical function optimization: Artificial bee colony (ABC) algorithm. Journal of Global Optimization. 2007;39(3):459–471.

[4] Goudos SK, Kalialakis C, and Mittra R. Evolutionary algorithms applied to antennas and propagation: A review of state of the art. International Journal of Antennas and Propagation. 2016;2016:12.

[5] Goudos SK. In: Khatib DM, editor. Evolutionary Algorithms for Wireless Communications – A Review of the State-of-the Art. Croatia: InTech; 2014. p. 1–24.

[6] Clerc M. The swarm and the queen: Towards a deterministic and adaptive particle swarm optimization. In: Proceedings of the 1999 Congress on Evolutionary Computation, 1999. CEC 99. vol. 3; 1999. p. 1951–1957.

[7] Baskar S, Alphones A, Suganthan PM, *et al.* Design of Yagi-Uda antennas using comprehensive learning particle swarm optimisation. IEE Proceedings: Microwaves, Antennas and Propagation. 2005;152:340–346.

[8] Goudos SK, Moysiadou V, Samaras T, *et al.* Application of a comprehensive learning particle swarm optimizer to unequally spaced linear array synthesis with sidelobe level suppression and null control. IEEE Antennas and Wireless Propagation Letters. 2010;9:125–129.

[9] Kennedy J and Eberhart RC. Discrete binary version of the particle swarm algorithm. In: IEEE International Conference on Systems, Man and Cybernetics. 1997;5:4104–4108.

[10] Mirjalili S and Lewis A. S-shaped versus V-shaped transfer functions for binary particle swarm optimization. Swarm and Evolutionary Computation. 2013;9:1–14.

[11] Marandi A, Afshinmanesh F, Shahabadi M, *et al.* Boolean particle swarm optimization and its application to the design of a dual-band dual-polarized planar antenna. In: IEEE Congress on Evolutionary Computation, CEC 2006; 2006. p. 3212–3218.

[12] Deligkaris KV, Zaharis ZD, Kampitaki DG, *et al.* Thinned planar array design using Boolean PSO with velocity mutation. IEEE Transactions on Magnetics. 2009;45(3):1490–1493.

[13] Zaharis ZD, Gotsis KA, and Sahalos JN. Adaptive beamforming with low side lobe level using neural networks trained by mutated Boolean PSO. Progress in Electromagnetics Research. 2012;127:139–154.

[14] Zaharis ZD, Goudos SK, and Yioultsis TV. Application of Boolean PSO with adaptive velocity mutation to the design of optimal linear antenna arrays excited by uniform amplitude current distribution. Journal of Electromagnetic Waves and Applications. 2011;25(10):1422–1436.

[15] Afshinmanesh F, Marandi A, and Shahabadi M. Design of a single-feed dual-band dual-polarized printed microstrip antenna using a Boolean particle swarm optimization. IEEE Transactions on Antennas and Propagation. 2008;56(7):1845–1852.

[16] da Silva Maximiano M, Vega-Rodríguez MA, Gómez-Pulido JA, *et al.* A new multiobjective artificial bee colony algorithm to solve a real-world frequency assignment problem. Neural Computing and Applications. 2013;22(7–8):1447–1459.

[17] Goudos SK, Siakavara K, Theopoulos A, *et al.* Application of Gbest-guided artificial bee colony algorithm to passive UHF RFID tag design. International Journal of Microwave and Wireless Technologies. 2016;8(3):537–545.

[18] Goudos SK, Siakavara K, and Sahalos JN. Novel spiral antenna design using artificial bee colony optimization for UHF RFID applications. IEEE Antennas and Wireless Propagation Letters. 2014;13:528–531.

[19] Yang J, Li WT, Shi XW, *et al.* A hybrid ABC-DE algorithm and its application for time-modulated arrays pattern synthesis. IEEE Transactions on Antennas and Propagation. 2013;61(11):5485–5495.

[20] Zhu G and Kwong S. Gbest-guided artificial bee colony algorithm for numerical function optimization. Applied Mathematics and Computation. 2010;217(7):3166–3173.

[21] Liang JJ, Qin AK, Suganthan PN, *et al.* Comprehensive learning particle swarm optimizer for global optimization of multimodal functions. IEEE Transactions on Evolutionary Computation. 2006;10(3):281–295.

[22] Zaharis ZD and Yioultsis TV. A novel adaptive beamforming technique applied on linear antenna arrays using adaptive mutated Boolean PSO. Progress in Electromagnetics Research. 2011;117:165–179.

[23] Dorigo M and Gambardella LM. Ant colonies for the travelling salesman problem. BioSystems. 1997;43(2):73–81.

[24] Dorigo M and Stutzle T. Ant Colony Optimization. Cambridge, MA: The MIT Press; 2004.

[25] Mirjalili S and Lewis A. The whale optimization algorithm. Advances in Engineering Software. 2016;95:51–67.

[26] Mirjalili S, Gandomi AH, Mirjalili SZ, *et al.* Salp swarm algorithm: A bio-inspired optimizer for engineering design problems. Advances in Engineering Software. 2017;114:163–191.

[27] Mirjalili S, Mirjalili SM, and Lewis A. Grey wolf optimizer. Advances in Engineering Software. 2014;69:46–61.

[28] Goudos SK, Tsoulos G, and Athanasiadou G. Artificial neural network optimal modelling of received signal strength in mobile communications using UAV measurements. In: 12th European Conference on Antennas and Propagation (EuCAP 2018); 2018. p. 1–4.

[29] Saxena P and Kothari A. Optimal pattern synthesis of linear antenna array using grey wolf optimization algorithm. International Journal of Antennas and Propagation. 2016;2016.

[30] Ozkaya U and Seyfi L. A comparative study on parameters of leaf-shaped patch antenna using hybrid artificial intelligence network models. Neural Computing and Applications. 2018;29(8):35–45.

[31] Emary E, Zawbaa HM, and Hassanien AE. Binary grey wolf optimization approaches for feature selection. Neurocomputing. 2016;172:371–381.

[32] Boursianis AD, Goudos SK, Yioultsis TV, *et al.* Low-cost dual-band E-shaped patch antenna for energy harvesting applications using grey wolf optimizer. In: 13th European Conference on Antennas and Propagation (EuCAP 2019); Krakow, Poland. 2019. p. 1–5.

[33] Boursianis AD, Koulouridis S, Georgoulas D, *et al.* Wearable 5-gigahertz Wi-Fi antenna design using whale optimization algorithm. In: 14th European Conference on Antennas and Propagation (EuCAP 2020); Copenhagen, Denmark. 2020, p. 1–4.

[34] Goudos SK. Joint power allocation and user association in non-orthogonal multiple access networks: An evolutionary approach. Physical Communication. 2019;37:100841.

[35] Boursianis AD, Goudos SK, Yioultsis TV, *et al.* MIMO antenna design for 5G communication systems using salp swarm algorithm. In: International Workshop on Antenna Technology (iWAT 2020); Bucharest, Romania. 2020, p. 1–3.

[36] Storn R and Price K. Differential evolution – A simple and efficient heuristic for global optimization over continuous spaces. Journal of Global Optimization. 1997;11(4):341–359.

[37] Storn R and Price K. Differential evolution – A simple and efficient heuristic for global optimization over continuous spaces. Journal of Global Optimization. 1997;11(4):341–359.

[38] Storn R. Differential evolution research – Trends and open questions. Studies in Computational Intelligence. 2008;143:1–31.

[39] Mezura-Montes E, Velázquez-Reyes J, and Coello Coello CA. A comparative study of differential evolution variants for global optimization. In: GECCO'06; 2006. p. 485–492.

[40] Brest J, Greiner S, Boskovic B, *et al.* Self-adapting control parameters in differential evolution: A comparative study on numerical benchmark problems. IEEE Transactions on Evolutionary Computation. 2006;10(6): 646–657.

[41] Goudos SK. Design of microwave broadband absorbers using a self-adaptive differential evolution algorithm. International Journal of RF and Microwave Computer-Aided Engineering. 2009;19(3):364–372.

[42] Dib NI, Goudos SK, and Muhsen H. Application of Taguchi's optimization method and self-adaptive differential evolution to the synthesis of linear antenna arrays. Progress in Electromagnetics Research. 2010;102: 159–180.

[43] Goudos SK, Siakavara K, Samaras T, *et al.* Self-adaptive differential evolution applied to real-valued antenna and microwave design problems. IEEE Transactions on Antennas and Propagation. 2011;59(4):1286–1298.

[44] Qin AK and Suganthan PN. Self-adaptive differential evolution algorithm for numerical optimization. In: Evolutionary Computation, 2005. The 2005 IEEE Congress on. vol. 2; 2005. p. 1785–1791.

[45] Huang VL, Qin AK, and Suganthan PN. Self-adaptive differential evolution algorithm for constrained real-parameter optimization. In: Evolutionary Computation, 2006. CEC 2006. IEEE Congress on; 2006. p. 17–24.

[46] Qin AK, Huang VL, and Suganthan PN. Differential evolution algorithm with strategy adaptation for global numerical optimization. IEEE Transactions on Evolutionary Computation. 2009;13(2):398–417.

[47] Goudos SK, Zaharis ZD, and Yioultsis TV. Application of a differential evolution algorithm with strategy adaptation to the design of multi-band microwave filters for wireless communications. Progress in Electromagnetics Research. 2010;109:123–137.

[48] Goudos SK, Siakavara K, Samaras T, *et al.* Sparse linear array synthesis with multiple constraints using differential evolution with strategy adaptation. IEEE Antennas and Wireless Propagation Letters. 2011;10:670–673.

[49] Wang Y, Cai Z, and Zhang Q. Differential evolution with composite trial vector generation strategies and control parameters. IEEE Transactions on Evolutionary Computation. 2011;15(1):55–66.

[50] Sotiroudis SP, Goudos SK, Gotsis KA, *et al.* Application of a composite differential evolution algorithm in optimal neural network design for propagation path-loss prediction in mobile communication systems. IEEE Antennas and Wireless Propagation Letters. 2013;12:364–367.

[51] Goudos SK. Shaped beam pattern synthesis of antenna arrays using composite differential evolution with eigenvector-based crossover operator. International Journal of Antennas and Propagation. 2015;2015. Article ID 29501.

[52] Omran MGH, Engelbrecht AP, and Salman A. Bare bones differential evolution. European Journal of Operational Research. 2009;196(1):128–139.

[53] Kennedy J. Bare bones particle swarms. In: 2003 IEEE Swarm Intelligence Symposium, SIS 2003 – Proceedings; 2013. p. 80–87.

[54] Wang H, Rahnamayan S, Sun H, *et al.* Gaussian bare-bones differential evolution. IEEE Transactions on Cybernetics. 2013;43(2):634–647.

[55] Doanis P, Boursianis AD, Huillery J, *et al.* Differential evolution in waveform design for wireless power transfer. Telecom. 2020;1(2):96–113.

[56] Guo S and Yang C. Enhancing differential evolution utilizing eigenvector-based crossover operator. IEEE Transactions on Evolutionary Computation. 2015;19(1):31–49.

[57] Demmel J and Veselić K. Jacobi's method is more accurate than QR. SIAM Journal on Matrix Analysis and Applications. 1992;13(4):1204–1245.

[58] Qin AK, Huang VL, and Suganthan PN. Differential evolution algorithm with strategy adaptation for global numerical optimization. IEEE Transactions on Evolutionary Computation. 2009;13(2):398–417.

[59] Zhang J and Sanderson AC. JADE: Adaptive differential evolution with optional external archive. IEEE Transactions on Evolutionary Computation. 2009;13(5):945–958.

[60] Wang L, Fu X, Mao Y, *et al.* A novel modified binary differential evolution algorithm and its applications. Neurocomputing. 2012;98:55–75.

[61] Goudos S. Antenna design using binary differential evolution: Application to discrete-valued design problems. IEEE Antennas and Propagation Magazine. 2017;59(1):74–93.

[62] Baluja S. Population-based incremental learning: A method for integrating genetic search based function optimization and competitive learning. Technical Report. 1994.

[63] Simon D. Biogeography-based optimization. IEEE Transactions on Evolutionary Computation. 2008;12(6):702–713.

[64] Ma H. An analysis of the equilibrium of migration models for biogeography-based optimization. Information Sciences. 2010;180(18):3444–3464.

[65] Guo W, Wang L, and Wu Q. An analysis of the migration rates for biogeography-based optimization. Information Sciences. 2014;254: 111–140.

[66] Saremi S, Mirjalili S, and Lewis A. Biogeography-based optimisation with chaos. Neural Computing and Applications. 2014;25(5):1077–1097.

[67] Yang XS. In: Watanabe O, Zeugmann T, editors. Firefly Algorithms for Multimodal Optimization. Berlin, Heidelberg: Springer Berlin Heidelberg; 2009. p. 169–178.

[68] Wang GG, Deb S, and Cui Z. Monarch butterfly optimization. Neural Computing and Applications. 2019;31:1995–2014.

[69] Wang GG. Moth search algorithm: A bio-inspired metaheuristic algorithm for global optimization problems. Memetic Computing. 2018;10(2): 151–164.

[70] Wang G, Deb S, and dos S Coelho L. Elephant herding optimization. In: 2015 3rd International Symposium on Computational and Business Intelligence (ISCBI); 2015. p. 1–5.

[71] Eusuff M, Lansey K, and Pasha F. Shuffled frog-leaping algorithm: A memetic meta-heuristic for discrete optimization. Engineering Optimization. 2006;38(2):129–154.

[72] Rashedi E, Nezamabadi-pour H, and Saryazdi S. GSA: A gravitational search algorithm. Information Sciences. 2009;179(13):2232–2248.

[73] Mirjalili S and Hashim SZM. A new hybrid PSOGSA algorithm for function optimization. In: 2010 International Conference on Computer and Information Application; 2010. p. 374–377.

[74] Bayraktar Z, Komurcu M, Bossard JA, *et al.* The wind driven optimization technique and its application in electromagnetics. IEEE Transactions on Antennas and Propagation. 2013;61(5):2745–2757.

[75] Rao RV, Savsani VJ, and Vakharia DP. Teaching-learning-based optimization: A novel method for constrained mechanical design optimization problems. CAD Computer Aided Design. 2011;43(3):303–315.

[76] Tsiflikiotis A, Goudos SK, and Karagiannidis GK. Hybrid teaching-learning optimization of wireless sensor networks. Transactions on Emerging Telecommunications Technologies. 2017;28(11):e3194.

[77] Rao RV. Jaya: A simple and new optimization algorithm for solving constrained and unconstrained optimization problems. International Journal of Industrial Engineering Computations. 2016;7(1):19–34.

[78] Geem ZW, Kim JH, and Loganathan GV. A new heuristic optimization algorithm: Harmony search. Simulation. 2001;76(2):60–68.

[79] Chandrasekaran K, Simon SP, and Padhy NP. Binary real coded firefly algorithm for solving unit commitment problem. Information Sciences. 2013;249:67–84.

[80] Crawford B, Soto R, Olivares-Suárez M, *et al.* A binary firefly algorithm for the set covering problem. In: Silhavy R, Senkerik R, Oplatkova ZK, *et al.*, editors. Modern Trends and Techniques in Computer Science: 3rd Computer Science On-line Conference 2014 (CSOC 2014). Cham: Springer International Publishing; 2014. p. 65–73.

[81] Boursianis AD, Goudos SK, Kantartzis NV, *et al.* Patch antenna design for C2C communication systems using monarch butterfly optimization. In: International Workshop on Antenna Technology (iWAT 2019); Miami, FL, USA. 2019, p. 239–241.

[82] Goudos SK. Cognitive radio engine design for IoT using monarch butterfly optimization and fuzzy decision making. Internet of Things; Springer Nature, Switzerland, 2020. p. 81–100.

[83] Stromberger I, Tuba E, Bacanin N, *et al.* Monarch butterfly optimization algorithm for localization in wireless sensor networks. In: 2018 28th International Conference Radioelektronika, RADIOELEKTRONIKA 2018; 2018. p. 1–6.

[84] Wang GG, Zhao X, and Deb S. A novel monarch butterfly optimization with greedy strategy and self-adaptive. In: Proceedings – 2015 2nd International Conference on Soft Computing and Machine Intelligence, ISCMI 2015; 2015. p. 45–50.

[85] Boursianis AD, Koulouridis S, Rocca P, *et al.* Modified patch antenna design using moth search algorithm for RF energy harvesting applications. In: International Workshop on Antenna Technology (iWAT 2020); Bucharest, Romania. 2020, p. 1–3.

[86] Boursianis A, Patakakis K, Goudos SK, *et al.* Phased antenna array design using shuffled frog-leaping algorithm. In: 2018 IEEE International Symposium on Antennas and Propagation USNC/URSI National Radio Science Meeting; 2018. p. 1845–1846.

[87] Makiabadi MH, Baghlani A, Rahnema H, *et al.* Optimal design of truss bridges using teaching-learning-based optimization algorithm. International Journal of Optimization in Civil Engineering. 2013;3(3):499–510.

[88] Krishnasamy U and Nanjundappan D. A refined teaching-learning based optimization algorithm for dynamic economic dispatch of integrated multiple fuel and wind power plants. Mathematical Problems in Engineering. 2014;2014. Article ID 956405.

[89] Tizhoosh HR. Opposition-based learning: A new scheme for machine intelligence. In: Proceedings – International Conference on Computational Intelligence for Modelling, Control and Automation, CIMCA 2005 and International Conference on Intelligent Agents, Web Technologies and Internet. vol. 1; 2005. p. 695–701.

[90] Tizhoosh HR. Reinforcement learning based on actions and opposite actions. In: Proceedings of the ICGST International Conference on Artificial Intelligence and Machine Learning; 2005. p. 94–98.

[91] Tizhoosh HR and Ventresca M. Oppositional concepts in computational intelligence. Oppositional Concepts in Computational Intelligence. New York: Springer; 2008.

[92] Mohseni Pour F and Gharaveisi AA. Opposition-based discrete action reinforcement learning automata algorithm case study: Optimal design of a PID controller. Turkish Journal of Electrical Engineering and Computer Sciences. 2013;21(6):1603–1614.

[93] Rahnamayan S, Tizhoosh HR, Salama MMA, *et al.* Quasi-oppositional differential evolution. In: 2007 IEEE Congress on Evolutionary Computation, Proceedings, vols. 1–10; 2007.

[94] Ergezer M, Simon D, Du D, *et al.* Oppositional biogeography-based optimization. In: IEEE International Conference on Systems Man and Cybernetics Conference Proceedings; 2009. p. 1009–1014.

[95] Wang H, Wu Z, Rahnamayan S, *et al.* Enhancing particle swarm optimization using generalized opposition-based learning. Information Sciences. 2011;181(20):4699–4714.

[96] Xu Q, Wang L, He B, *et al.* Modified opposition-based differential evolution for function optimization. Journal of Computational Information Systems. 2011;7:1582–1592.

[97] Rahnamayan S, Jesuthasan J, Bourennani F, *et al.* Computing opposition by involving entire population. In: 2014 IEEE Congress on Evolutionary Computation; 2014.

[98] Seif Z and Ahmadi MB. An opposition-based algorithm for function optimization. Engineering Applications of Artificial Intelligence. 2015;37:293–306.

[99] Rahnamayan RS, Tizhoosh HR, and Salama MMA. Opposition-based differential evolution. IEEE Transactions on Evolutionary Computation. 2008;12(1):64–79.

[100] Ergezer M and Simon D. Oppositional biogeography-based optimization for combinatorial problems. In: 2011 IEEE Congress of Evolutionary Computation, CEC 2011; 2011. p. 1496–1503.

[101] Ergezer M, Simon D, and Du D. Oppositional biogeography-based optimization. In: Conference Proceedings – IEEE International Conference on Systems, Man and Cybernetics; 2009. p. 1009–1014.

[102] Goudos SK. A novel generalized oppositional biogeography-based optimization algorithm: Application to peak to average power ratio reduction in OFDM systems. Open Mathematics. 2016;14(1):705–722.

[103] Deb K, Pratap A, Agarwal S, *et al.* A fast and elitist multiobjective genetic algorithm: NSGA-II. IEEE Transactions on Evolutionary Computation. 2002;6(2):182–197.

[104] Deb K and Agrawal RB. Simulated binary crossover for continuous search space. Complex Systems. 1995;9(2):115–148.

[105] Deb K and Jain H. An evolutionary many-objective optimization algorithm using reference-point-based nondominated sorting approach, Part I: Solving problems with box constraints. IEEE Transactions on Evolutionary Computation. 2014;18(4):577–601.

[106] Deb K, Thiele L, Laumanns M, *et al.* In: Abraham A, Jain L, Goldberg R, editors. Scalable Test Problems for Evolutionary Multiobjective Optimization. London: Springer London; 2005. p. 105–145.

[107] Kukkonen S and Lampinen J. GDE3: The third evolution step of generalized differential evolution. In: Proceedings of the 2005 IEEE Congress on Evolutionary Computation (CEC 2005). vol. 1; 2005. p. 443–450.

[108] Kukkonen S and Lampinen J. Performance assessment of generalized differential evolution 3 (GDE3) with a given set of problems. In: Proceedings of the IEEE Congress on Evolutionary Computation, 2007. CEC 2007; 2007. p. 3593–3600.

[109] Tan KC. CEC 2007 conference report. IEEE Computational Intelligence Magazine. 2008;3(2):72–73.

[110] Goudos SK, Siakavara K, Vafiadis EE, *et al.* Pareto optimal Yagi-Uda antenna design using multi-objective differential evolution. Progress in Electromagnetics Research. 2010;105:231–251.

[111] Goudos SK and Sahalos JN. Pareto optimal microwave filter design using multiobjective differential evolution. IEEE Transactions on Antennas and Propagation. 2010;58(1):132–144.

[112] Goudos SK, Gotsis KA, Siakavara K, *et al.* A multi-objective approach to subarrayed linear antenna arrays design based on memetic differential evolution. IEEE Transactions on Antennas and Propagation. 2013;61(6):3042–3052.

[113] Deb K, Thiele L, Laumanns M, *et al.* Evolutionary multiobjective optimization: Theoretical advances and applications. In: Abraham A, Jain L, Goldberg R, editors. Scalable Test Problems for Evolutionary Multiobjective Optimization. London: Springer; 2005. p. 105–145.

[114] Kukkonen S and Deb K. A fast and effective method for pruning of nondominated solutions in many-objective problems. Lecture Notes in Computer Science. 2006;4193:553–562.

[115] Kukkonen S and Lampinen J. GDE3: The third evolution step of generalized differential evolution. In: 2005 IEEE Congress on Evolutionary Computation. vol. 1; 2005. p. 443–450.

[116] Kukkonen S and Lampinen J. Performance assessment of generalized differential evolution 3 with a given set of constrained multi-objective test problems. In: 2009 IEEE Congress on Evolutionary Computation; 2009. p. 1943–1950.

[117] Nebro AJ, Durillo JJ, Garcia-Nieto J, *et al.* SMPSO: A new PSO-based metaheuristic for multi-objective optimization. In: IEEE Symposium on Computational Intelligence in Multi-Criteria Decision-Making (MCDM); 2009. p. 66–73.

[118] Kukkonen S and Deb K. A fast and effective method for pruning of nondominated solutions in many-objective problems. In: Runarsson TP, Beyer HG, Burke E, *et al.*, editors. Parallel Problem Solving from Nature – PPSN IX. Berlin, Heidelberg: Springer Berlin Heidelberg; 2006. p. 553–562.

[119] Goudos SK, Plets D, Liu N, *et al.* A multi-objective approach to indoor wireless heterogeneous networks planning based on biogeography-based optimization. Computer Networks. 2015;91:564–576.

Chapter 3

Antenna array design using EAs

This chapter presents several antenna array design cases by using different evolutionary algorithms (EAs) and comparing results.

The synthesis of antenna arrays plays a very important role in communication systems. Array synthesis is a classic and challenging optimization problem, which has been extensively studied using several analytical or stochastic methods [1–7]. The increased use of such arrays creates more challenges upon the antenna engineers. More requirements, such as pattern shaping, low profile, wideband/narrowband, and interference cancellation; and more limitations such as power dissipation and antenna size, lead to the urgent need for simple, time saving, and efficient optimization techniques. Common optimization goals in array synthesis are the sidelobe level (SLL) suppression and the matching of the mainlobe to a desired pattern. Thus, the optimization problem is usually to find a set of element excitations and/or positions that closely match a desired pattern. The desired pattern shape can vary widely depending on the application. Several new synthesis and optimization techniques have emerged in the last two decades that mimic biological evolution, brain function, or the way biological entities communicate in nature. Several of these methods have been applied to the array design problem.

Additionally, EAs like genetic algorithms (GAs), particle swarm optimization (PSO), and differential evolution (DE) [8], which is a population-based stochastic global optimization algorithm, have been applied to a variety of array design problems [3,9–16].

3.1 Linear-array design

We consider an N-element linear-array symmetrically placed along the x-axis (Figure 3.1). When N is even, the 0 element does not exist. The array factor in the $x-z$ plane is expressed as

$$AF(\vartheta,\bar{x},\bar{\phi},\bar{I}) = \sum_{n=1}^{N} I_n e^{j((2\pi/\lambda)x_n \sin\vartheta+\phi_n)} \tag{3.1}$$

where λ is the wavelength, ϑ is the steering angle measured with respect to the positive direction of the z-axis, x_n, I_n, and ϕ_n are the position, amplitude, and the phase of

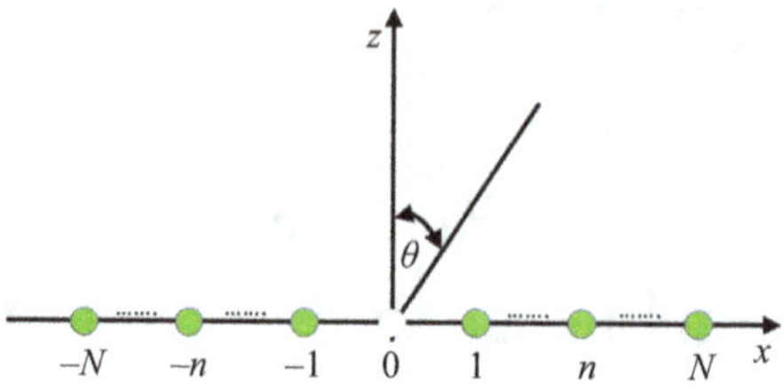

Figure 3.1 Geometry of a linear array

the nth element, respectively, and $\bar{x}, \bar{I}$, and $\bar{\phi}$ are the corresponding vectors. For a symmetrically excited array, (3.1) becomes

$$AF(\vartheta, \bar{x}, \bar{\phi}, \bar{I}) = 2\sum_{n=1}^{M} I_n \cos\left[\frac{2\pi}{\lambda} x_n \sin\vartheta + \phi_n\right] + I_0 \tag{3.2}$$

where $M = \lfloor N/2 \rfloor$ is the largest integer less than or equal to $N/2$. For an even number of elements $I_0 = 0$, and for an odd number of elements we set $I_0 = 1$. The element positions in this array satisfy the following constraints:

$$x_1 \geq \frac{d_{\min}}{2}, \quad x_1 \leq \frac{d_{\max}}{2}, \qquad \text{when } N = 2M \tag{3.3}$$

$$|x_0 - x_1| \geq d_{\min}, \quad |x_0 - x_1| \leq d_{\max}, \qquad \text{when } N = 2M + 1 \tag{3.4}$$

$$\left.\begin{array}{l} |x_i - x_{i+1}| \geq d_{\min} \\ |x_i - x_{i+1}| \leq d_{\max} \end{array}\right\}, 1 \leq i \leq M - 1$$

where $d_{\min}$ and $d_{\max}$ are the minimum and maximum distance between two adjacent elements, respectively. In the case of a uniformly excited linear array symmetrically placed along the x-axis, the array factor in the x–z plane is then expressed as

$$AF(\vartheta, \bar{y}, \bar{\phi}) = 2\sum_{m=1}^{M} \cos\left[\frac{2\pi}{\lambda} y_m \sin\vartheta + \phi_m\right] \tag{3.5}$$

Equation (3.5) expressed in dB is written as

$$AF_{\mathrm{dB}}\left(\vartheta, \bar{x}, \bar{\phi}\right) = 20 \times \log_{10}\left|\frac{AF\left(\vartheta, \bar{x}, \bar{\phi}\right)}{AF\left(\vartheta_o, \bar{x}, \bar{\phi}\right)}\right| \tag{3.6}$$

where ϑ_o is the direction of the maximum. The optimization goal is the SLL suppression. In the case of position or phase-only synthesis, this is achieved by finding the optimum element positions and phases. This problem is defined by the minimization of the objective function:

$$F_1(\bar{y}, \bar{\phi}) = \max_{\vartheta \in T}\{AF_{\mathrm{dB}}(\vartheta, \bar{y}, \bar{\phi})\} \tag{3.7}$$

where T is the set of theta angles that are outside the angular range of the mainlobe. As in [11] the linear-array synthesis design cases are executed 50 times for each algorithm. The algorithms selected for this case are different self-adaptive DE algorithms, which were presented in Chapter 2. More specifically, we select for comparison the algorithms jDE [17], barebones DE (BBDE) [18], Gaussian barebones DE (GBDE) [19], composite DE (CoDE) [20], and JADE [21].

We examine different design options for uniformly excited arrays, which include position-only, phase-only, and position–phase synthesis. For position- and phase-only synthesis the number of unknowns is N, while for position–phase synthesis this number is $2N$. We set $d_{\min} = 0.5\lambda$ for all cases, where $d_{\min}$ is the minimum interelement distance.

3.1.1 Position-only optimization

The first design case is that of a 32-element array. We consider first that $d_{\max} = 0.7\lambda$, and we apply all algorithms to position-only synthesis. The optimization process uses the distances d_n between two adjacent elements as the unknown position variables instead of using the absolute positions x_n.

Table 3.1 holds the comparative results for this case. We notice that for this case all algorithms perform similarly. Additionally, all algorithms have found the same best value result. The boxplot of the algorithms results is depicted in Figure 3.2. We notice

Table 3.1 Position-only synthesis with 32-elements $d_{max} = 0.7\lambda$ (comparative results)

Algorithm	Best	Worst	Mean	Median	St. Dev.
BBDE	−18.78	−18.13	−18.59	−18.63	1.51E−01
GBDE	−18.78	−18.78	−18.78	−18.78	0.00E+00
CoDE	−18.78	−18.78	−18.78	−18.78	0.00E+00
JADE	−18.78	−18.78	−18.78	−18.78	1.71E−05
jDE	−18.78	−18.78	−18.78	−18.78	2.40E−09

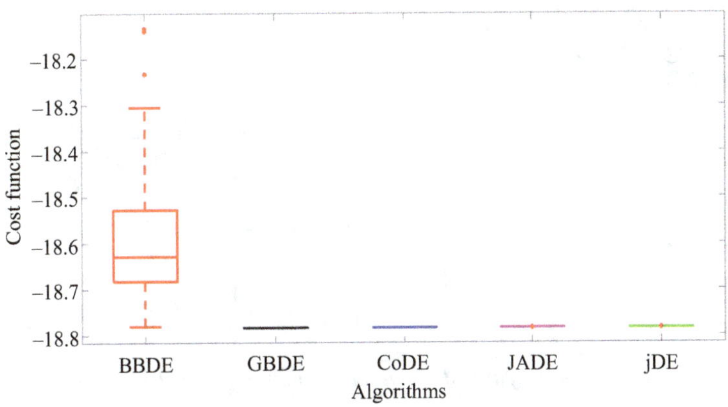

Figure 3.2 Position-only synthesis with 32-element $d_{max} = 0.7\lambda$. Boxplot of all algorithms results

that only the results obtained by BBDE present a large dispersion of values. The corresponding convergence rate plot is shown in Figure 3.3. From this plot, we can conclude that all algorithms converge at similar rate. Figure 3.4 shows the obtained radiation pattern for this case. The element positions of the best obtained results are shown in Table 3.2.

The next design case is that of position-only synthesis, where we set $d_{max} = \lambda$. Table 3.3 reports the algorithms comparative results. For this case, we notice that CoDE obtained the best results in terms of all statistical measures. This is also evident in the results boxplot that is depicted in Figure 3.5. The convergence speed is shown

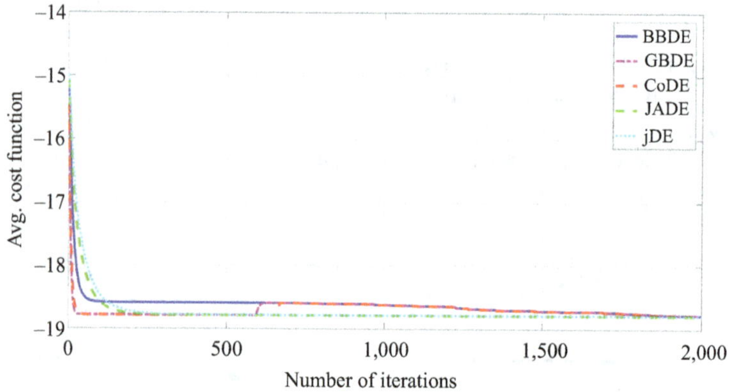

Figure 3.3 Position-only synthesis with 32-element $d_{max} = 0.7\lambda$. Convergence rate graph

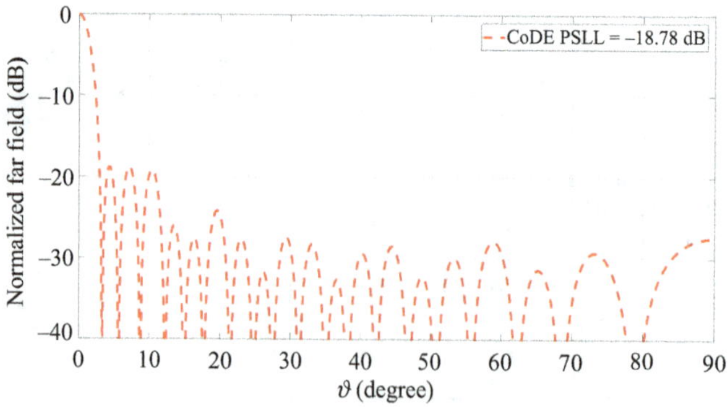

Figure 3.4 Position-only synthesis with 32-element $d_{max} = 0.7\lambda$. Best radiation pattern found

in Figure 3.6. All algorithms converge at similar speed except BBDE, which seems to be slower. Figure 3.7 holds the best radiation patterns obtained by all algorithms, while the corresponding interelement positions for the best result are reported in Table 3.4.

Table 3.2 Position-only synthesis with 32-element $d_{max} = 0.7\lambda$ (element positions of the best obtained result)

n	x_n	n	x_n
1	0.250	9	4.553
2	0.750	10	5.094
3	1.250	11	5.662
4	1.750	12	6.490
5	2.250	13	7.352
6	2.750	14	8.240
7	3.297	15	9.238
8	3.801	16	10.064

Table 3.3 Position-only synthesis with 32-element $d_{max} = \lambda$ (comparative results)

Algorithm	**Best**	**Worst**	**Mean**	**Median**	**St. Dev.**
BBDE	−22.20	−19.09	−20.84	−20.86	6.05E−01
GBDE	−22.60	−21.95	−22.42	−22.45	1.59E−01
CoDE	−22.64	−22.56	−22.63	−22.64	2.11E−02
JADE	−22.16	−22.07	−22.13	−22.15	2.84E−02
jDE	−22.60	−22.05	−22.18	−22.12	1.61E−01

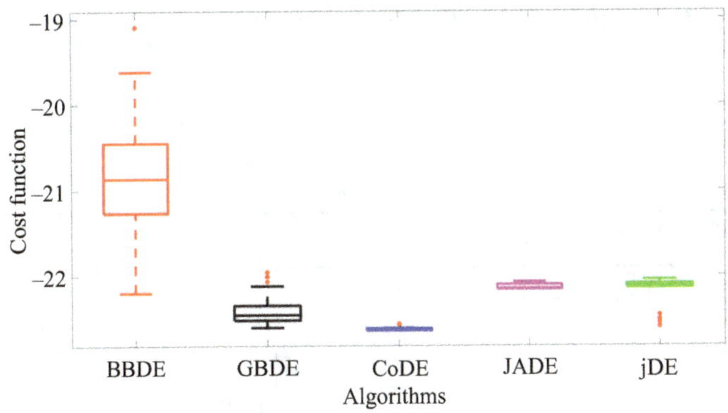

Figure 3.5 Position-only synthesis with 32-element $d_{max} = \lambda$. Boxplot of all algorithms results

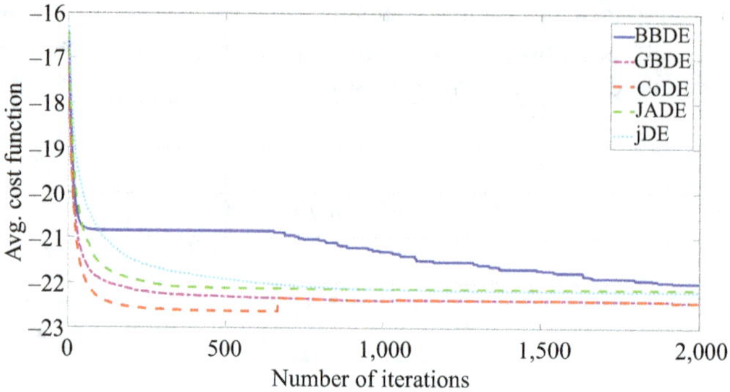

Figure 3.6 Position-only synthesis with 32-element $d_{max} = \lambda$. Convergence rate graph

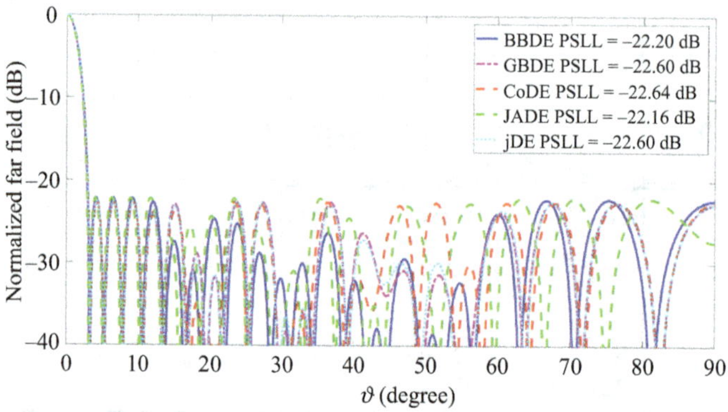

Figure 3.7 Position-only synthesis with 32-element $d_{max} = \lambda$. Best radiation patterns found by all algorithms

Table 3.4 Position-only synthesis with 32-element $d_{max} = \lambda$ (element positions of the best obtained result)

n	x_n	n	x_n
1	0.250	9	4.572
2	0.750	10	5.098
3	1.250	11	5.634
4	1.750	12	6.472
5	2.250	13	7.385
6	2.750	14	8.239
7	3.288	15	9.179
8	3.788	16	10.102

3.1.2 Phase-only optimization

In this section, we apply phase-only optimization, i.e., all element amplitudes are set to one and all interelement positions are set to $d_{\max} = 0.5\lambda$. The only array feature that is varied in the optimization is the array phases, which can be from 0° to 359°. Here, we consider the design case of a 60-element array. We apply all algorithms to phase-only synthesis. Moreover, it is set $d_{\max} = 0.7\lambda$. Table 3.5 holds all the comparative results. In this case the jDE algorithm clearly outperforms the other algorithms. JADE has found the second best results. This fact is also shown in a graphical manner in boxplot graph of Figure 3.8.

Moreover, Figure 3.9 shows that all algorithms converge at similar speed, except BBDE and jDE that seem to be slower. Figure 3.10 presents the three best radiation patterns obtained by jDE, JADE, and GBDE. The configuration of the best array obtained by jDE is reported in Table 3.6.

Table 3.5 Phase-only synthesis with 60-element $d_{max} = 0.7\lambda$ (comparative results)

Algorithm	Best	Worst	Mean	Median	St. Dev.
BBDE	−18.03	−14.81	−17.01	−17.2	7.99E−01
GBDE	−19.52	−18.5	−18.92	−18.95	2.08E−01
CoDE	−18.83	−18.48	−18.65	−18.65	8.19E−02
JADE	−19.38	−18.69	−19.07	−19.05	1.38E−01
jDE	−19.56	−18.51	−19.14	−19.19	2.59E−01

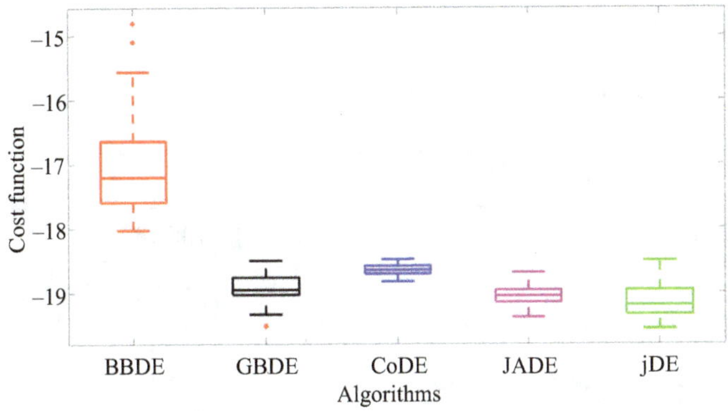

Figure 3.8 Phase-only synthesis with 60-element $d_{max} = 0.7\lambda$. Boxplot of all algorithms results

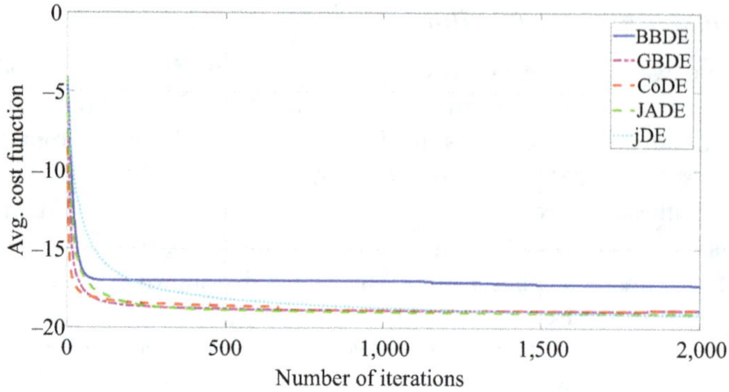

Figure 3.9 Phase-only synthesis with 60-element $d_{max} = 0.7\lambda$. Convergence rate graph

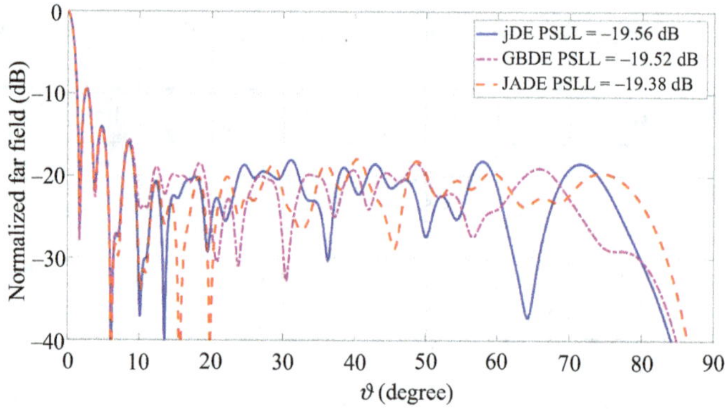

Figure 3.10 Phase-only synthesis with 60-element $d_{max} = 0.7\lambda$. Best radiation patterns found

Table 3.6 Phase-only synthesis with 60-element $d_{max} = 0.7\lambda$ (element phases)

n	φ_n°	n	φ_n°	n	φ_n°	n	φ_n°
1	151.17	9	138.90	17	170.93	25	169.95
2	157.60	10	132.92	18	162.57	26	162.90
3	195.29	11	136.79	19	174.07	27	169.65
4	117.75	12	216.68	20	155.84	28	162.65
5	74.74	13	142.82	21	158.99	29	163.43
6	172.56	14	204.10	22	173.80	30	159.60
7	166.67	15	199.66	23	158.50		
8	311.21	16	168.28	24	165.47		

3.1.3 Position and phase optimization

In order to further evaluate the algorithms results, we apply them to a more complex design case of both position and phase optimization. We consider two cases.

The first design case is that of a 32-element array. We assume first that $d_{max} = 0.6\lambda$ and we apply all algorithms to position-phase optimization. Table 3.7 holds the comparative results. In this case, we notice that the algorithms differences are small. jDE and JADE obtained the same best result; however, GBDE obtained the best mean and median values. The pattern obtained by GBDE is very close to the ones obtained by jDE and JADE. CoDE also obtained comparative results and the smaller standard deviation value. Figure 3.11 shows the boxplot for this case. The convergence plot is shown in Figure 3.12. One may notice that GBDE and CoDE converge at higher speed.

Figure 3.13 depicts the radiation pattern of the three best patterns obtained by jDE, GBDE, and JADE. We notice that the main lode beamwidth is almost identical

Table 3.7 Phase–position synthesis with 32-element $d_{max} = 0.6\lambda$ (comparative results)

Algorithm	Best	Worst	Mean	Median	St. Dev.
BBDE	−18.65	−15.11	−17.52	−17.45	6.22E−01
GBDE	−19.32	−18.72	−19.11	−19.14	1.49E−01
CoDE	−19.21	−18.96	−19.08	−19.07	4.59E−02
JADE	−19.35	−18.48	−18.81	−18.80	1.65E−01
jDE	−19.35	−18.18	−18.99	−19.00	2.44E−01

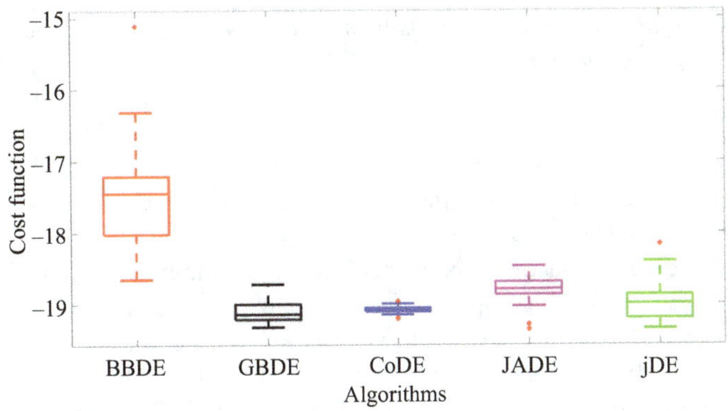

Figure 3.11 Phase–position synthesis with 32-element $d_{max} = 0.6\lambda$. Boxplot of all algorithms results

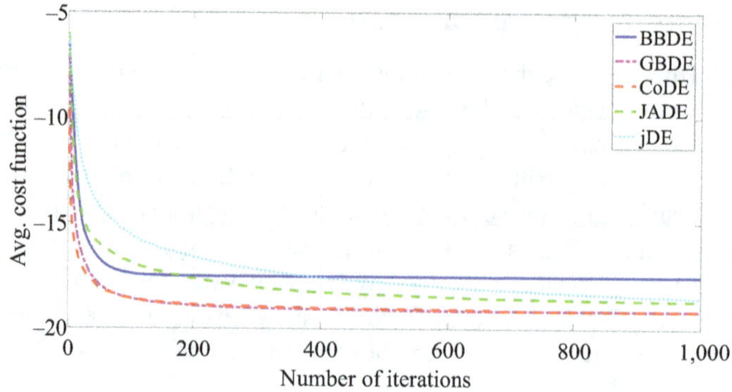

Figure 3.12 Phase–position synthesis with 32-element $d_{max} = 0.6\lambda$. Convergence rate graph

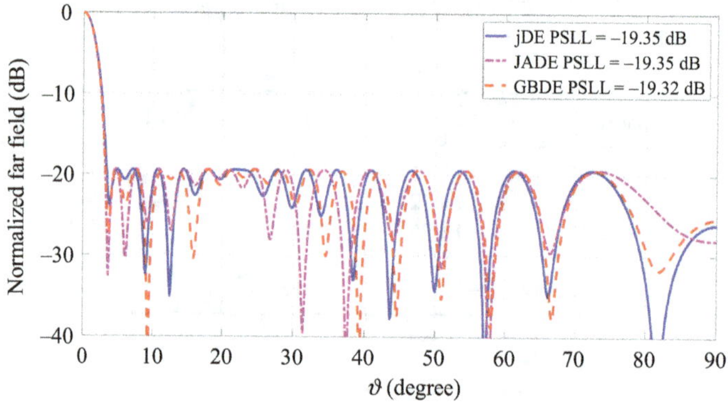

Figure 3.13 Phase–position synthesis with 32-element $d_{max} = 0.6\lambda$. Best radiation pattern found

in all patterns. The values of positions and phases of the best obtained pattern are reported in Table 3.8.

The next case is that of 32 elements but the maximum interelement distance is now set to $d_{\max} = \lambda$. Table 3.9 reports the comparative results. jDE obtained the best results in terms of best and mean values. GBDE, CoDE, and JADE obtained similar best patterns, but the mean values are less than that of jDE. This indicates that jDE can produce good results with fewer algorithm runs. The boxplot is presented in Figure 3.14, where we can see the dispersion of values. CoDE results have the smaller standard deviation as it is shown from the boxplot. Figure 3.15 shows the convergence rate graph. GBDE, CoDE, and JADE converge at similar speed, while

Table 3.8 Phase–position synthesis with 32-element $d_{max} = 0.6\lambda$ (element positions and phases of the best obtained pattern)

n	x_n	$\varphi_n{}^\circ$	n	x_n	$\varphi_n{}^\circ$
1	0.250	161.198	9	4.455	157.317
2	0.751	160.504	10	5.054	160.018
3	1.252	158.759	11	5.654	145.344
4	1.752	158.198	12	6.253	165.219
5	2.252	159.750	13	6.853	112.864
6	2.754	159.849	14	7.423	242.935
7	3.272	157.265	15	8.023	199.848
8	3.860	154.244	16	8.623	168.333

Table 3.9 Phase–position synthesis with 32-element $d_{max} = \lambda$ (comparative results)

Algorithm	**Best**	**Worst**	**Mean**	**Median**	**St. Dev.**
BBDE	−22.04	−16.03	−19.69	−20.03	1.38E+00
GBDE	−23.60	−20.67	−22.99	−23.00	4.60E−01
CoDE	−23.00	−22.54	−22.76	−22.74	1.14E−01
JADE	−23.43	−22.10	−23.13	−23.37	3.20E−01
jDE	−23.66	−22.74	−23.27	−23.32	1.90E−01

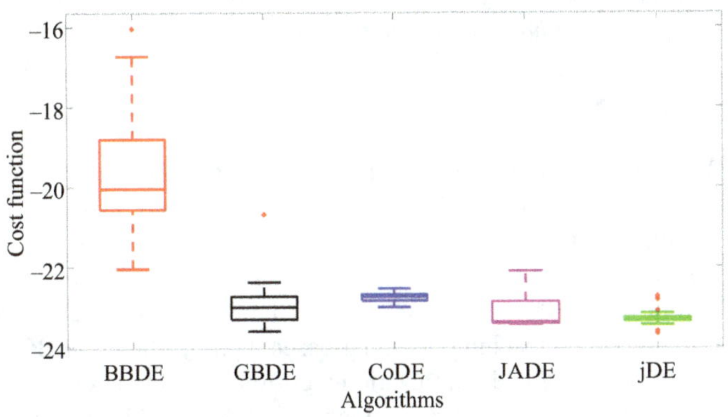

Figure 3.14 Phase–position synthesis with 32-element $d_{max} = \lambda$. Boxplot of all algorithms results

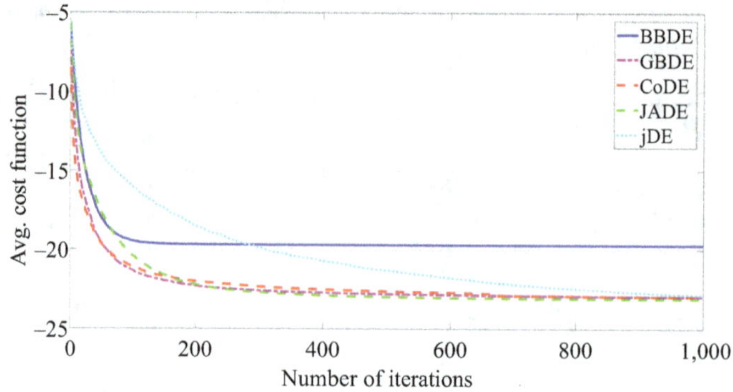

Figure 3.15 Phase–position synthesis with 32-element $d_{max} = \lambda$. Convergence rate graph

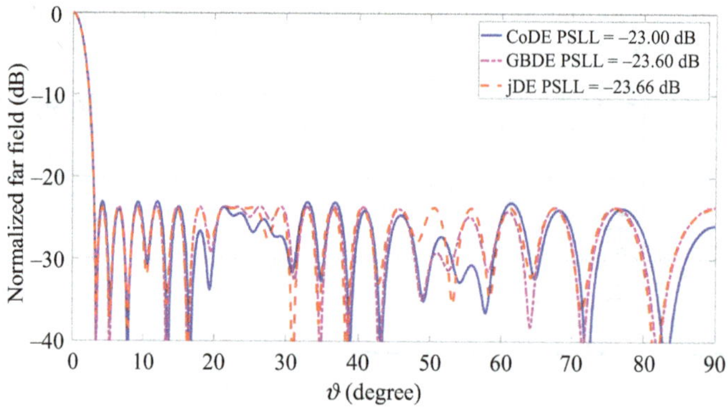

Figure 3.16 Phase–position synthesis with 32-element $d_{max} = \lambda$. Best radiation patterns found by all algorithms

jDE and BBDE converge more slowly. Figure 3.16 illustrates the radiation patterns of the three best obtained results. Table 3.10 reports the values of positions and phases of the best obtained pattern.

Next, a design case with 60 elements and maximum interelement distance at $d_{\max} = 0.7\lambda$ follows. In this case, we apply BBDE and emerging nature-inspired algorithms, namely, the teaching–learning-based optimization (TLBO) [22], the Jaya algorithm [23], the whale optimization algorithm (WOA) [24], the salp swarm algorithm (SSA) [25], and the grey wolf optimizer (GWO) [26].

Table 3.11 holds the comparative results. We notice that GWO obtained the best value, the best mean values, and the best median value. This is also evident from the boxplot in Figure 3.17. However, GWO, Jaya, and BBDE obtained results with a large

Table 3.10 Phase–position synthesis with 32-element $d_{max} = \lambda$ (element positions and phases of the best obtained result)

n	x_n	$\varphi_n{}^\circ$	n	x_n	$\varphi_n{}^\circ$
1	0.250	156.221	9	4.546	158.296
2	0.750	158.217	10	5.088	156.851
3	1.250	158.053	11	5.760	159.751
4	1.751	157.234	12	6.589	148.336
5	2.252	160.483	13	7.451	156.334
6	2.783	158.618	14	8.318	186.150
7	3.323	156.363	15	9.253	125.566
8	3.899	159.985	16	10.142	155.963

Table 3.11 Phase–position synthesis with 60-element $d_{max} = 0.7\lambda$ (comparative results)

Algorithm	**Best**	**Worst**	**Mean**	**Median**	**St. Dev.**
GWO	−22.05	−19.00	−20.98	−21.37	9.61E−01
SSA	−21.28	−16.84	−17.67	−17.48	9.55E−01
WOA	−19.78	−17.52	−18.57	−18.53	5.23E−01
Jaya	−21.42	−16.43	−18.95	−18.92	1.15E+00
TLBO	−20.88	−18.57	−19.59	−19.60	4.17E−01
BBDE	−19.13	−14.90	−17.22	−17.40	1.05E+00

dispersion of values. Figure 3.18 portrays the convergence rate plot. All algorithms seem to converge at similar speed. The radiation patterns of the three best results are shown in Figure 3.19. The GWO result is clearly the best, while the Jaya and SSA results are very similar. Table 3.12 reports the values of positions and phases of the best obtained pattern by GWO.

3.1.4 Amplitude-only optimization

The amplitude optimization problem is the most common in array design [27–30]. If we consider a broadside array, where all the array phases are set to zero, then the decision variables are the unknown array amplitudes. In order to optimize the amplitudes, the remaining control parameters x_n and ϕ_n are fixed, where ϕ_n is taken as zero and the spacing between adjacent elements is taken as $(\lambda/2), n = 1 \ldots N$. Assuming that the first element is placed at $x_1 = \lambda/4$, the array factor can be simplified as

$$AF(\vartheta) = 2\sum_{n=1}^{N} I_n \cos\left[(n - 0.5)\pi \sin(\vartheta)\right] \tag{3.8}$$

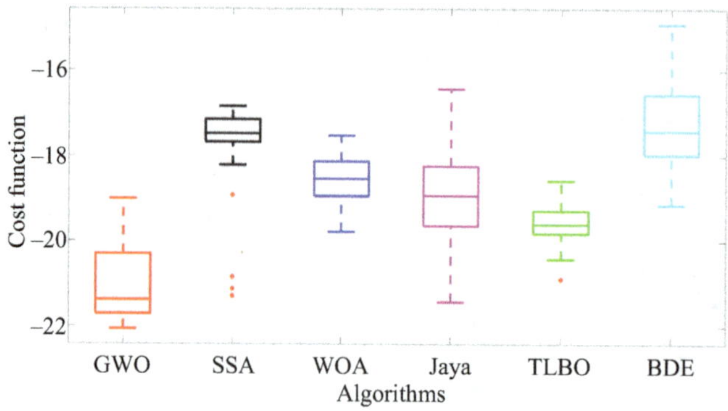

Figure 3.17 Phase–position synthesis with 60-element $d_{max} = 0.7\lambda$. Boxplot of all algorithms results

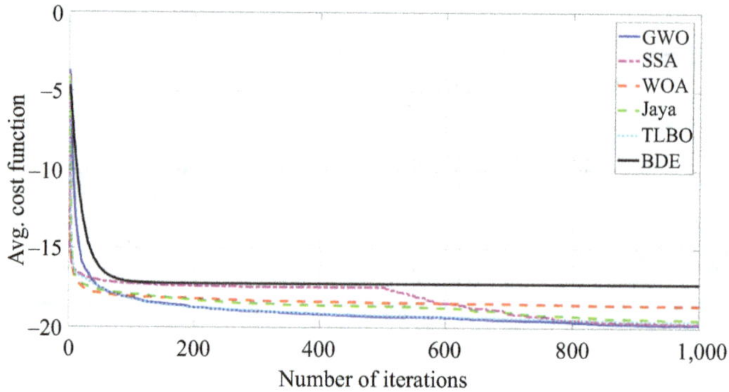

Figure 3.18 Phase–position synthesis with 60-element $d_{max} = 0.7\lambda$. Convergence rate graph

The excitation amplitudes of the $2N$-element array will be optimized in the range [0, 1]. For this case we consider the self-adaptive DE algorithms used in the previous sections, which are the jDE [17], BBDE [18], GBDE [19], CoDE [20], and JADE [21]. Moreover, we also apply two swarm intelligence (SI) algorithms: the constriction factor PSO (CFPSO) [31] and the gbest-guided ABC (GABC) [32].

Table 3.13 reports the comparative results for this case. We notice that CoDE, jDE, and JADE obtained almost the same best result. However, JADE obtained the same result in each algorithm run and has the best standard deviation value. The two SI algorithms obtained results slightly worse than the DE algorithms. Figure 3.20 shows the dispersion of values for this case. Moreover, Figure 3.21 illustrates the convergence speed of the algorithms. The JADE, CoDE, GBDE, and CFPSO converge at similar

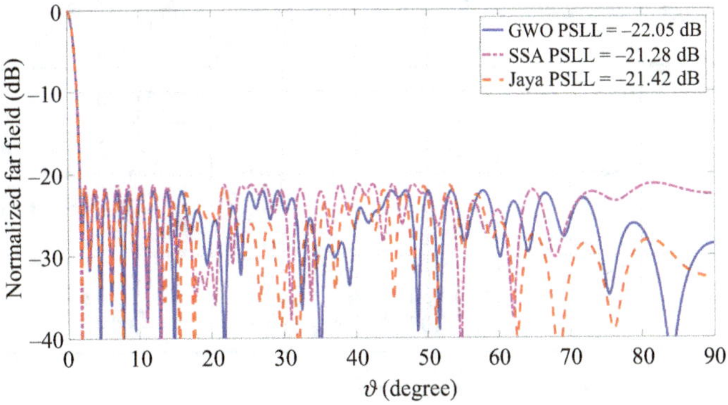

Figure 3.19 Phase–position synthesis with 60-element $d_{max} = 0.7\lambda$. Best radiation patterns found

Table 3.12 Phase–position synthesis with 60-element $d_{max} = 0.7\lambda$ (element positions and phases)

n	x_n	$\varphi_n{}^\circ$	n	x_n	$\varphi_n{}^\circ$
1	0.250	180.245	16	8.734	191.755
2	0.750	177.881	17	9.342	173.065
3	1.256	176.724	18	9.947	141.936
4	1.825	179.277	19	10.539	200.348
5	2.330	181.129	20	11.175	195.207
6	2.852	184.953	21	11.873	175.026
7	3.403	182.861	22	12.474	97.036
8	3.908	185.995	23	13.118	181.036
9	4.507	182.018	24	13.818	183.837
10	5.106	176.790	25	14.383	286.044
11	5.623	194.714	26	15.011	174.824
12	6.230	180.928	27	15.703	212.203
13	6.798	166.319	28	16.401	178.438
14	7.435	183.536	29	17.062	155.196
15	8.075	173.074	30	17.661	190.306

speed. Figure 3.22 depicts the three best obtained radiation patterns in comparison with a uniform array of 24 elements. We notice the three best patterns found by CoDE, jDE, and JADE are identical. Table 3.14 holds the amplitude configuration for the best obtained pattern.

Table 3.13 Amplitude-only synthesis with 24 elements (comparative results)

Algorithm	Best	Worst	Mean	Median	St. Dev.
BBDE	−33.51	−26.87	−31.12	−31.26	2.04E+00
GBDE	−35.07	−33.92	−34.77	−34.91	4.05E−01
CoDE	−35.22	−35.17	−35.19	−35.19	1.43E−02
JADE	−35.23	−35.23	−35.23	−35.23	7.49E−15
jDE	−35.23	−35.21	−35.22	−35.22	8.63E−03
CFPSO	−35.05	−34.54	−34.83	−34.82	1.54E−01
GABC	−34.56	−33.08	−33.98	−33.97	4.19E−01

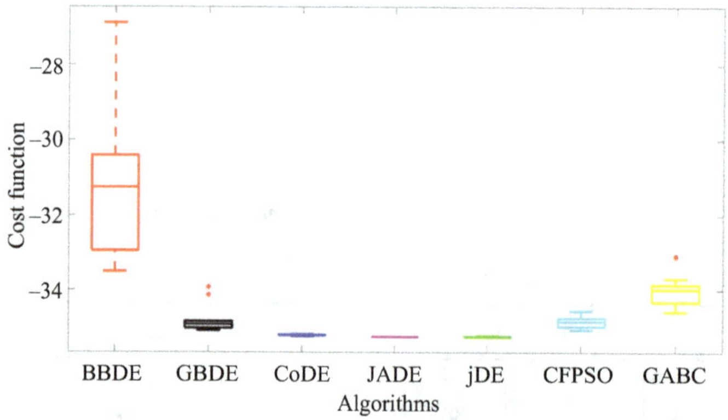

Figure 3.20 Amplitude-only synthesis with 24 elements. Boxplot of all algorithms results

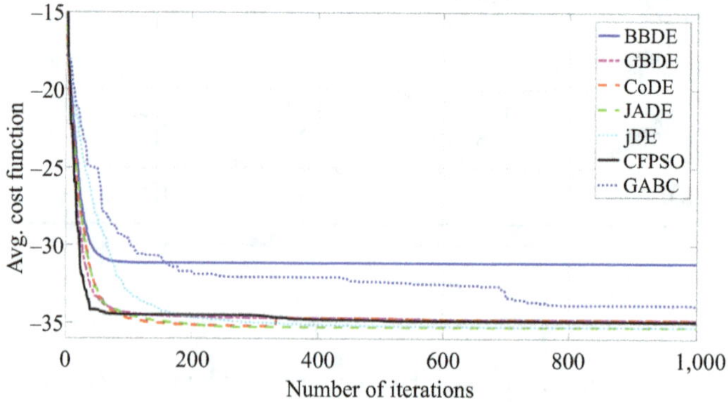

Figure 3.21 Amplitude-only synthesis with 24 elements. Convergence rate graph

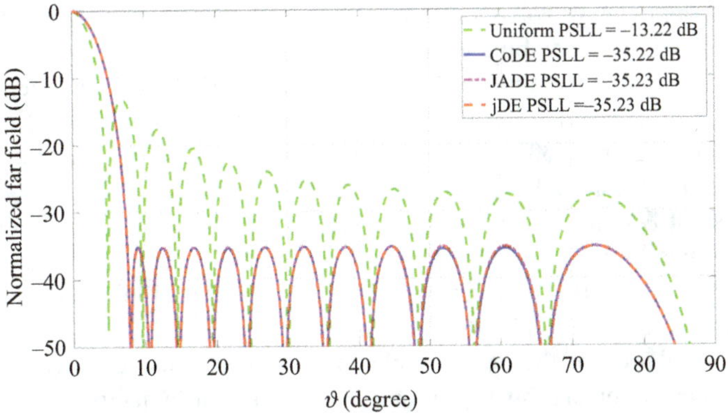

Figure 3.22 Amplitude-only synthesis with 24 elements. Best radiation patterns found

Table 3.14 Amplitude-only synthesis with 24 elements (element amplitudes of the best obtained pattern)

n	I_n	n	I_n
1	1	7	0.574
2	0.975	8	0.471
3	0.927	9	0.371
4	0.858	10	0.278
5	0.773	11	0.196
6	0.677	12	0.206

3.2 Thinned-array design

Array thinning is defined as the removal/extraction of a specific number of radiating elements from a periodic antenna array. This approach aims to create arrays with lower SLL, compared to the SLL of uniform excitation. Moreover, fill factor percentage pf is defined as the ratio percentage of the powered elements (elements in "on" state) to the total number of elements (elements in both "on" and "off" state) in an array.

These elements (powered "on") are connected to the feed network, whereas the extracted from the array elements (powered "off") are connected to a matched load. Array thinning delivers several advantages, including the cost and weight reduction. The array thinning optimization problem aims to obtain the best possible combination of powered "on" and powered "off" elements, in order to satisfy the objective function of the problem, which is usually the SLL minimization.

An apparent way to solve the array thinning problem is to perform a thorough search of all possible combinations. However, in most cases, this is not feasible, because the computational time increases exponentially with the array size. Therefore, array thinning is classified to the category of discrete, combinatorial NP complete optimization problems [33]. Any possible solution vector $\vec{x}$ in an thinned array optimization problem is a binary string.

The use of EAs in the array thinning problem is common in the literature. This problem has been solved in the literature using several EAs like binary GAs [3], PSO [34], ant colony optimization (ACO) [35], binary DE (BDE) [36], and biogeography-based optimization (BBO) [37,38].

The formulation of the design problem follows. Let us consider a K-element broadside linear antenna array of isotropic sources. The array factor can be written as

$$AF(v, \bar{I}) = \sum_{k=1}^{K} I_k e^{j(2\pi/\lambda)\,kdu} \tag{3.9}$$

where λ denotes the wavelength, I_k is the complex amplitude of the kth element, $\bar{I}$ is the corresponding vector of element amplitudes, d is the distance between two adjacent elements, $v = \sin(\vartheta)$ is the direction cosine, and ϑ is the steering angle calculated from broadside of the array. In a thinned-array case, the $\bar{I}$ is a binary string consisting of ones (element excitation) or zeros (element non-excitation). For the case of a symmetrically excited array, (3.9) becomes

$$AF(v, \bar{I}) = 2\sum_{n=1}^{M} I_n \cos\left[\frac{2\pi}{\lambda} dnv\right] + I_0 \tag{3.10}$$

where $M = \lfloor K/2 \rfloor$ is the largest integer less than or equal to $K/2$. For an even number of elements $I_0 = 0$, and for an odd number of elements, we set $I_0 = 1$. For all simulations, we assume that the interelement distance is $d = 0.5\lambda$. We define the optimization objective as the peak SLL (PSLL) minimization, which can be achieved by obtaining the best possible binary string of element excitations. In this context, we define the antenna array thinning synthesis problem as the minimization problem of the following objective function:

$$F_1(\bar{v}, \bar{I}) = \max_{v \in S_v}\{AF_{dB}(\bar{v}, \bar{I})\} \tag{3.11}$$

where S_v is the set of direction cosines that are not including in the angular range of the mainlobe.

In this subsection, we compare the BBO algorithm [39] with the two binary PSO (BPSO) algorithms (BPSO [40] and Boolean PSO (BoolPSO) [41]), the two binary-coded GWO algorithms (bGWO1), (bGWO2) proposed in [42], binary TLBO (BTLBO) that was described in the previous chapter, and the novel BDE (NBDE) algorithm [43]. For all the previously mentioned algorithms, the population size is set to 100 and the maximum number of iterations is set to 500.

The first case is a 150-element linear array with symmetrical excitation. Due to array symmetry, the dimensionality of the problem is equal to 75. Table 3.15

lists the comparative results of all algorithms. It is clear that NBDE exceeds the other algorithms in terms of best, worst, and mean values. BTLBO obtained the second best result, which is slightly worse than the best. BBO also obtained completive results. This is also evident from the algorithms boxplot, which is presented in Figure 3.23. We notice that NBDE and BTLBO obtained the best results in terms of small dispersion of values. On the other hand, the BoolPSO obtained results with a larger dispersion of values. Figure 3.24 depicts a convergence rate graph for this case. We notice that BBO presents the fastest convergence speed, while NBDE and BTLBO converge more slowly. The best-obtained results achieve a PSLL value of −22.61 dB and a fill factor of 72%, which is shown in Figure 3.25.

To evaluate the algorithms performance in a more complex problem, we consider a synthesis case of a 150-element asymmetrical array. This array lacks of symmetry and therefore the dimensionality of the problem is doubled (150), when compared to a symmetrical array. We increase the difficulty level for all algorithms by keeping the

Table 3.15 Comparative results for the case of 150-element symmetrically thinned array

Algorithm	Best	Worst	Mean	Median	St. Dev.
NBDE	−22.61	−21.89	−22.12	−22.10	1.38E−01
BPSO	−21.08	−19.83	−20.24	−20.19	3.22E−01
BoolPSO	−21.66	−20.15	−21.17	−21.30	4.98E−01
BBO	−22.24	−21.67	−21.94	−21.93	1.42E−01
bGWO1	−21.56	−19.98	−20.86	−20.93	3.18E−01
bGWO2	−22.09	−20.91	−21.62	−21.63	3.10E−01
BTLBO	−22.32	−21.84	−22.08	−22.09	1.17E−01

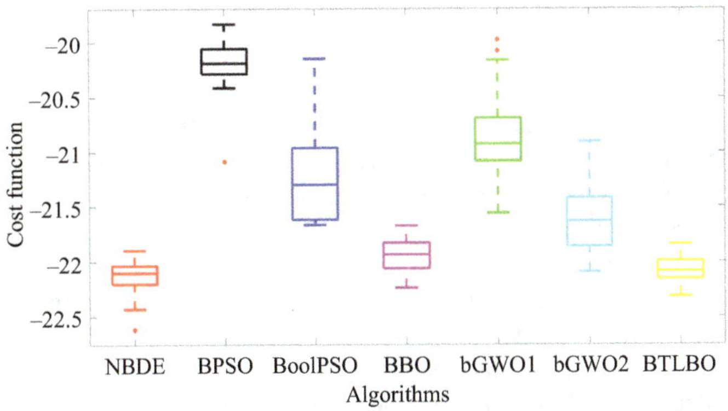

Figure 3.23 A 150-element symmetrical array. Boxplot of all algorithms results

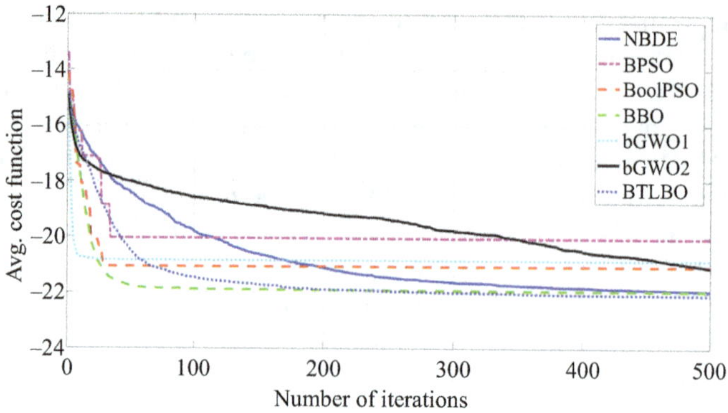

Figure 3.24 A 150-element symmetrical array. Convergence rate graph

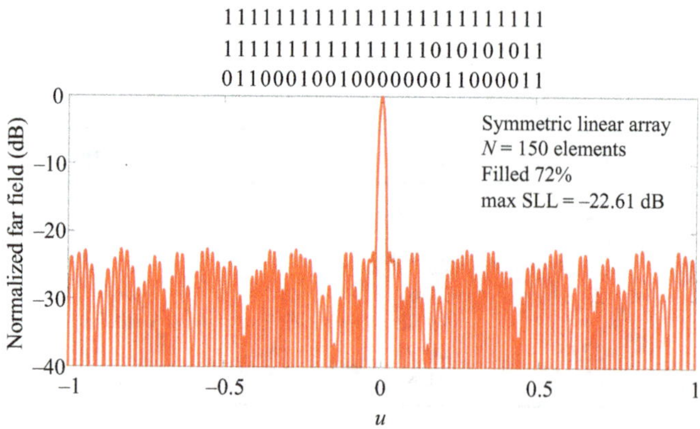

Figure 3.25 A 150-element symmetrical array. Best radiation patterns found by all algorithms

same population size and the maximum number of iterations as in the corresponding symmetrical array. Although the dimensionality of the problem is doubled, the total number of objective function evaluations for both use-cases (symmetrical and asymmetrical array) remains the same.

Table 3.16 reports the comparative results for this case. We notice that for this case BBO obtains the best result in terms of best, worst, mean and median values. For this case, BTLBO and bGWO2 also obtain very completive results close to that of BBO. Overall, the algorithms performance is different than in the previous case. Figure 3.26 depicts the algorithms boxplot for this case. It is obvious that NBDE, BBO, and BTLBO also obtain results with a small dispersion of values. The convergence rate graph for this is illustrated in Figure 3.27. BBO and BoolPSO converse quickly

Table 3.16 Comparative results for the case of 150-element asymmetrically thinned array

Algorithm	Best	Worst	Mean	Median	St. Dev.
NBDE	−23.02	−22.25	−22.62	−22.61	1.58E−01
BPSO	−20.13	−18.83	−19.3	−19.29	3.17E−01
BoolPSO	−22.52	−19.94	−21.32	−21.36	5.81E−01
BBO	−23.72	−22.81	−23.37	−23.40	1.91E−01
bGWO1	−22.72	−20.98	−21.79	−21.78	3.29E−01
bGWO2	−23.61	−19.93	−22.31	−22.54	9.22E−01
BTLBO	−23.42	−22.77	−23.05	−23.05	1.60E−01

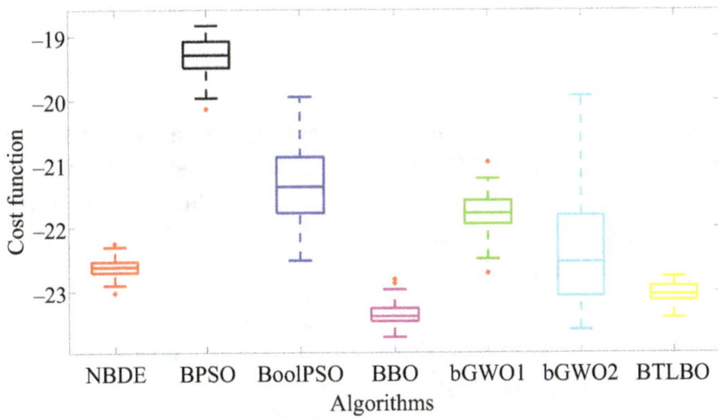

Figure 3.26 A 150-element asymmetrical array. Boxplot of all algorithms results

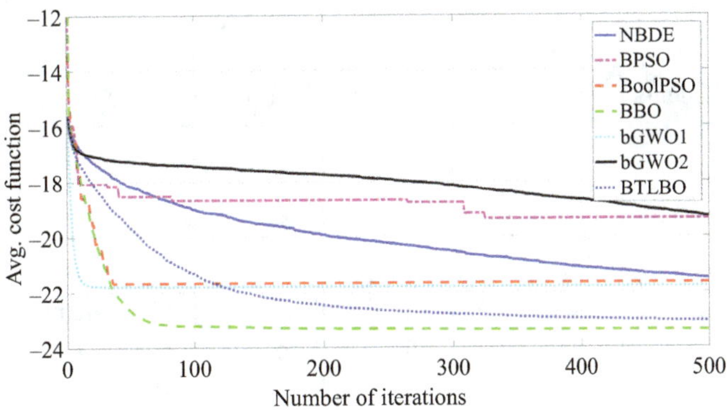

Figure 3.27 A 150-element asymmetrical array. Convergence rate graph

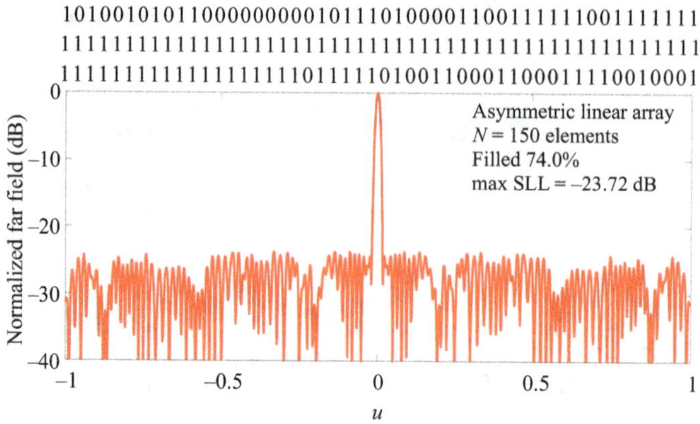

Figure 3.28 A 150-element asymmetrical array. Best radiation patterns found by all algorithms

at less than 100 iterations. However, BBO converges at a lower value than all the other algorithms. NBDE, BTLBO, and bGWO2 converge more slowly and require the maximum number of iterations to reach the final value. The radiation pattern of the best result obtained by BBO is depicted in Figure 3.28. This has a PSLL value of −23.72 dB and a fill factor of 74%.

3.3 Shaped beam synthesis

The shaped-beam pattern synthesis problem has received wide attention over the years, and several methods or techniques have been reported in the literature [44–51]. Among others, these methods include tabu search [45], ACO [44], and linear programming [47]. Moreover, in [51] a new hybrid DE algorithm is proposed named CoDE-EIG, and it is applied successfully in shaped beam synthesis.

The problem formulation follows. We consider the same linear array of K isotropic elements to be equally spaced, but with different phase values. In this case (3.9) becomes

$$AF(u,\bar{I},\bar{\varphi}) = \sum_{k=1}^{K} I_k e^{j(2\pi/\lambda)\,kdu+\varphi_k} \tag{3.12}$$

where I_k and φ_k are the amplitude and the phase of the kth element, respectively, $\bar{I}$ and $\bar{\varphi}$ are the corresponding vectors. For a symmetrically excited array, (3.12) becomes

$$AF(u,\bar{I},\bar{\varphi}) = 2\sum_{k=1}^{M} I_k \cos\left[\frac{2\pi}{\lambda}kdu + \varphi_k\right] + I_0 \tag{3.13}$$

The shaped beam design cases are chosen as the desired pattern is a cosecant beam, and outside the main beam the PSLL should not exceed predefined values as in [44,45,47,51]. We consider case 1, where the desired pattern is given by the following expression:

$$AF_{desired}(u) = \begin{cases} \text{cosecant}(u) & \text{for } 0.1 \le u \le 0.5 \\ 0 & \text{elsewhere} \end{cases} \tag{3.14}$$

where $PSLL_d$ is the desired peak sidelobe level set with the constraint $PSLL_d(u) \le -30$ dB. One may observe that the $PSLL_d$ is set to -30 dB instead of -25 dB in [44,47]. The second case uses the same cosecant main beam in the sector of $u \in [0.1, 0.5]$ as in (3.14), while the $PSLL_d$ is given by [44,45,47,51]

$$PSLL_d(u) = \begin{cases} -35\,\text{dB} & \text{for } -1 \le u \le -0.35 \\ -25\,\text{dB} & \text{for } -0.35 \le u \le 0 \\ -40\,\text{dB} & \text{for } 0.6 \le u \le 1 \end{cases} \tag{3.15}$$

where $PSLL_d$ is the desired peak sidelobe level. One may observe that the $PSLL_d$ is set to -30 dB instead of -25 dB in [44,47]. The second case uses the same cosecant main beam in the sector of $u \in [0.1, 0.5]$ as in (3.14), while the $PSLL_d$ is given by [44,45,47,51]. The requirement of matching a desired shaped beam pattern can be expressed as follows:

$$F(\bar{x}) = \sum_{n=1}^{N} w(u_n)\,[|AF_{desired}(u_n)| - |AF_{calc}(u_n, \bar{x})|]^2 \tag{3.16}$$

$$w(u) = \begin{cases} 500 & \text{for } 0.1 < u \le 0.5 \\ 1{,}000 & \text{for } u = 0.1 \\ 300 & \text{if } PSLL_c(u) > PSLL_d(u) \\ 1 & \text{otherwise} \end{cases} \tag{3.17}$$

where $PSLL_c$ is the calculated peak sidelobe level, N is the number of pattern samples taken, and $\bar{x}$ is the vector of the unknown variables (the array amplitudes and phases).

The previous weight values are obtained in [51] using a trial-and-error procedure. In [51] the population size is set to 100, and the maximum number of generations is set to 1,000 for all linear array design cases. The CoDE-EIG is applied in all cases. In [51] CoDE-EIG is compared with other popular algorithms. These algorithms include the original CoDE algorithm [52], the popular *DE/rand/1/bin* strategy (binomial crossover), the *DE/rand/1/eig* strategy (eigenvector-based mixed with binomial crossover), the CFPSO [31], and a real-coded GA.

The first example case is that of a symmetrically excited array of 20 and 18 elements. Thus, the total number of unknowns is 20 and 18 respectively. The desired pattern is that of case 1 given in (3.14). The best radiation patterns obtained by CoDE-EIG in [51] for this case are depicted in Figure 3.29. One may notice that the obtained patterns are in good agreement with the desired one. The results show that for the

20-element array a PSLL value of −30 dB was found by CoDE-EIG, while for the 18-element array, a PSLL value of −29.96 dB was obtained.

The second example case is again a symmetrically excited array of 20 and 18 elements. However, the desired pattern is that of case 2, expressed with (3.15). Figure 3.30 shows the patterns of the best designs obtained by CoDE-EIG for both $K = 20$ and $K = 18$ elements [51]. It is clear that the derived patterns closely match the desired one, while the 20-element array is a better approximation than the 18-element one. The maximum SLL of the 20-element case is −25.09 dB while for the 18-element case the PSLL value is −24.94 dB.

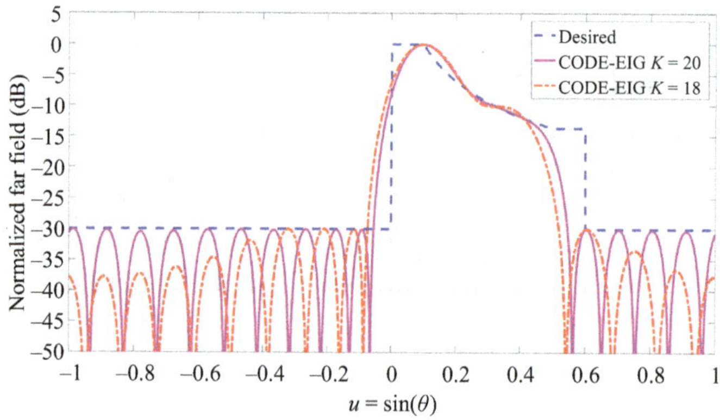

Figure 3.29 Shaped beam Case 1. Radiation patterns obtained for K = 20 and K = 18 symmetrically excited isotropic elements [51]

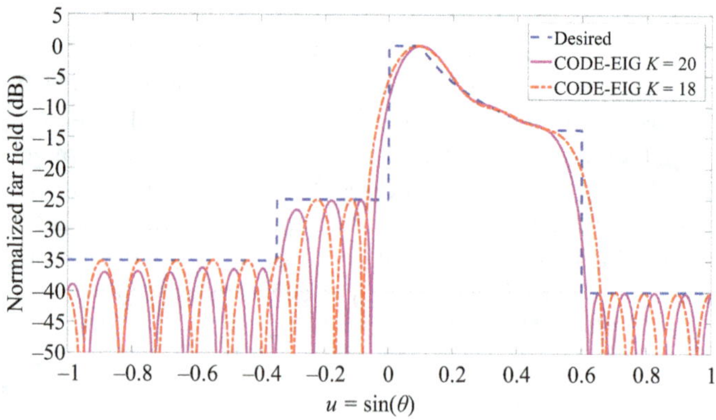

Figure 3.30 Shaped beam Case 2. Radiation patterns obtained for K = 20 and K = 18 symmetrically excited isotropic elements [51]

3.4 Planar thinned-array design

In this section, we evaluate the performance of different binary EAs on planar thinned-array design. More specifically, we compare the two BPSO algorithms (BPSO [40] and BoolPSO [41]), the two binary-coded GWO algorithms (bGWO1), (bGWO2) proposed in [42], the BBO algorithm [39], and the NBDE algorithm [43].

We assume a $2L \times 2K$ planar array that is placed on the x–y plane. We can then express the array factor by

$$AF(u, v, \bar{I}) = 4\sum_{l=1}^{L}\sum_{k=1}^{K} I_{kl} \cos\left[\pi(2k-1)d_y u\right] \cos\left[\pi(2l-1)d_x v\right], \tag{3.18}$$

where d_x and d_y are the distances between two adjunct elements in the x and y directions accordingly, and $u = \sin\vartheta\cos\phi$, $v = \sin\vartheta\sin\phi$ are the direction cosines.

The purpose of the objective function is to suppress the PSLL at two different ϕ planes ($\phi = 0°$, $\phi = 90°$). Moreover, a desired fill factor percentage is defined in the objective function. As a result, the objective function can be formulated as [36]

$$F_{planar}(\bar{I}) = \max\{PSLL_{dB}^{\phi=0^\circ}(\bar{I}), PSLL_{dB}^{\phi=90^\circ}(\bar{I})\} + \Xi \times \max\{0, pf_c - pf_d\}$$

where Ξ is a very large number, pf_c is the calculated fill factor percentage, and pf_d is the desired fill factor percentage, respectively, $PSLL_{dB}^{\phi=0^\circ}$ and $PSLL_{dB}^{\phi=90^\circ}$ are the computed PSLLs at $\phi = 0°$ and $\phi = 90°$ planes, accordingly.

We evaluate the binary algorithms at a 20×10 planar array with $d_x = d_y = 0.5\lambda$ as in the literature [3,33,36]. We select a population of 200 vectors for all algorithms. The total number of generations is set to 500. All algorithms are evaluated for 50 independent trials.

Table 3.17 lists the comparative results for the first design case of 54% desired fill factor. We notice that NBDE, BoolPSO, and bGWO2 obtain the same best result, where NBDE also achieves the same best result in every algorithm run. This is also

Table 3.17 Comparative results for the case of a 200-element planar (fill factor=54%)

Algorithm	Best	Worst	Mean	Median	St. Dev.
NBDE	−25.07	−25.07	−25.07	−25.07	1.08E−14
BPSO	−24.45	−21.07	−22.68	−22.55	8.51E−01
BoolPSO	−25.07	−20.67	−23.68	−24.06	1.09E+00
BBO	−25.07	−24.06	−24.84	−24.88	3.01E−01
bGWO1	−23.64	−16.05	−20.05	−20.06	1.50E+00
bGWO2	−25.07	−22.42	−24.17	−23.83	5.44E−01

evident from Figure 3.31, where we notice that bGWO1 has obtained the larger dispersion of values. Figure 3.32 shows the convergence rate graph, one may conclude that BBO, bGWO1, BPSO, and BoolPSO converge very quickly at less than 100 iterations. However, NBDE and bGWO2 converge more slowly but at lower objective function value.

Figures 3.33 and 3.34 present the 3D radiation pattern of the best derived result at two different views (perspective and top view). Figure 3.35 depicts the array factor of the best result for the two main ϕ planes ($\phi = 0°$ and $\phi = 90°$). We notice that the best pattern achieved has a PSLL of -25.07 dB for $\phi = 0°$, and a PSLL value of -26.09 dB for $\phi = 90°$.

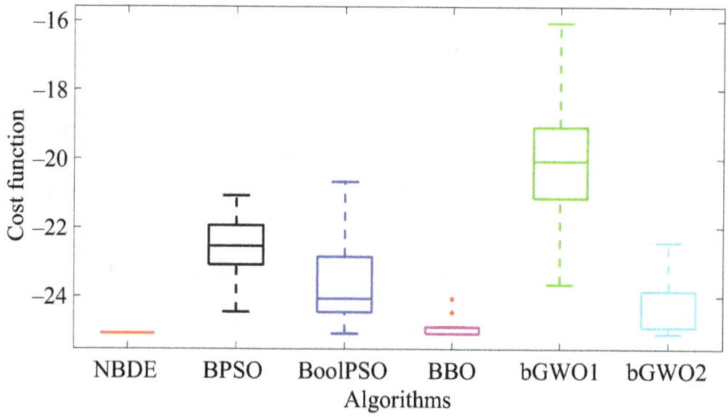

Figure 3.31 200-Element planar thinned-array case with a fill factor of 54%. Boxplot of all algorithms results

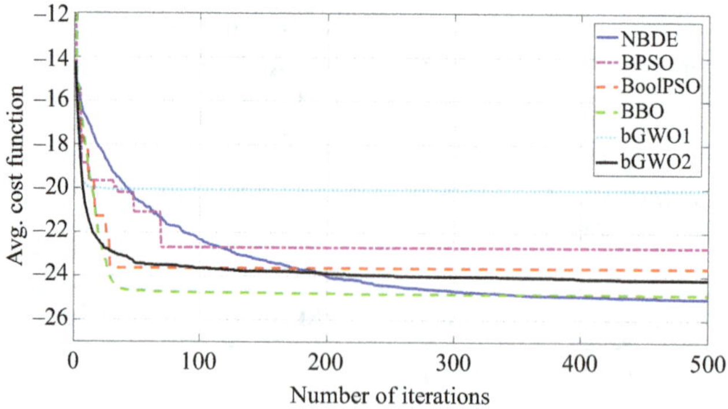

Figure 3.32 200-Element planar thinned-array case with a fill factor of 54%. Convergence rate graph

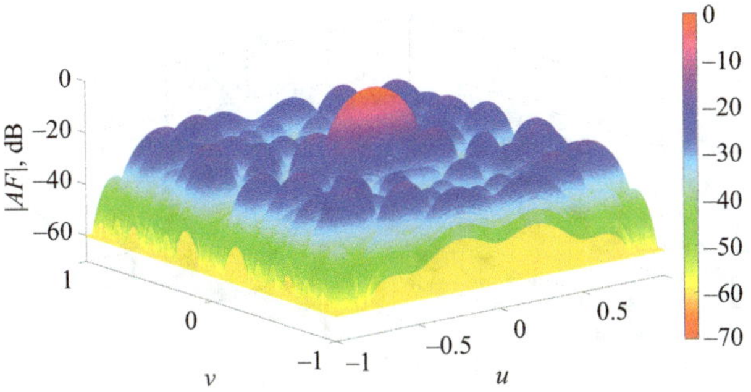

Figure 3.33 200-Element planar thinned-array case with a fill factor of 54%. 3D radiation pattern (perspective view)

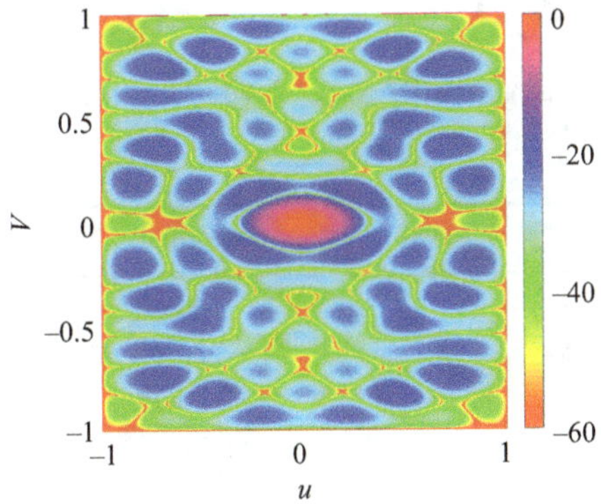

Figure 3.34 200-Element planar thinned-array case with a fill factor of 54%. 3D radiation pattern (top view)

Next, another design case with 58% desired fill factor follows. The comparative results for this case are reported in Table 3.18. For this case, NBDE and BBO achieve the same best result. However, NBDE outperforms the other algorithms in terms of worst, mean, and median values. The boxplot of the algorithms results in Figure 3.36 shows that the NBDE achieved the smaller dispersion of values, where almost all values are below −25 dB. On the contrary, bGWO2 obtained the larger dispersion of values. Figure 3.37 plots the convergence rate graph. This is similar to the previous case, where NBDE converges more slowly but at a lower objective function value.

Figures 3.38 and 3.39 depict the 3D radiation pattern of the best derived result at two different views (perspective and top view). Figure 3.40 shows the radiation patterns of the best derived result for the two main ϕ cuts ($\phi = 0°$ and $\phi = 90°$). We notice that the best pattern obtained by NBDE has a PSLL value of −26.58 dB for $\phi = 0°$, and a PSLL value of −28.32 dB for $\phi = 90°$.

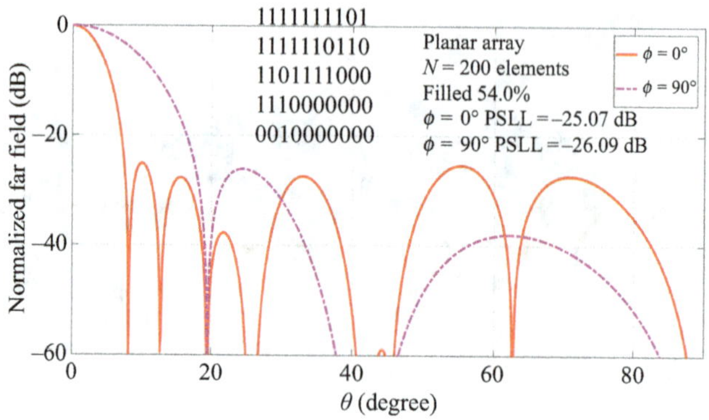

Figure 3.35 Planar thinned-array case with a fill factor of 54%. Far-field radiation patterns at $\phi = 0°$ and $\phi = 90°$ planes

Table 3.18 Comparative results for the case of 200-element planar (fill factor=58%)

Algorithm	Best	Worst	Mean	Median	St. Dev.
NBDE	−26.58	−25.07	−25.46	−25.12	6.35E−01
BPSO	−24.45	−21.41	−23.02	−23.01	7.65E−01
BoolPSO	−25.07	−21.79	−23.82	−23.86	6.12E−01
BBO	−26.58	−24.16	−25.03	−25.07	3.71E−01
bGWO1	−23.98	−17.49	−20.50	−20.70	1.79E+00
bGWO2	−26.02	−22.32	−24.15	−23.83	7.31E−01

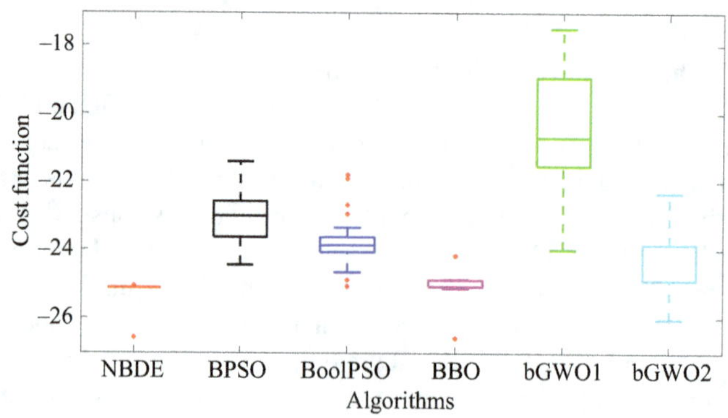

Figure 3.36 200-Element planar thinned-array case with a fill factor of 58%. Boxplot of all algorithms results

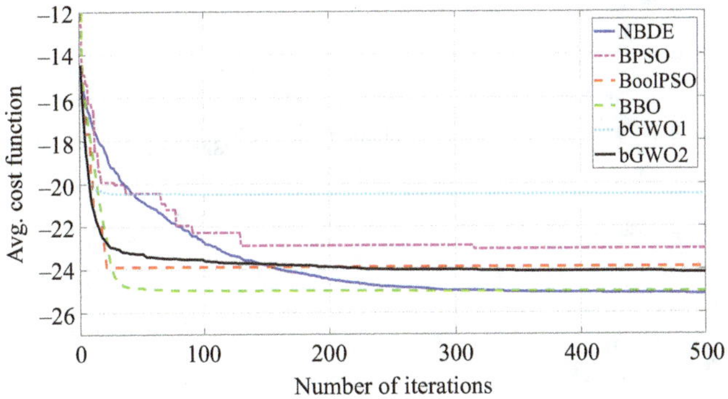

Figure 3.37 200-Element planar thinned-array case with a fill factor of 58%. Convergence rate graph

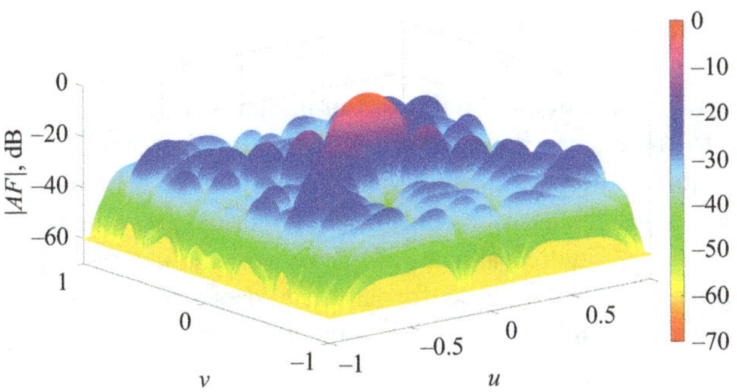

Figure 3.38 200-Element planar thinned-array case with a fill factor of 58%. 3D radiation pattern (perspective view)

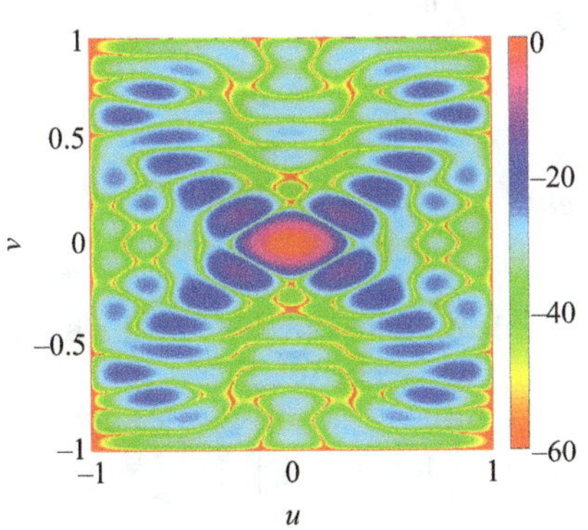

Figure 3.39 200-Element planar thinned-array case with a fill factor of 58%. 3D radiation pattern (top view)

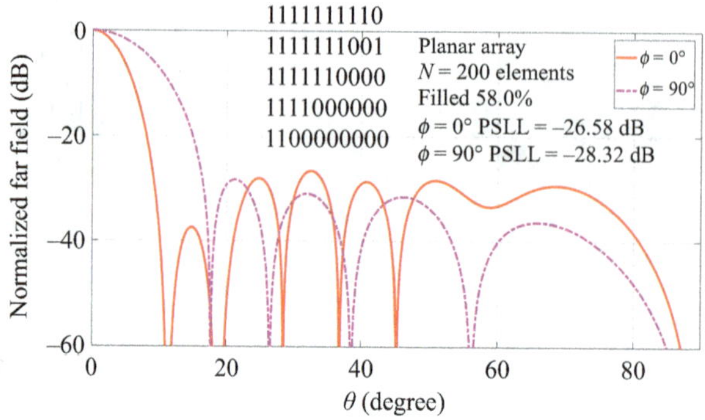

Figure 3.40 Planar thinned-array case with a fill factor of 58%. Far-field radiation patterns at $\phi = 0°$ and $\phi = 90°$ planes

3.5 Conformal array design

There are several applications in wireless commutations that require antenna arrays that conform to some surface. For example, if an omnidirectional pattern in the ϕ direction is required then the array can be conformed to a cylinder surface. This section presents a design case of a conformal cylindrical array. Such a case has been solved in [1] using simulated annealing. Moreover, the author in [47] uses linear programming to solve the same conformal array design problem. In [51] a design case of a conformal cylindrical array is optimized using CoDE-EIG.

The problem formulation is given later. We assume a cylindrical array consisting of $Q \times N$ elements in the ϕ and z directions, respectively. For this case we set $Q = N = 8$. The radius R_0 of the cylindrical surface is 5λ, while the cylinder height is $h_0 = 3.5\lambda$. The angular region of the array extends from $-\phi_0$ to ϕ_0, where $\phi_0 = 20°$. The array elements are uniformly spaced by 0.5λ. The far-field radiation pattern of the conformal cylindrical array is given by

$$FF(\vartheta, \phi, \bar{I}, \bar{\varphi})$$
$$= \sum_{q=1}^{Q}\sum_{n=1}^{N} I_{qn} \times el(\vartheta, \phi) \times e^{j\left[\varphi_{qn} + \left(2\pi/\lambda\left(x_{qn}\sin\vartheta\cos\phi + y_{qn}\sin\vartheta\sin\phi + z_{qn}\cos\vartheta\right)\right)\right]} \tag{3.19}$$

where q and n are the element indices in the ϕ and z directions, respectively, and $el(\vartheta, \phi)$ is the element pattern given by [47,51]

$$el(\vartheta, \phi) = \begin{cases} \sin\vartheta\cos\phi & \text{for } |\phi - \phi_q| \leqslant \dfrac{\pi}{2} \\ 0 & \text{otherwise} \end{cases} \tag{3.20}$$

where ϕ_q is the element azimuthal position. The positions of the array elements are

$$\begin{aligned} x_{qn} &= R_0 \cos \phi_q \\ y_{qn} &= R_0 \sin \phi_q \\ z_{qn} &= -\frac{h_0}{2} + n \times 0.5\lambda \end{aligned} \tag{3.21}$$

The desired far-field pattern, FF_d, has a cosecant square dependency in the $\phi = 0°$ plane for an elevation angle ranging between $\vartheta_{start}^{\cos}$ and $\vartheta_{end}^{\cos}$. The setting of the half-power beamwidth (HPBW) in the $\vartheta = 90°$ plane to a desired value, $HPBW_d^{\vartheta=90°}$, within a tolerance percentage is also a requirement. Additionally, the PSLLs outside the shaped beam region must remain below a predefined desired value, $PSLL_d$. For this case, the objective function is formulated as [51]

$$\begin{aligned} F(\bar{I},\bar{\varphi}) = & \frac{1}{K_s}\sum_{k=1}^{K_1}\left[\left|FF_d(\vartheta_k,0,\bar{I},\bar{\varphi})\right| - \left|FF_c(\vartheta_k,0,\bar{I},\bar{\varphi})\right|\right]^2 \\ & + \Xi \times \max\{0, PSLL_d - PSLL_c\} \\ & + \Xi \times \max\left\{0, \frac{\left|HPBW_c^{\vartheta=90°} - HPBW_d^{\vartheta=90°}\right|}{HPBW_d^{\vartheta=90°}} - T_{HPBW}\right\} \end{aligned} \tag{3.22}$$

where FF_c is the calculated far-field pattern in the $\phi = 0°$ plane, K_s is the number of pattern samples taken for the $\phi = 0°$ plane, $PSLL_c$ is the calculated peak sidelobe level in a set of $\phi -$ cuts, $HPBW_c^{\vartheta=90°}$ is the calculated half-power beamwidth in the $\vartheta = 90°$ plane, and $T_{HPBW}$is the desired beamwidth tolerance.

The design parameters we have selected for this case are the same as in [51]. These are $\vartheta_{start}^{\cos} = 100°$, $\vartheta_{end}^{\cos} = 140°$, $HPBW_d^{\vartheta=90°} = 16°$, $T_{HPBW} = 0.01$, and $PSLL_d = -27$ dB (instead of -25 dB in [1,47]). We solve this problem using a low complexity algorithm, namely, the TLBO. As in [51] we select the population size equal to 100 vectors, and the maximum number of iterations is chosen to be 3,000.

Figures 3.41 and 3.42 plot the best achieved radiation pattern with TLBO at two ϕ cuts, at $\phi = 0°$, $\vartheta = 90°$, respectively. The 3D radiation pattern is depicted in Figure 3.43. From the $\phi = 0°$ plane pattern a PSLL value of -28.14 dB is derived, which is about 4 dB below the one from [1], while it is about 0.5 dB less than the one from [47] and 1 dB higher than the one in [51]. Moreover, the $\vartheta = 90°$ plane pattern has a HPBW equal to 16.2°. Additionally, the PSLL value in this pattern is about 1 dB less than the corresponding values from previous works in [1,47,51]. The amplitudes and phases of the best design achieved by TLBO are reported in Table 3.19.

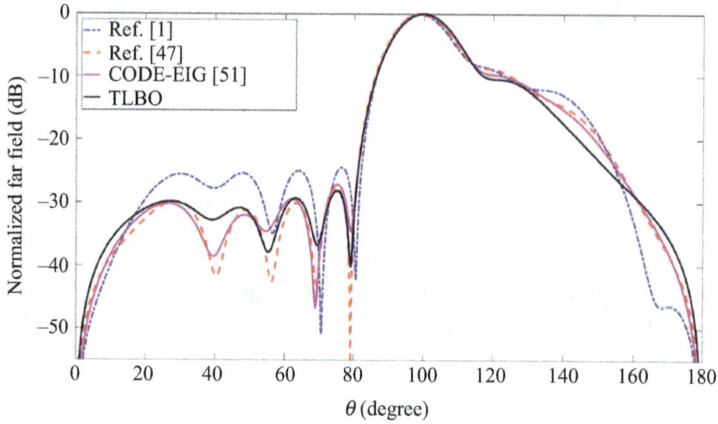

Figure 3.41 Conformal array radiation far-field patterns in $\vartheta = 90°$ plane

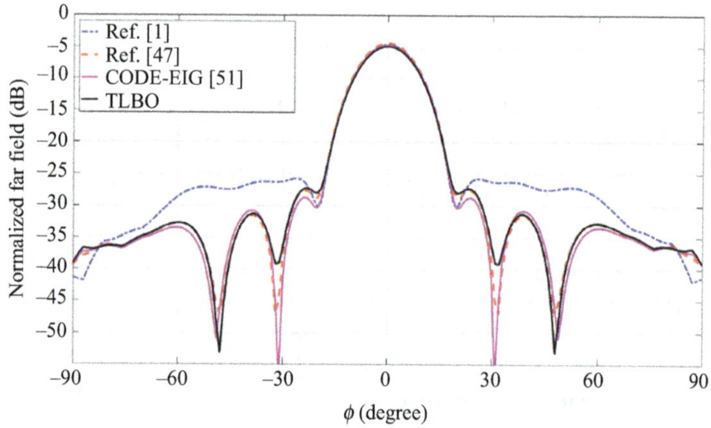

Figure 3.42 Conformal array radiation far-field patterns in $\phi = 0°$ plane

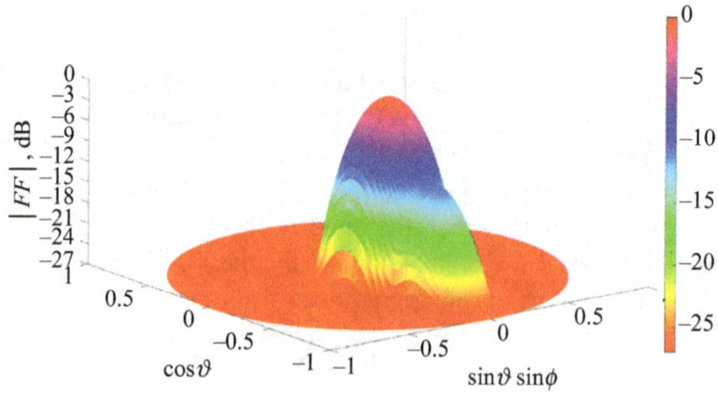

Figure 3.43 Conformal array radiation far-field pattern in 3D (perspective view)

Table 3.19 Excitation coefficients for the best conformal design obtained by TLBO

m	1,8		2,7		3,6		4,5	
n	I_{qn}	φ_{qn}	I_{qn}	φ_{qn}	I_{qn}	φ_{qn}	I_{qn}	φ_{qn}
1	0.41	121.89	0.16	−108.55	0.86	63.20	0.60	20.93
2	0.49	−107.02	0.87	102.02	0.18	148.29	0.77	101.22
3	0.19	177.89	0.19	−113.44	0.61	95.06	0.93	71.72
4	0.32	−154.72	0.80	143.17	1.15	120.2	1.10	115.63
5	0.69	−107.64	1.08	−157.66	0.99	178.72	1.41	135.14
6	0.75	−66.26	0.24	−119.57	1.17	−147.99	1.04	−162.19
7	0.35	10.32	0.57	−118.18	0.54	−72.17	0.60	−135.34
8	0.08	78.54	0.36	−22.97	0.28	−167.65	0.31	−79.49

The phase values are in degrees.

3.6 Reducing the number of elements in array design

The reduction of array elements plays an important role in cases where the cost, weight, and size of antennas are limited. Such cases include antennas in radar systems and satellite communications. Thus, one common objective in antenna array design is to create a desired pattern using the minimum number of elements. Additional constraints could include the minimum and maximum distance between two adjacent elements.

The problem of reducing the number of elements in antenna arrays has been addressed in the literature using the matrix pencil method (MPM) [53–55], sparseness constrained optimization (SCO) [56], and a hybrid MoM/GA approach [57]. Moreover, in [58] BBO is applied to the abovementioned problem.

EAs can be utilized to synthesize a desired pattern using a minimum number of elements. In this case, the decision variables are the optimum element amplitudes, phases, and positions. The abovementioned design problems can be expressed with an objective function formulation based on mean square error (MSE) metric subject to constraints. Thus, the optimization objective is to obtain an array with the minimum number of elements that matches a desired pattern. This requirement of matching a desired pattern can be formulated mathematically using the MSE as

$$F\left(\bar{x},\bar{\phi},\bar{I}\right) = \frac{1}{K}\sum_{k=1}^{K}\left[|AF_d(\vartheta_k)| - \left|AF\left(\vartheta_k,\bar{x},\bar{\phi},\bar{I}\right)\right|\right]^2 \tag{3.23}$$

where $AF_d(\vartheta) = \sum_{n=1}^{N'} I'_n e^{j\left[(2\pi/\lambda)x'_n \sin\vartheta + \phi'_n\right]}$, $N' > N$ is the desired pattern, N' is the number of elements of the desired pattern, K is the number of pattern samples taken, and x'_n, I'_n, and ϕ'_n are the position, amplitude, and the phase of the nth element of the desired pattern, respectively. Therefore, the optimization goal is to find the element positions, phases, and amplitudes that minimize (3.23). Equation (3.23) can also be used as a performance metric of how close a desired pattern is matched.

The linear-array synthesis design cases in this section are each executed 20 times. The best results are reported. We examine the synthesis of different patterns taken from the literature. We assume that $d_{\min} = 0.5\lambda$, and $d_{\max} = \lambda$ for all cases. We consider a design scheme that optimizes element positions and amplitudes (assuming that element phases are zero). The objective function in (3.23) has been calculated using $K = 901$ pattern samples from 0° to 90° with steps of 0.1°. The population size and the number of generations is set to 100 and 2,000, respectively. The number of unknown variables is $2M$. In order to compare with results from the literature, we calculate the MSE value using the same number of pattern samples. We apply to all design cases the algorithms, the TLBO [22], the Jaya algorithm [23], the WOA [24], the SSA [25], and the GWO [26].

3.6.1 20-Element Chebyshev array

The desired pattern in the first example is that of a 20-element Chebyshev array uniformly spaced with $d = 0.5\lambda$, $SLL_d = -30$ dB, and $BW_d = 17.2°$. The authors in [55] use a non-iterative synthesis method that is based on the MPM. They first sample the desired radiation pattern to form a discrete pattern data set. Then they organize the discrete data set in the form of a Hankel matrix and perform the singular value decomposition of the matrix. By discarding the non-principal singular values, they obtain an optimal lower rank approximation of the Hankel matrix, which corresponds to fewer antenna elements. Finally, MPM is used to reconstruct the excitation and position distributions from the approximated matrix.

We run the algorithm for two cases with $N = 12$ and $N = 14$ elements, respectively. Tables 3.20 and 3.21 hold the corresponding comparative results for both design cases. We notice that adding two more elements improves the MSE by about two orders of magnitude. For $N = 12$ all the algorithms obtain similar results, SSA found the best result, very close to that of Jaya. This indicates that for this type of problem any of these algorithms can be used. We can deduce that similar conclusion for the $N = 14$ case. All algorithms perform quite similarly with the exception of WOA that obtained the worst result. TLBO obtained the slightly better than the others result.

Table 3.20 Comparative results for the case of a 20-element Chebyshev pattern with SLL = −30 dB with N = 12

Algorithm	Best	Worst	Mean	Median	St. Dev.
GWO	5.62E−06	1.59E−03	3.25E−04	2.42E−04	3.87E−04
WOA	8.65E−06	5.16E−04	2.85E−04	3.37E−04	1.72E−04
SSA	5.59E−06	5.72E−03	1.15E−03	3.16E−04	1.67E−03
TLBO	5.60E−06	3.11E−04	2.32E−04	2.96E−04	1.18E−04
Jaya	5.59E−06	4.97E−04	3.64E−05	5.59E−06	9.71E−05

Table 3.21 Comparative results for the case of a 20-element Chebyshev pattern with SLL = −30 dB with N = 14

Algorithm	Best	Worst	Mean	Median	St. Dev.
GWO	9.66E−08	2.60E−03	1.21E−04	1.82E−05	3.79E−04
WOA	2.68E−06	4.83E−04	1.33E−04	7.63E−05	1.47E−04
SSA	7.30E−08	4.98E−03	8.12E−04	3.56E−05	1.49E−03
TLBO	3.27E−08	5.29E−05	5.77E−06	1.52E−06	1.34E−05
Jaya	8.61E−08	3.82E−04	3.13E−05	2.97E−06	7.49E−05

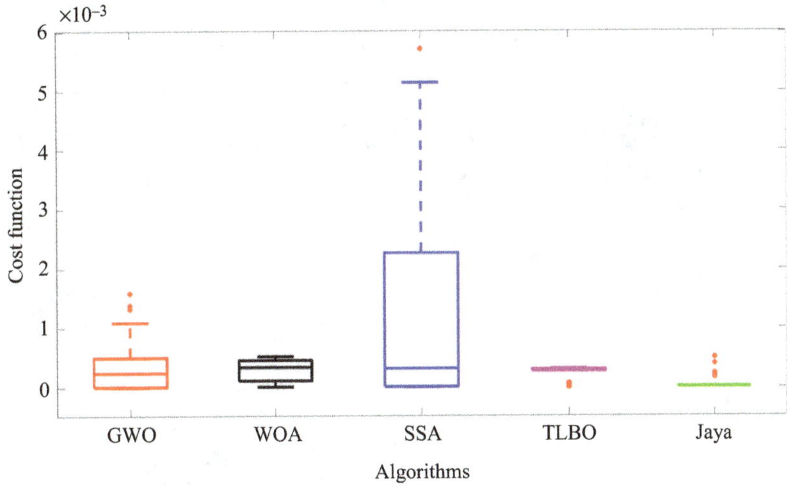

Figure 3.44 Synthesis of a 20-element Chebyshev pattern with SLL = −30 dB with N = 12. Boxplot of all algorithms results

Figure 3.44 shows the boxplot of the algorithms results for $N = 12$. We notice that the best results in terms of smaller dispersion of values are derived by Jaya and TLBO. SSA although obtained the best result is the worst in terms of standard deviation values. Thus, Jaya and TLBO would be a better option for this type of problem. The boxplot for the $N = 14$ case is depicted in Figure 3.45. In this case all algorithms obtain smaller dispersion of values, except SSA. The corresponding convergence rate graphs are plotted in Figures 3.46 and 3.47. Both of them show that TLBO and Jaya converge faster than the other algorithms.

Table 3.22 shows the radiation pattern characteristics. We notice that the new obtained results by SSA and TLBO have improved the MSE compared with the results

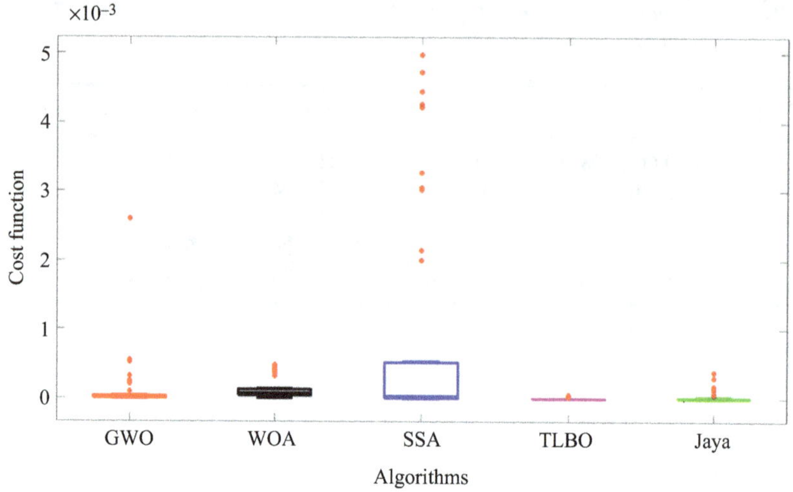

Figure 3.45 Synthesis of a 20-element Chebyshev pattern with SLL= −30 dB with N=14. Boxplot of all algorithms results

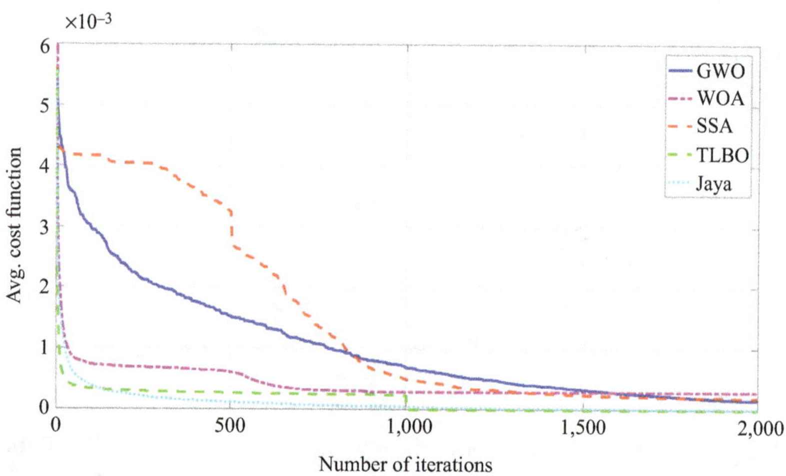

Figure 3.46 Synthesis of a 20-element Chebyshev pattern with SLL= −30 dB with N=12. Convergence rate graph

from [55]. The desired mainlobe beamwidth of 17.2° is obtained by both patterns. The SSA obtained pattern with $N = 12$ slightly deviates from the desired and has a PSLL value of −29.02 dB. The TLBO synthesized pattern with $N = 14$ matches closely the desired one with $SLL = -29.88$ dB.

The best reconstructed patterns are shown in Figure 3.48 in comparison with the case from [55]. We may notice that the MSE difference between the new reconstructed

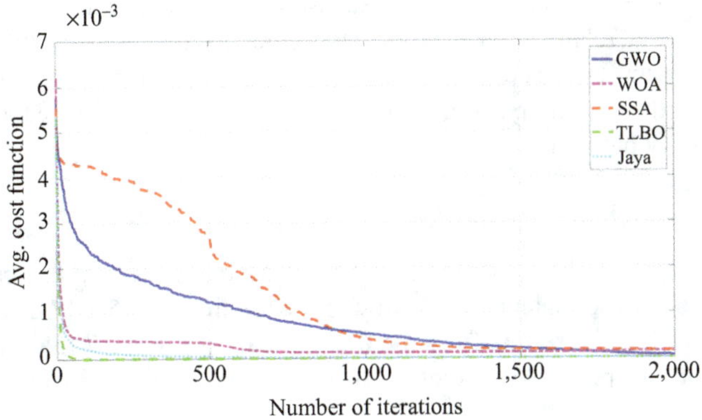

Figure 3.47 Synthesis of a 20-element Chebyshev pattern with SLL= −30 dB with N=14. Convergence rate graph

Table 3.22 Radiation pattern characteristics for the 20-element Chebyshev array design case

	Desired	**MPM (N = 12 [55])**	**SSA (N = 12)**	**TLBO (N = 14)**
SLL (dB)	−30	−29.05	−29.02	−29.88
BW (degree)	17.20	17.20	17.20	17.20
MSE	N/A	7.91E−06	5.69E−06	9.54E−08

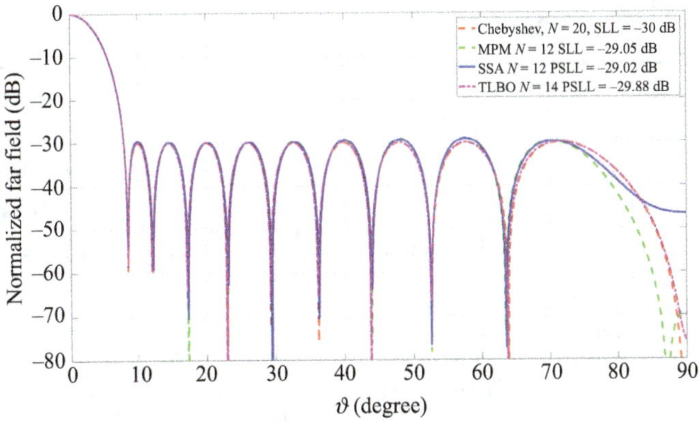

Figure 3.48 Synthesis of a 20-element Chebyshev pattern with SLL= −30 dB with N=12 and N=14. Radiation patterns of the best results found

patterns with $N = 12$ and $N = 14$ is because of the patterns difference at angles greater than about 75°. If lower MSE values are needed then we have to increase the number of elements by 2. The best amplitudes and positions are reported in Table 3.23. We notice that the element positions obtained in both cases exceed the total length of the desired array pattern.

3.6.2 A 29-element Taylor–Kaiser array

The next example is taken from [55,56,58]. The desired pattern in this case is a 29-element Taylor–Kaiser antenna uniformly spaced with $d = 0.5\lambda$, $SLL_d = -25$ dB, and $BW_d = 11.8°$. The results from the literature are for $N = 17$ [55], and $N = 18$ [56], elements, respectively. Therefore, we obtain reconstructed patterns for both cases.

Table 3.24 holds the comparative results of all the algorithms for $N = 17$. SSA obtained the best result. Jaya obtained the best mean and median value. Moreover, Table 3.25 reports the corresponding results for $N = 18$. In this case Jaya found the best result in terms of best, mean, and median values. GWO derived the second best result. The boxplot for the first case of $N = 17$ is depicted in Figure 3.49. This shows

Table 3.23 Positions and amplitudes of the best reconstructed patterns found by SSA and TLBO for the 20-element Chebyshev array design case

	SSA (N = 12)			**TLBO (N = 14)**	
n	x_n	I_n	n	x_n	I_n
1	0.426	1.000	1	0.357	1.000
2	1.278	0.914	2	1.078	0.957
3	2.127	0.759	3	1.814	0.856
4	2.971	0.566	4	2.561	0.698
5	3.804	0.371	5	3.307	0.511
6	4.637	0.268	6	4.033	0.328
			7	4.718	0.267

Table 3.24 Comparative results for the case of a 29-element Taylor–Kaiser pattern with SLL = −25 dB with N = 17

Algorithm	**Best**	**Worst**	**Mean**	**Median**	**St. Dev.**
GWO	9.49E−07	7.62E−04	1.48E−04	7.14E−05	2.19E−04
WOA	9.89E−06	1.25E−03	3.36E−04	2.50E−04	2.66E−04
SSA	6.50E−07	7.60E−04	1.65E−04	1.07E−04	2.08E−04
TLBO	7.20E−05	9.94E−05	9.20E−05	9.25E−05	5.26E−06
Jaya	3.08E−06	3.57E−04	5.07E−05	2.37E−05	6.71E−05

Table 3.25 Comparative results for the case of a 29-element Taylor–Kaiser pattern with SLL = −25 dB with N = 18

Algorithm	Best	Worst	Mean	Median	St. Dev.
GWO	9.47E−07	6.40E−04	1.01E−04	8.95E−05	1.28E−04
WOA	9.27E−06	6.65E−04	2.09E−04	1.36E−04	1.78E−04
SSA	4.47E−06	1.15E−03	1.68E−04	9.01E−05	1.97E−04
TLBO	7.07E−06	6.74E−05	2.18E−05	1.51E−05	1.53E−05
Jaya	4.97E−07	1.01E−04	2.02E−05	6.42E−06	2.98E−05

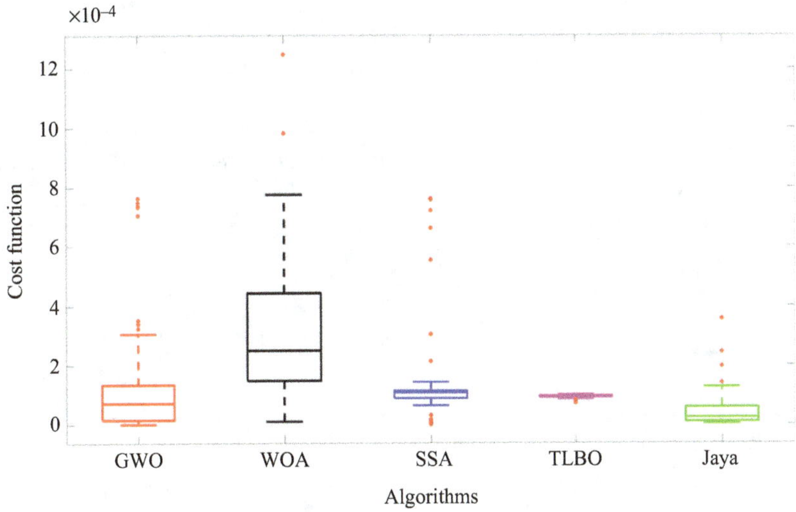

Figure 3.49 Synthesis of a 29-element Taylor–Kaiser pattern with SLL= −25 dB N= 17. Boxplot of all algorithms results

that SSA, TLBO, and Jaya obtained results with the smaller dispersion of values. WOA obtained results with the larger dispersion of values. Figure 3.50 shows the results for $N = 18$. Again Jaya and TLBO obtained the best results in terms of dispersion of values. The corresponding converge rate graphs are plotted in Figures 3.51 and 3.52. For both cases TLBO and Jaya converge faster than the other algorithms. We also notice that GWO converges slower than the other algorithms and requires the maximum number of iterations to reach the final value.

The radiation pattern characteristics are reported in Table 3.26. We notice that SSA found slightly better results than [55] for the 17-element case. The PSLL is

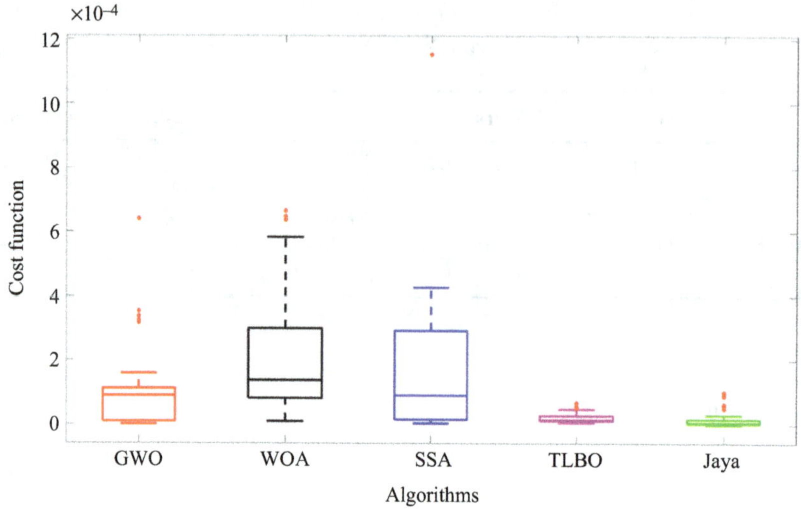

Figure 3.50 Synthesis of a 29-element Taylor–Kaiser pattern with SLL= −25 dB N = 18. Boxplot of all algorithms results

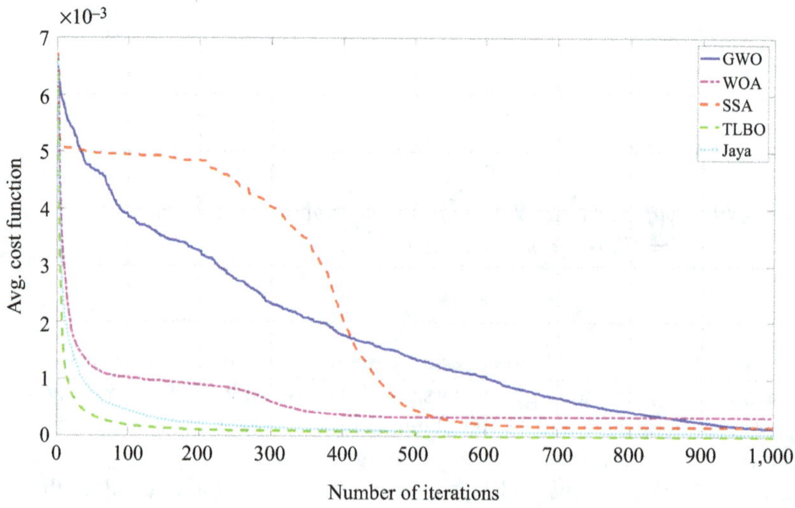

Figure 3.51 Synthesis of a 29-element Taylor–Kaiser pattern with SLL= −25 dB, N = 17. Convergence rate graph

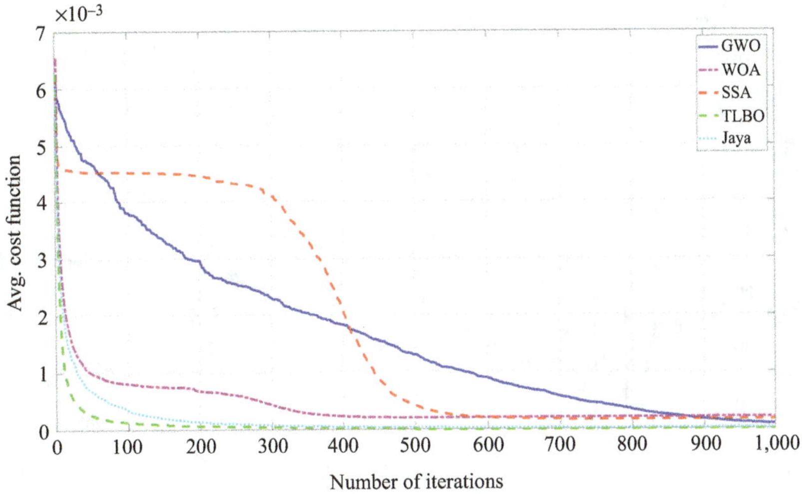

Figure 3.52 Synthesis of a 29-element Taylor–Kaiser pattern with SLL= −25 dB, N= 18. Convergence rate graph

Table 3.26 Radiation pattern characteristics for the 29-element Taylor–Kaiser array design case

	Desired	**MPM (*N* = 17 [55])**	**SSA (*N* = 17)**	**SCO (*N* = 18 [56])**	**Jaya (*N* = 18)**
SLL (dB)	−26.1	−25.96	−25.98	−25.57	−26.01
BW (degree)	11.80	11.80	11.80	12.00	11.80
MSE	N/A	9.59E−07	6.50E−07	2.65E−05	4.97E−07

better while the beamwidth is the same as that of the desired pattern. The value of MSE is also better. For the 18-element case, Jaya clearly outperforms [56]. The SLL is improved and the beamwidth exactly matches the desired pattern. The MSE is improved by two orders of magnitude compared with [56].

Figure 3.53 shows the two reconstructed patterns compared with the desired pattern and the pattern from [55]. We notice again that the reconstructed patterns deviate from the desired pattern at angles greater than 75°. For a better match, the array should include more elements. The best amplitudes and positions are reported in Table 3.27. Again, our results show that the position of the last element is less than 7λ, which is the last element position of the desired pattern.

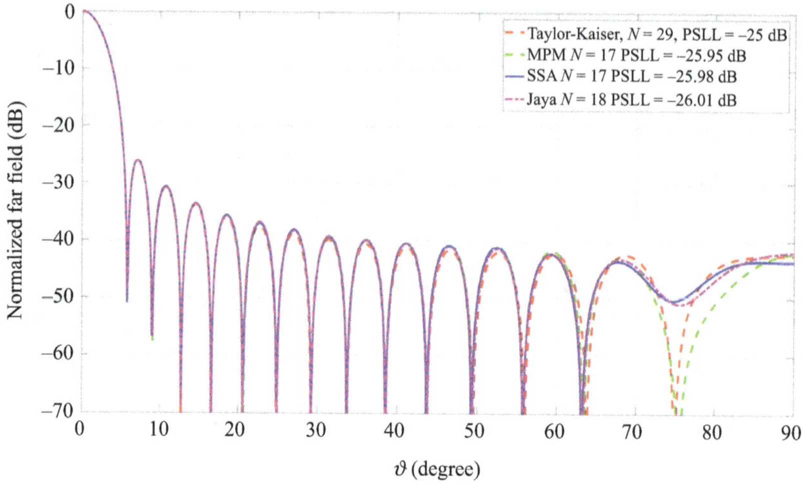

Figure 3.53 Synthesis of a 29-element Taylor–Kaiser pattern with SLL= −25 dB, N=17, and N = 18. Radiation patterns of the best results found

Table 3.27 Positions and amplitudes of the best reconstructed patterns found by SSA and Jaya for the 29-element Taylor–Kaiser array design case

	SSA (N = 17)			**Jaya (N = 18)**	
n	x_n	I_n	n	x_n	I_n
1	0.879	0.979	1	0.379	1.000
2	1.758	0.919	2	1.149	1.000
3	2.636	0.823	3	1.946	0.967
4	3.512	0.701	4	2.767	0.887
5	4.383	0.563	5	3.603	0.769
6	5.245	0.419	6	4.445	0.624
7	6.086	0.282	7	5.285	0.468
8	6.873	0.159	8	6.109	0.317
			9	6.880	0.180

References

[1] Ferreira JA and Ares F. Pattern synthesis of conformal arrays by the simulated annealing technique. Electronics Letters. 1997;33(14):1187–1189.

[2] Gomez NG, Rodriguez JJ, Melde KL, *et al.* Design of low-sidelobe linear arrays with high aperture efficiency and interference nulls. IEEE Antennas and Wireless Propagation Letters. 2009;8:607–610.

[3] Haupt RL. Thinned arrays using genetic algorithms. IEEE Transactions on Antennas and Propagation. 1994;42(7):993–999.

[4] Hooker JW and Arora RK. Optimal thinning levels in linear arrays. IEEE Antennas and Wireless Propagation Letters. 2010;9:771–774.

[5] Isernia T, Ares Pena FJ, Bucci OM, *et al.* A hybrid approach for the optimal synthesis of pencil beams through array antennas. IEEE Transactions on Antennas and Propagation. 2004;52(11):2912–2918.

[6] Oliveri G, Caramanica F, Fontanari C, *et al.* Rectangular thinned arrays based on McFarland difference sets. IEEE Transactions on Antennas and Propagation. 2011;59(5):1546–1552.

[7] Oliveri G, Donelli M, and Massa A. Linear array thinning exploiting almost difference sets. IEEE Transactions on Antennas and Propagation. 2009;57(12):3800–3812.

[8] Storn R and Price K. Differential evolution – A simple and efficient heuristic for global optimization over continuous spaces. Journal of Global Optimization. 1997;11(4):341–359.

[9] Goudos SK, Gotsis KA, Siakavara K, *et al.* A multi-objective approach to subarrayed linear antenna arrays design based on memetic differential evolution. IEEE Transactions on Antennas and Propagation. 2013;61(6):3042–3052.

[10] Goudos SK, Moysiadou V, Samaras T, *et al.* Application of a comprehensive learning particle swarm optimizer to unequally spaced linear array synthesis with sidelobe level suppression and null control. IEEE Antennas and Wireless Propagation Letters. 2010;9:125–129.

[11] Goudos SK, Siakavara K, Samaras T, *et al.* Self-adaptive differential evolution applied to real-valued antenna and microwave design problems. IEEE Transactions on Antennas and Propagation. 2011;59(4):1286–1298.

[12] Goudos SK, Siakavara K, Samaras T, *et al.* Sparse linear array synthesis with multiple constraints using differential evolution with strategy adaptation. IEEE Antennas and Wireless Propagation Letters. 2011;10:670–673.

[13] Haupt RL. Optimized weighting of uniform subarrays of unequal sizes. IEEE Transactions on Antennas and Propagation. 2007;55(4):1207–1210.

[14] Khodier MM and Christodoulou CG. Linear array geometry synthesis with minimum sidelobe level and null control using particle swarm optimization. IEEE Transactions on Antennas and Propagation. 2005;53(8 II):2674–2679.

[15] Kurup DG, Himdi M, and Rydberg A. Synthesis of uniform amplitude unequally spaced antenna arrays using the differential evolution algorithm. IEEE Transactions on Antennas and Propagation. 2003;51(9):2210–2217.

[16] Lin C, Qing A, and Feng Q. Synthesis of unequally spaced antenna arrays by using differential evolution. IEEE Transactions on Antennas and Propagation. 2010;58(8):2553–2561.

[17] Brest J, Greiner S, Boskovic B, *et al.* Self-adapting control parameters in differential evolution: A comparative study on numerical benchmark problems. Evolutionary Computation, IEEE Transactions on. 2006;10(6):646–657.

[18] Omran MGH, Engelbrecht AP, and Salman A. Bare bones differential evolution. European Journal of Operational Research. 2009;196(1):128–139.

[19] Wang H, Rahnamayan S, Sun H, *et al.* Gaussian bare-bones differential evolution. IEEE Transactions on Cybernetics. 2013;43(2):634–647.

[20] Wang Y, Cai Z, and Zhang Q. Differential evolution with composite trial vector generation strategies and control parameters. Evolutionary Computation, IEEE Transactions on. 2011;15(1):55–66.

[21] Zhang J and Sanderson AC. JADE: Adaptive differential evolution with optional external archive. IEEE Transactions on Evolutionary Computation. 2009;13(5):945–958.

[22] Rao RV, Savsani VJ, and Vakharia DP. Teaching-learning-based optimization: A novel method for constrained mechanical design optimization problems. CAD Computer Aided Design. 2011;43(3):303–315.

[23] Rao RV. Jaya: A simple and new optimization algorithm for solving constrained and unconstrained optimization problems. International Journal of Industrial Engineering Computations. 2016;7(1):19–34.

[24] Mirjalili S and Lewis A. The whale optimization algorithm. Advances in Engineering Software. 2016;95:51–67.

[25] Mirjalili S, Gandomi AH, Mirjalili SZ, *et al.* Salp swarm algorithm: A bio-inspired optimizer for engineering design problems. Advances in Engineering Software. 2017;114:163–191.

[26] Mirjalili S, Mirjalili SM, and Lewis A. Grey wolf optimizer. Advances in Engineering Software. 2014;69:46–61.

[27] Sartori D, Oliveri G, Manica L, *et al.* Hybrid design of non-regular linear arrays with accurate control of the pattern sidelobes. IEEE Transactions on Antennas and Propagation. 2013;61(12):6237–6242.

[28] Oliveri G, Caramanica F, and Massa A. Hybrid ADS-based techniques for radio astronomy array design. IEEE Transactions on Antennas and Propagation. 2011;59(6 PART 1):1817–1827.

[29] Zaharis ZD, Goudos SK, and Yioultsis TV. Application of Boolean PSO with adaptive velocity mutation to the design of optimal linear antenna arrays excited by uniform amplitude current distribution. Journal of Electromagnetic Waves and Applications. 2011;25(10):1422–1436.

[30] Dib NI, Goudos SK, and Muhsen H. Application of Taguchi's optimization method and self-adaptive differential evolution to the synthesis of linear antenna arrays. 2010;102:159–180.

[31] Clerc M. The swarm and the queen: Towards a deterministic and adaptive particle swarm optimization. In: Proceedings of the 1999 Congress on Evolutionary Computation, 1999. CEC 99. vol. 3; 1999. p. 1951–1957.

[32] Zhu G and Kwong S. Gbest-guided artificial bee colony algorithm for numerical function optimization. Applied Mathematics and Computation. 2010;217(7):3166–3173.

[33] Jain R and Mani GS. Solving "antenna array thinning problem" using genetic algorithm. Applied Computational Intelligence and Soft Computing. 2012;2012:14.

[34] Jin N and Rahmat-Samii Y. Advances in particle swarm optimization for antenna designs: Real-number, binary, single-objective and multiobjective

implementations. IEEE Transactions on Antennas and Propagation. 2007; 55(3 I):556–567.

[35] Quevedo-Teruel O and Rajo-Iglesias E. Ant colony optimization in thinned array synthesis with minimum sidelobe level. IEEE Antennas and Wireless Propagation Letters. 2006;5(1):349–352.

[36] Goudos S. Antenna design using binary differential evolution: Application to discrete-valued design problems. IEEE Antennas and Propagation Magazine. 2017;59(1):74–93.

[37] Goudos SK and Sahalos JN. Design of large thinned arrays using different biogeography-based optimization migration models. International Journal of Antennas and Propagation. vol. 2016, 11 pages, 2016. https://doi.org/10.1155/2016/5359298.

[38] Goudos SK, Siakavara K, and Sahalos JN. A comparative study of different biogeography based optimization migration models performance on antenna array thinning problems. International Conference on High Performance Computing and Simulation, HPCS 2016; 2016. p. 270–277.

[39] Simon D. Biogeography-based optimization. IEEE Transactions on Evolutionary Computation. 2008;12(6):702–713.

[40] Kennedy J and Eberhart RC. Discrete binary version of the particle swarm algorithm. In: IEEE International Conference on Systems, Man and Cybernetics. 1997;5:4104–4108.

[41] Marandi A, Afshinmanesh F, Shahabadi M, *et al.* Boolean particle swarm optimization and its application to the design of a dual-band dual-polarized planar antenna. In: 2006 IEEE Congress on Evolutionary Computation, CEC 2006; 2006. p. 3212–3218.

[42] Emary E, Zawbaa HM, and Hassanien AE. Binary grey wolf optimization approaches for feature selection. Neurocomputing. 2016;172: 371–381.

[43] Wang L, Fu X, Mao Y, *et al.* A novel modified binary differential evolution algorithm and its applications. Neurocomputing. 2012;98: 55–75.

[44] Akdagli AA, Guney K, and Karaboga D. Touring ant colony optimization algorithm for shaped-beam pattern synthesis of linear antenna. Electromagnetics. 2006;26(8):615–628.

[45] Akdagli A and Guney K. Shaped-beam pattern synthesis of equally and unequally spaced linear antenna arrays using a modified tabu search algorithm. Microwave and Optical Technology Letters. 2003;36(1):16–20.

[46] Buckley MJ. Synthesis of shaped beam antenna patterns using implicitly constrained current elements. IEEE Transactions on Antennas and Propagation. 1996;44(2):192–197.

[47] Fuchs B. Shaped beam synthesis of arbitrary arrays via linear programming. IEEE Antennas and Wireless Propagation Letters. 2010;9:481–484.

[48] Ho SL and Yang S. Multiobjective synthesis of antenna arrays using a vector tabu search algorithm. IEEE Antennas and Wireless Propagation Letters. 2009;8:947–950.

[49] Ho SL and Yang S. Multiobjective optimization of inverse problems using a vector cross entropy method. IEEE Transactions on Magnetics. 2012;48(2): 247–250.

[50] Isernia T, Bucci OM, and Fiorentino N. Shaped beam antenna synthesis problems: Feasibility criteria and new strategies. Journal of Electromagnetic Waves and Applications. 1998;12(1):103–138.

[51] Goudos SK. Shaped beam pattern synthesis of antenna arrays using composite differential evolution with eigenvector-based crossover operator. International Journal of Antennas and Propagation. 2015;2015:Article ID 295012.

[52] Iorio AW and Li X. Solving rotated multi-objective optimization problems using differential evolution. Lecture Notes in Artificial Intelligence (Subseries of Lecture Notes in Computer Science). 2004;3339:861–872.

[53] Yang S, Liu Y, and Liu QH. Combined strategies based on matrix pencil method and tabu search algorithm to minimize elements of non-uniform antenna array. Progress in Electromagnetics Research B. 2009;18:259–277.

[54] Yanhui L, Qing Huo L, and Zaiping N. Reducing the number of elements in the synthesis of shaped-beam patterns by the forward-backward matrix pencil method. IEEE Transactions on Antennas and Propagation. 2010;58(2): 604–608.

[55] Yanhui L, Zaiping N, and Qing Huo L. Reducing the number of elements in a linear antenna array by the matrix pencil method. IEEE Transactions on Antennas and Propagation. 2008;56(9):2955–2962.

[56] Wenji Z, Lian L, and Fang L. Reducing the number of elements in linear and planar antenna arrays with sparseness constrained optimization. IEEE Transactions on Antennas and Propagation. 2011;59(8):3106–3111.

[57] Hussein AH, Abdullah HH, Salem AM, *et al.* Optimum design of linear antenna arrays using a hybrid MoM/GA algorithm. IEEE Antennas and Wireless Propagation Letters. 2011;10:1232–1235.

[58] Goudos SK, Baltzis KB, Siakavara K, *et al.* Reducing the number of elements in linear arrays using biogeography-based optimization. In: 2012 6th European Conference on Antennas and Propagation (EUCAP); 2012. p. 1615–1618.

Chapter 4
Microstrip patch antenna design

This chapter describes patch design cases for different applications. The use of microstrip patch antennas in wireless communication systems provides several advantages like low profile, low cost, and ease of fabrication. Moreover, microstrip patch antennas can provide a possible solution for fifth generation (5G) antenna design. Different antenna shapes can be fabricated using the rectangular patch as an initial step.

This type of antenna design requires the simultaneous optimization of several different geometrical parameters. An optimization algorithm or techniques is a suitable approach for solving this problem. In the literature, there are several examples of patch antenna design and optimization using different evolutionary algorithms (EAs) [1–5]. These approaches include genetic algorithms [6], particle swarm optimization (PSO) [7–9], differential evolution (DE) [10–13], teaching–learning-based optimization (TLBO)[14], Jaya [15], and a hybrid Jaya–GWO algorithm [16].

4.1 E-shaped patch antenna design

E-shaped patch antennas extend the patch functionality and bandwidth [7,10,14]. The E-shape is achieved by incorporating two parallel slots into the rectangular patch that introduce multiple resonances. Such antenna shape is suitable for dual-band or wideband designs. The geometry of an E-shaped patch antenna with coaxial feed is given in Figure 4.1. The design of such antennas requires the determination of the geometrical parameters of the antenna that satisfy the design requirements at the desired frequencies. PSO and DE have been in several occasions [7,10,12] applied to E-shaped patch antenna design.

The E-shaped patch of Figure 4.1 requires the setting of six design parameters $(W, L, W_s, L_s, P_s, P_x)$ if we assume that the ground plane size is fixed. The decision variables for this case are the patch width W, the patch length L, the slot width W_s, the slot length L_s, the slot position P_s, and the feed position P_x. A possible objective function for dual-band operation in two frequencies f_1 and f_2 is to minimize the S_{11} magnitude in these frequencies, which can be expressed as

$$F_1(\bar{x}) = \max\left(S_{11,f_1}, S_{11,f_2}\right) \quad (4.1)$$

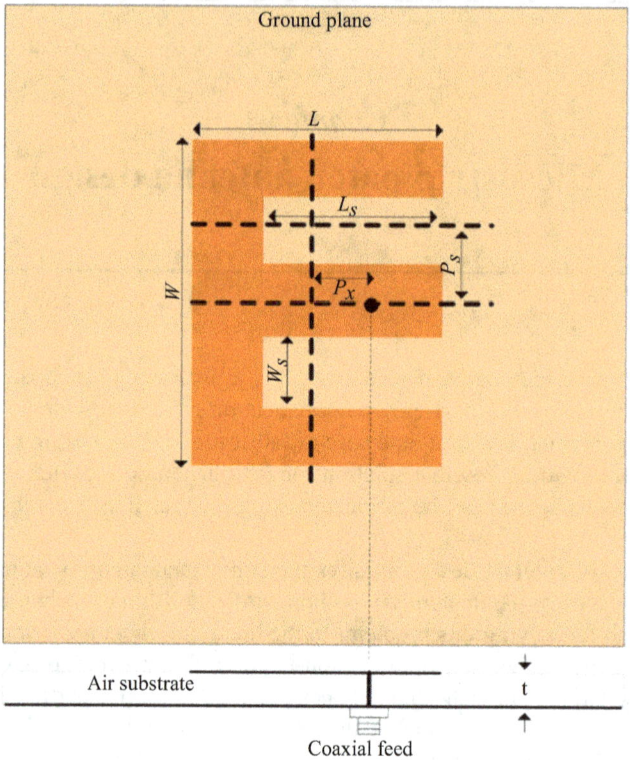

Figure 4.1 Geometry of an E-shaped dual-band patch antenna with coaxial feed

where $\bar{x} = (W, L, W_s, L_s, P_s, P_x)$ is the vector of the antenna geometry. In order to maintain the E-shape, additional restrictions apply to the design parameters. These are [7,12]

$$
\begin{aligned}
&L_s < L \text{ The slot cannot cross the patch}\\
&P_s > \frac{W_s}{2} \text{ The central stub must exist}\\
&P_s + \frac{W_s}{2} < \frac{W}{2} \text{ The top and bottom stubs must exist}\\
&|P_x| < \frac{L}{2} \text{ The feed cannot go beyond the patch}
\end{aligned}
\qquad (4.2)
$$

The antenna geometry complexity makes it difficult or even impossible to estimate the effect of each design parameter in order to achieve the desired antenna performance. Therefore, an optimization technique is suitable to address this design problem. The computation of such an objective function like (4.1) requires the use

Algorithm 1: Generic antenna design procedure

1: For each new geometry vector $\bar{x}$
2: Generate an HFSS Visual Basic script file from MATLAB
3: Run HFSS with the newly created script and generate antenna geometry
4: Calculate the metrics of interest (e.g., S_{11}) at desired frequency or frequencies
5: Generate an output text file with the results from the EM software
6: Read the output file and calculate the objective function value
7: Return result to evolutionary algorithm

of a full-wave numerical method. The E-shaped patch antenna can be modeled in commercial full-wave electromagnetic (EM) software. Some possible software to use suitable for this task is ANSYS HFSS [17]. Moreover, the integration of the in-house source code of an EA with such an EM solver requires the use of a wrapper program. The implementation of a separate optimizer that calls external software to do the simulations for the evaluation of the objective function can also be found in [18]. In order to integrate the source code of an EA written in MATLAB® with ANSYS HFSS, the wrapper program can use the HFFS MATLAB API [19]. The whole process can be described in Algorithm 1.

4.1.1 Frequency-independent design procedure

In the previous example, we need to set the upper and lower limits for every geometrical parameter. This is quite complex, and it is frequency dependent. If set the limits of the geometrical parameters are valid only for the specific design that will operate in specific frequencies. The question is how to optimize another antenna with the same shape that would operate at different frequencies. One would have to find the appropriate limits again for every geometrical parameter. This procedure is more time-consuming and one has to perform several tests to find the most suitable limits. However, a more generic frequency-independent design procedure can be followed. The basic idea is to set the first two initial parameters, i.e., the patch width W and the patch length L, with appropriate limits for specific frequency and then set the other parameters as a perchance of these two. Then the limits for the other unknowns will be within the range of [0, 1]. For example, the six optimization variables $(x_1, x_2, x_3, x_4, x_5, x_6)$ in the above-described problem can be expressed as

$$
\begin{aligned}
L &= x_1 \\
W &= x_2 \\
L_s &= x_3 \times L \\
W_s &= x_4 \times W \\
P_s &= x_5 \times W \\
P_x &= x_6 \times (L/2)
\end{aligned}
\tag{4.3}
$$

That way if we need to make another design only, the limits of the first two optimization variables have to be determined for the new frequencies. Setting the limits of an optimization variable within [0, 1] decreases the search space, and it is easier for an EA to obtain results. Moreover, some EAs operate better if the search space is within [0, 1].

4.1.2 Dual-band 5G antenna design

Future 5G framework of cellular systems will use millimeter-wave frequencies and is expected to offer extremely wide spectrum and multi-gigabit-per-second (Gbps) data rates to mobile communications. Antenna design for the new mobile devices seems to be a challenging task.

A possible solution for this case is expected to be the use of microstrip patch antennas. Shaped apertures have also been proposed to feed the antenna elements, providing dual-band operation, high Gain, wide bandwidth, and dual polarization characteristics to the antenna [20–23]. Several design efforts of patch antennas exist in the literature that obtain good performance in mm-wave frequency bands [24,25]. The E-shaped patch antennas are good candidates for mm-wave frequency bands design, since as stated previously, they are well suited for dual- or wideband operation. The design in mm-wave frequencies could be employed using aperture-coupled feeding [26]. In order to enable dual-band operation, such a feeding technique requires the modification of the aperture shape. In [14] this is accomplished using an H-shaped aperture that resonates in two operating frequencies. The geometrical parameters of both the patch and the aperture need to be determined in order to satisfy the performance requirements at the desired frequencies.

In this case, we will use as in [14] TLBO algorithm [27]. The main benefit of this approach is that TLBO is a parameter-free algorithm.

Figure 4.2 shows the geometry of the aperture-coupled E-shaped patch antenna. We notice that the antenna has an H-shaped aperture that is used to introduce two

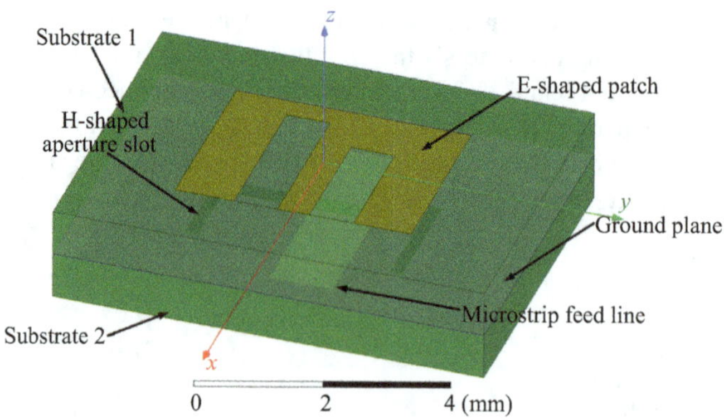

Figure 4.2 Geometry of an aperture-coupled E-shaped dual-band patch antenna

possible lengths to generate different resonant frequencies. The total number of design parameters for this case is 15. The geometrical parameters to be optimized are depicted in Figures 4.3–4.5. The antenna is fabricated in two substrate layers that consist of the same dielectric. This is an Arlon substrate with $\varepsilon_r = 3.38$ mi and 30 mi thick (0.762 mm).

The antenna is designed for dual-band operation in two 5G frequency bands. Thus, the design objective for this case is to minimize the S_{11} magnitude in both frequencies, in 25 GHz and 37 GHz and to reduce voltage standing wave ratio (VSWR) below

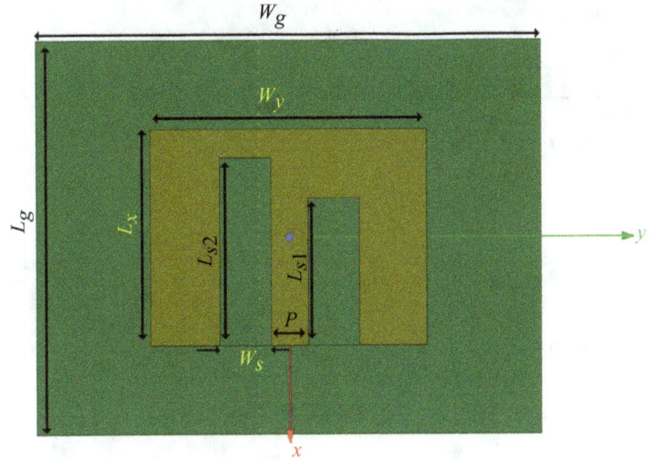

Figure 4.3 Geometry of an aperture-coupled E-shaped dual-band patch antenna. Patch plane

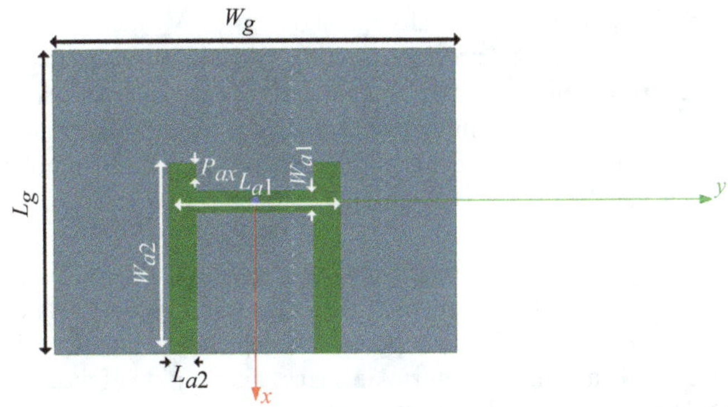

Figure 4.4 Geometry of an aperture-coupled E-shaped dual-band patch antenna. Ground plane with aperture

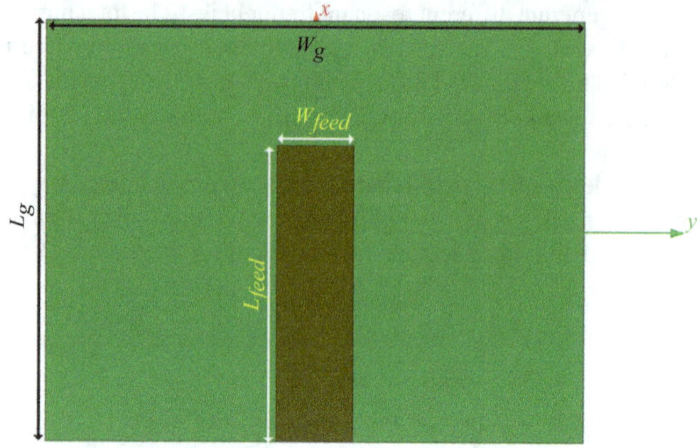

Figure 4.5 Geometry of an aperture-coupled E-shaped dual-band patch antenna. Feed line plane

2 in both frequencies. We define this design problem by the following objective function [14]:

$$F(\bar{x}) = \max\left(S_{11,25\,\text{GHz}}(\bar{x}), S_{11,37\,\text{GHz}}(\bar{x})\right) + VSWR(\bar{x})$$

$$y = \max\left(VSWR_{25\,\text{GHz}}(\bar{x}), VSWR_{37\,\text{GHz}}(\bar{x})\right) \tag{4.4}$$

$$VSWR(\bar{x}) = \begin{cases} \Xi \times (y-2) & \text{if } y > 2 \\ 0 & \text{otherwise} \end{cases}$$

where $\bar{x}$ is the vector of the antenna geometry design variables, $S_{11,\,25\,\text{GHz}}(\bar{x}), S_{11,\,37\,\text{GHz}}(\bar{x})$ are S_{11} magnitude, and $\left(VSWR_{25\,\text{GHz}}(\bar{x}), VSWR_{37\,\text{GHz}}(\bar{x})\right)$ is the VSWR in 25 GHz and 37 GHz, respectively, and Ξ is a penalty factor.

As was explained in the previous section, this kind of optimization problem is solved in conjunction with commercial EM solver software, which is ANSYS HFSS in our case. Similarly with [14] the TLBO algorithm is applied. We select the population size equal to 20, and the maximum number of iterations is selected equal to 100. After ten algorithm runs we select the best design found.

Figures 4.6 and 4.7 illustrate the 3D radiation patterns at both frequencies. The corresponding surface current distributions are shown in Figures 4.8 and 4.9. Moreover, the radiation patterns at different planes are depicted in Figures 4.10–4.12. We notice that at 25 GHz an almost uniform, in space, distribution of the radiated power is obtained as the maximum values of the surface current are substantially limited to a small area of the patch. However, at 37 GHz, the current spreads over larger areas appearing two separate regions of maximum and thus the distribution of the radiated

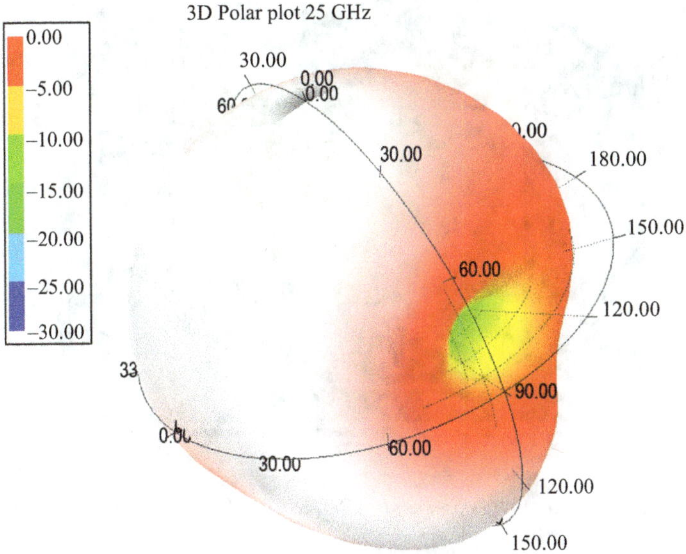

Figure 4.6 3D Gain pattern of the best design found by TLBO at 25 GHz for 5G operation

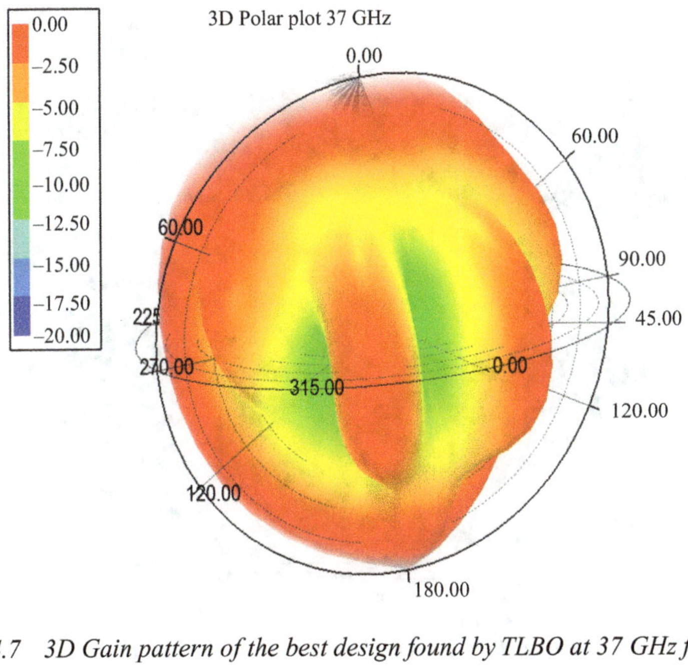

Figure 4.7 3D Gain pattern of the best design found by TLBO at 37 GHz for 5G operation

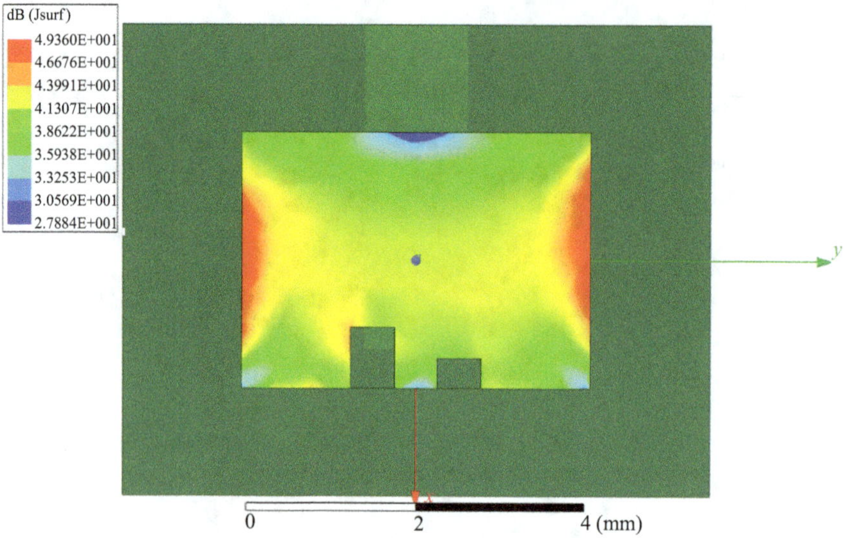

Figure 4.8 Simulated surface current distribution of the best design at 25 GHz for 5G operation

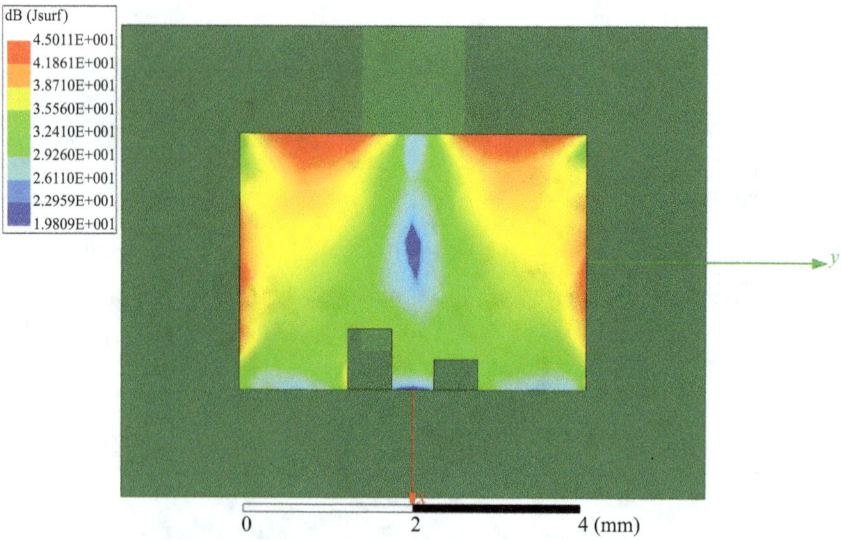

Figure 4.9 Simulated surface current distribution of the best design at 37 GHz for 5G operation

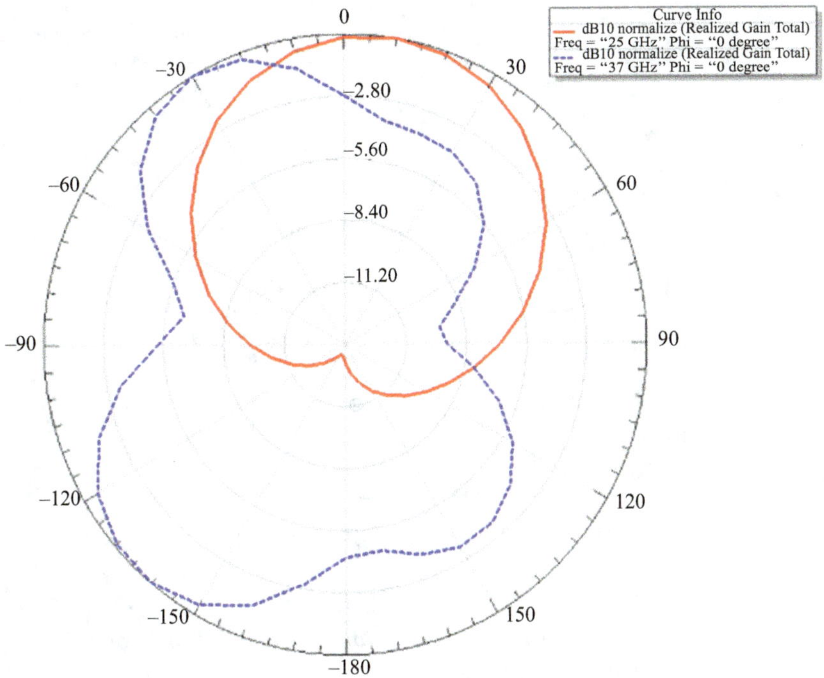

Figure 4.10 Radiation pattern of the best obtained design by TLBO for 5G operation ($\phi = 0°$)

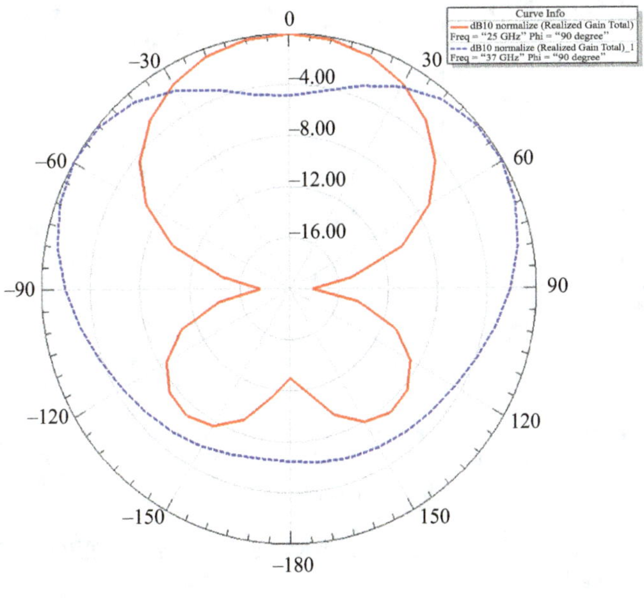

Figure 4.11 Radiation pattern of the best obtained design by TLBO for 5G operation ($\phi = 90°$)

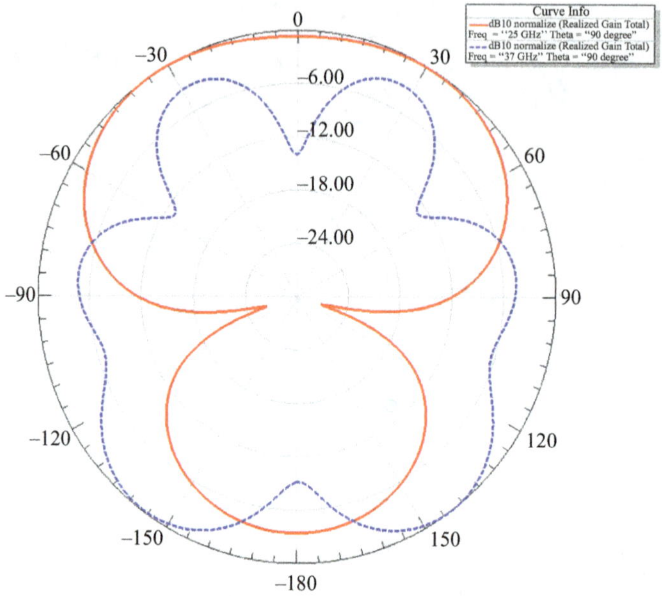

Figure 4.12 Radiation pattern of the best obtained design by TLBO for 5G operation ($\theta = 90°$)

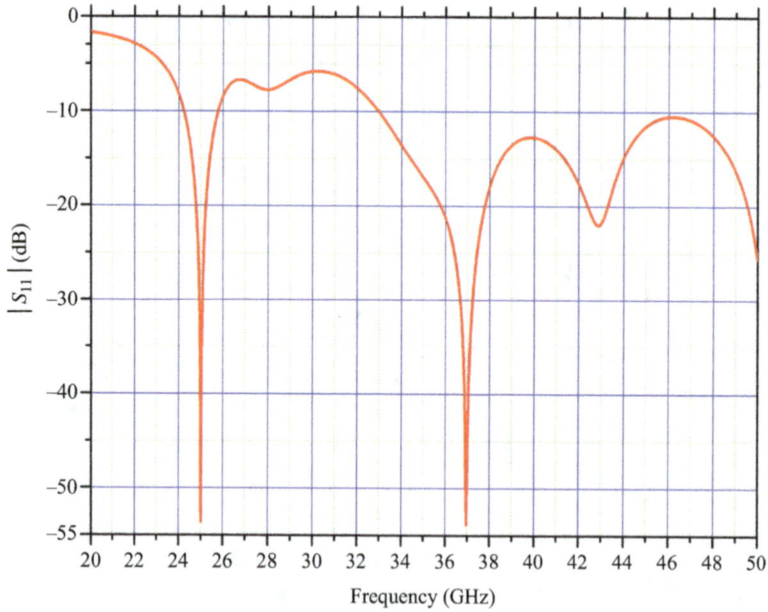

Figure 4.13 S_{11} plot of the best antenna best design found by TLBO for 5G operation

power is not uniform any more. This performance at 37 GHz is due to the increase of electrical size of the patch.

Figure 4.13 shows the frequency response of the best design found. It is clear that two distinct modes of operation exist, one at 25 GHz and the other at 37 GHz. Therefore, TLBO achieved the design goal of the optimization process. Moreover, the antenna is wide band with a large $S_{11} < -10$ dB bandwidth.

Figure 4.14 plots the VSWR versus frequency over a 50-Ω resistance. We notice that this plot is in accordance with the variation of S_{11} versus frequency in Figure 4.13. Additionally, the VSWR plot is in good agreement with the design constraint, namely, the VSWR to be lower than two around at both frequencies of interest.

Table 4.1 holds the antenna performance indices at both frequencies. We notice that they exhibit very satisfactory low values both of S_{11} and the bandwidth, especially at the upper frequency band. The Gain value is sufficiently high at 25 GHz; however, one can observe a degradation 37 GHz.

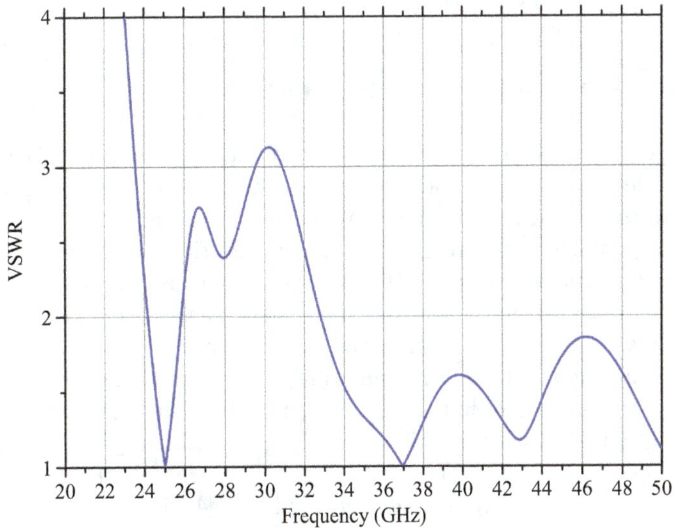

Figure 4.14 VSWR plot of the best design found by TLBO for 5G operation

Table 4.1 Operation indices values of the best antenna obtained by TLBO

Frequency (GHz)	Gain (dBi)	S_{11} (dB)	VSWR	Bandwidth (%)
25	6.09	−43.00	1.01	6.6
37	3.34	−47.15	1.01	>46.4

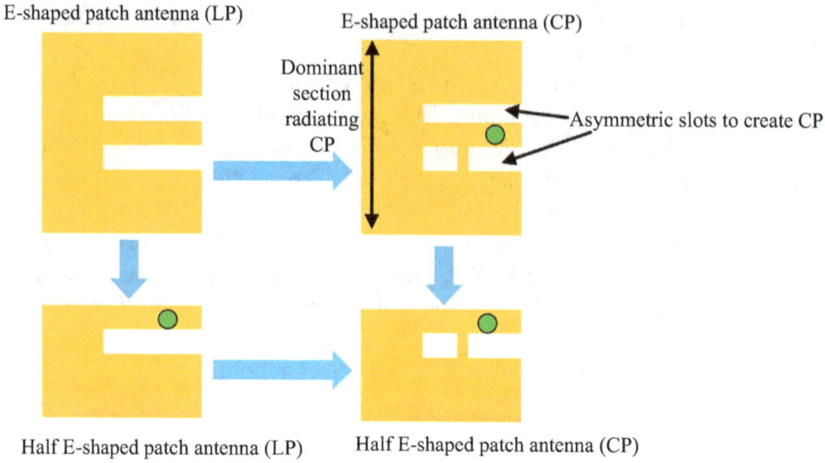

Figure 4.15 The evolution of shape from the LP E-shaped antenna to CP half E-shaped antenna [28]

4.2 Half E-shaped patch antenna design

The E-shaped patch antennas are wide band and linearly polarized (LP). One of the many possible applications of patch antennas is high-speed short-distance communication. Moreover, such communication systems have usually a key requirement to operate using circular polarization (CP). The main benefit of CP is the fact disengages the receiver from the dependency of the direction of arrival in an incident EM wave. CP eliminates the mismatch between the transmitter and the receiver due to misalignment and thus has the potential to improve the efficiency of the antenna. Therefore, it is sometimes important for some applications to design CP antennas. In order to design a CP antenna based on the E-shaped patch the authors in [28] design a half E-shaped antenna that operates at 2.4 GHz. The proposed design in [28] consists of two layers, and it is probe fed. The half E-shaped CP antenna can be derived from the E-shaped using the shape evolution depicted in Figure 4.15.

From these figures we may observe that the half E-shaped CP antenna requires the addition of a shorting bar of width W_b. This shorting bar induces currents in the y-direction. Thus, it allows the CP operation of the antenna. We may notice that one important factor that affects the axial ratio (AR) bandwidth is the shorting bar position. Thus, the suitable shorting bar position is another unknown variable.

4.2.1 Wireless LAN antenna design

The geometry of a modified half E-shaped patch antenna with coaxial feed is given in Figures 4.16 and 4.17. One may notice that the antenna geometry is quite complex. It consists of 11 different geometrical design parameters. Therefore, it is very difficult or even impossible to estimate the effect of each design parameter in order to achieve

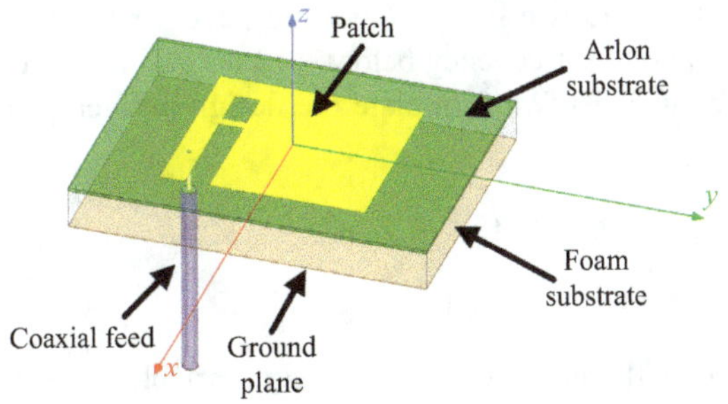

Figure 4.16 Antenna geometry for wireless LAN operation

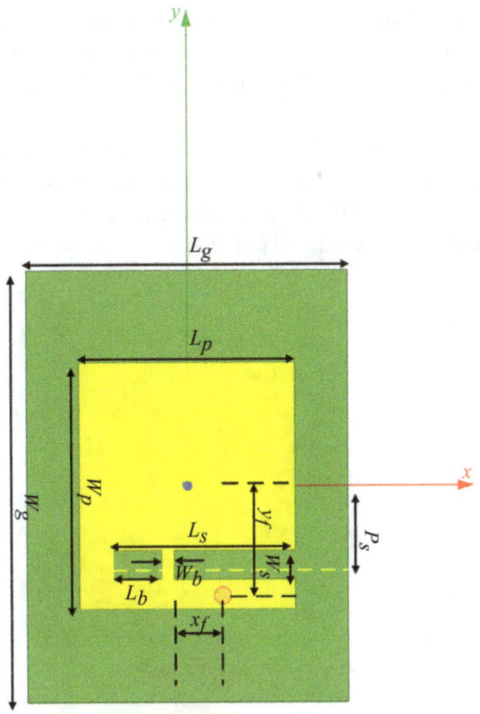

Figure 4.17 Antenna geometry top view for wireless LAN operation

the desired antenna performance. Thus, an optimization technique is the obvious solution to this design problem.

We design a half E-shaped patch antenna for operation in 2.45 GHz for Wi-Fi applications that uses a coaxial feed. The antenna for this case is fabricated on two layers. The first layer is an Arlon substrate ($\varepsilon_r = 3.38$) with 0.762-mm thickness, and the second one is foam with ($\varepsilon_r = 1.05$) with 10-mm thickness.

In this case there are two design objectives, the first is to minimize the S_{11} magnitude at the operating frequency below −10 dB and the second objective is to reduce the AR below 3 dB. We formulate the design problem using the following expression [16]:

$$F(\bar{u}) = S_{11}(\bar{u}) + \xi \times \left| |S_{11}(\bar{u})| - |L_{dB}| \right| + \xi \times \left| AR(\bar{u}) - L_{AR} \right| \tag{4.5}$$

where $\bar{u}$ is the vector of the antenna geometry design variables, S_{11} is the S_{11} magnitude, and AR is the axial ratio at the design frequency, respectively, L_{dB} is the S_{11} dB limit, L_{AR} is the AR dB limit, and ξ is a penalty factor. For both design cases, we select $L_{dB} = -10$ dB and $L_{AR} = 3$ dB.

In this case, we select the jDE [29] algorithm, which is parameter free. The use of parameter-free algorithms for this type of problems reduces the complexity and saves time spent on proper control parameters fine-tuning. We set the population size to 20 and the maximum number of iterations to 200. The best antenna out of ten independent runs is selected.

Figure 4.18 shows the 3D radiation pattern at 2.45 GHz of the best design obtained by jDE. Moreover, the surface current distribution at the same frequency is plotted in Figure 4.19. One may see the radiation pattern is almost uniform in space. This distribution of the radiated power is obtained as the maximum values of the surface current

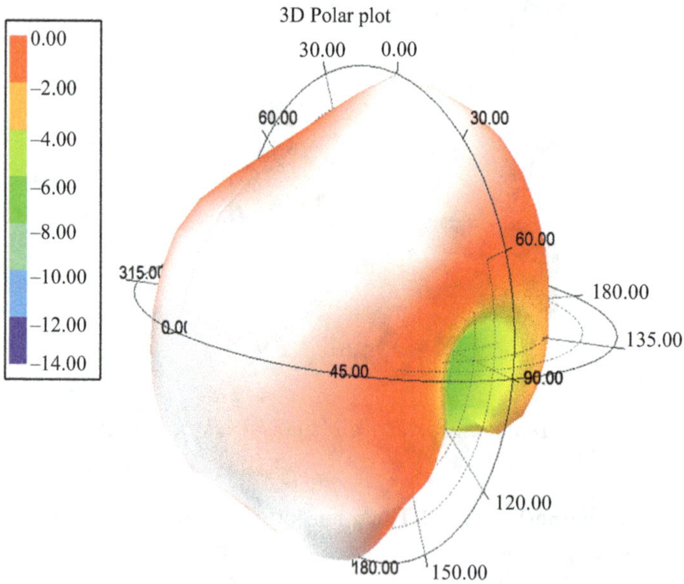

Figure 4.18 3D Gain pattern of the best design found by jDE at 2.45 GHz for wireless LAN operation

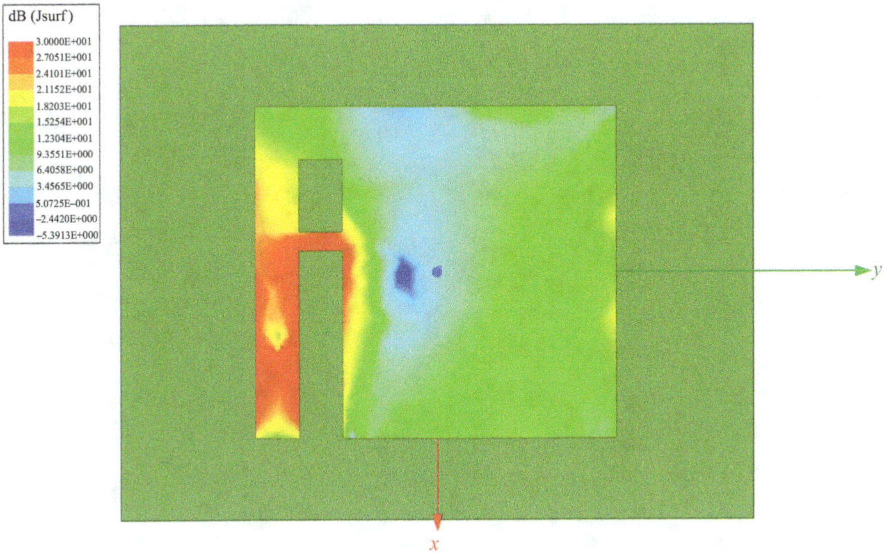

Figure 4.19 Simulated surface current distribution of the best design for wireless LAN operation at 2.45 GHz

are substantially limited to a small area of the patch and particularly near the shorting bar. The corresponding pattern in different planes are depicted in Figures 4.20–4.22.

Figure 4.23 shows both the S_{11} and the AR versus magnitudes versus frequency. The antenna has a resonance of -59.71 dB at 2.449 GHz, while a smaller resonance also exists at 2.05 GHz with -42.32 dB. We notice that the antenna exhibits a wideband behavior, since the ($S_{11} < -10$ dB) bandwidth is about 350 MHz and the bandwidth percentage is about 14%. However, the S_{11}–AR bandwidth ($AR < 3$ dB and $S_{11} < -10$ dB) is more narrow, extending from about 2.40 GHz to 2.51 GHz, which corresponds to a bandwidth of 4.63%. Figure 4.24 shows the VSWR versus frequency graph that shows similar behavior with the previous plot. The Gain versus frequency plot is depicted in Figure 4.25. We notice that up to 2.5 GHz the Gain values are above 8 dBi, and at a design frequency of 2.45 GHz, the Gain value is about 8.56 dBi, which is very satisfactory. Overall, the optimization process has derived an antenna that satisfies the design requirements.

4.2.2 5G antenna design

In this subsection, we design an antenna for 5G communication systems. This antenna is designed for operation at 26 GHz. A similar design case was presented in [16] using hybrid Jaya-GWO.

The antenna geometry for this case is depicted in Figures 4.26 and 4.27. The main differences with the previous design case is that in this case the antenna is fabricated on a single layer of dielectric and the feeding is accomplished using 50-Ω feed line.

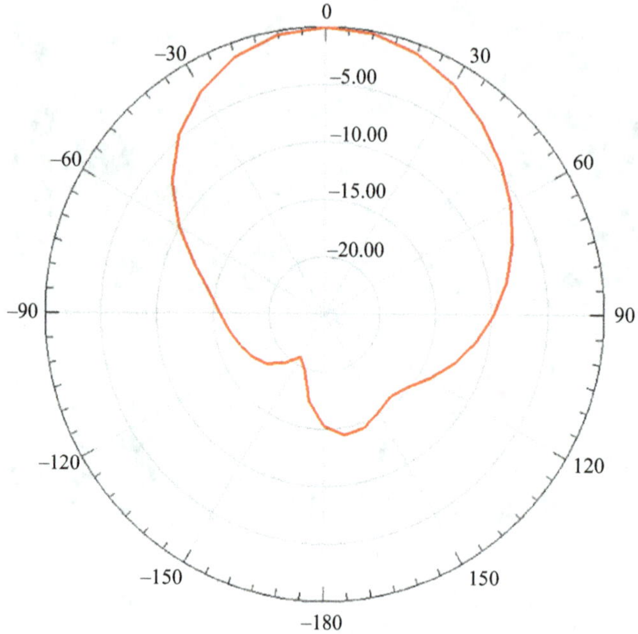

Figure 4.20 Radiation pattern of the best obtained design at 2.45 GHz for wireless LAN operation ($\phi = 0°$)

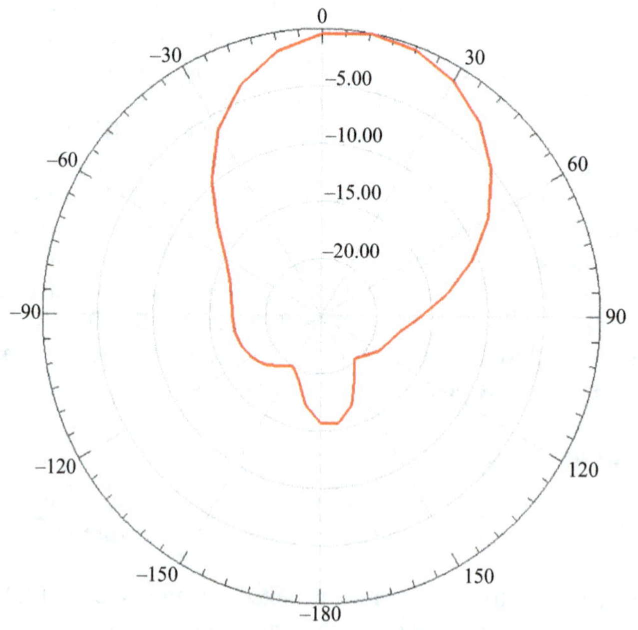

Figure 4.21 Radiation pattern of the best obtained design at 2.45 GHz for wireless LAN operation ($\phi = 90°$)

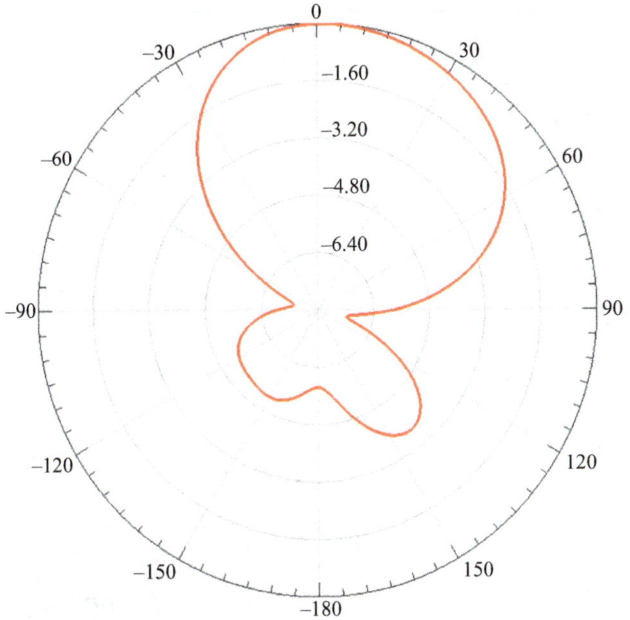

Figure 4.22 Radiation pattern of the best obtained design at 2.45 GHz for wireless LAN operation ($\theta = 90°$)

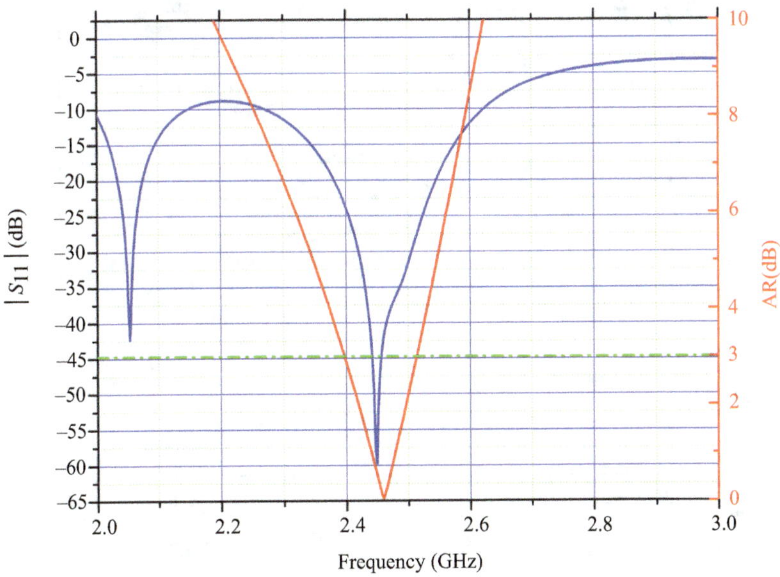

Figure 4.23 S_{11}–AR plot of the best antenna best design found by jDE for wireless LAN operation at 2.45 GHz

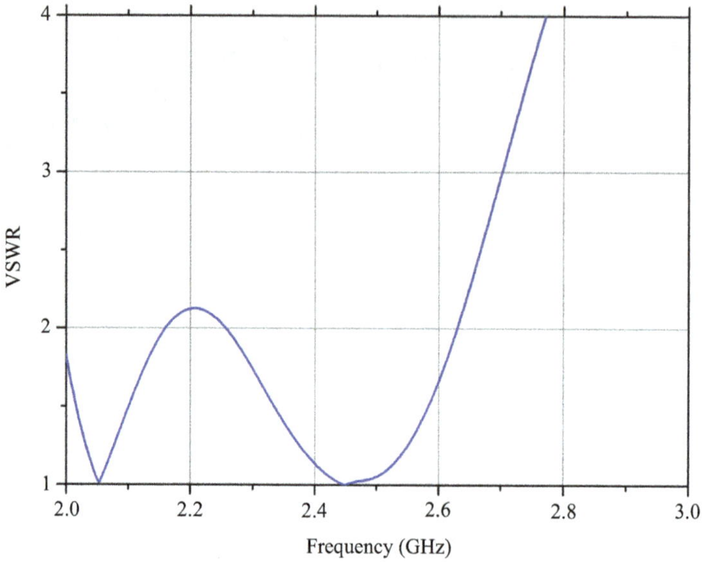

Figure 4.24 VSWR plot of the best design found by jDE for wireless LAN operation at 2.45 GHz

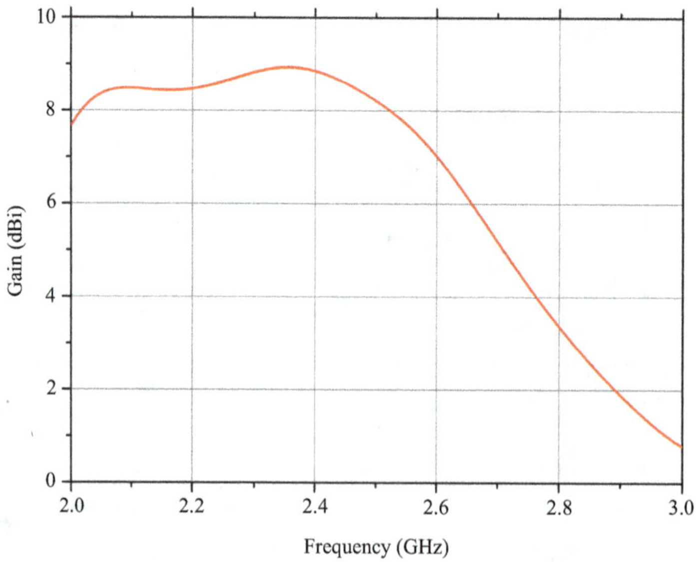

Figure 4.25 Gain plot of the best design found by jDE for wireless LAN operation at 2.45 GHz

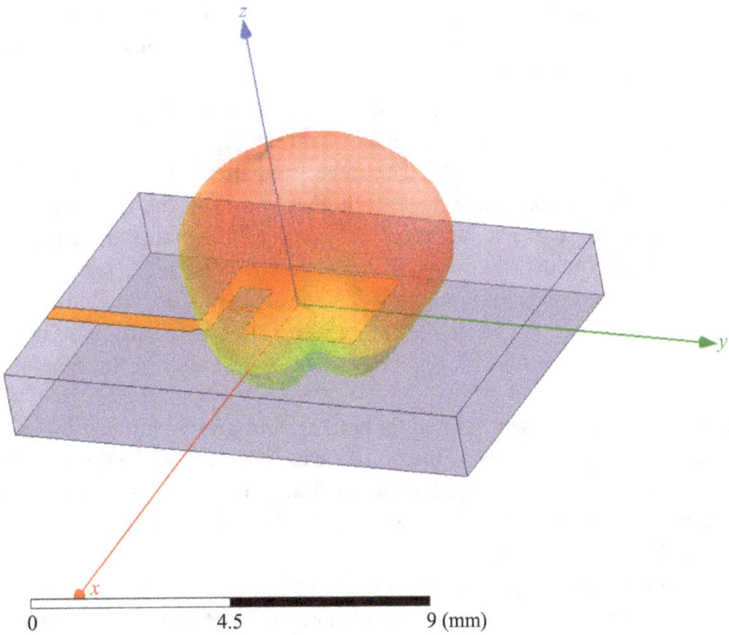

Figure 4.26 Antenna geometry for 5G at 26 GHz

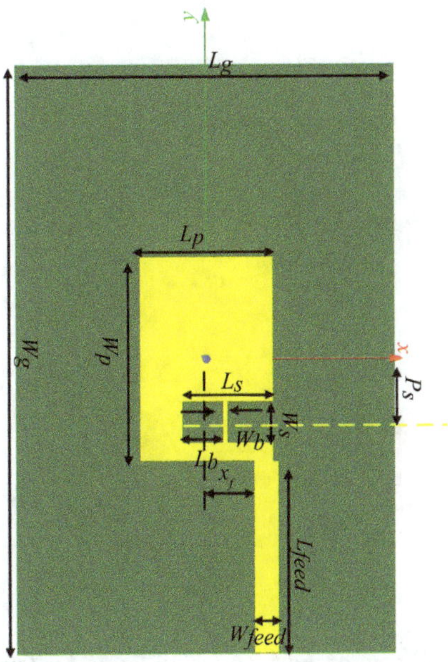

Figure 4.27 Antenna geometry top view for 5G at 26 GHz

The dielectric material selected is Taconic ($\varepsilon_r = 2.21$) with 1.58-mm thickness. This type of antenna requires the setting of 12 geometrical parameters. In this case, we optimize the antenna using the Jaya algorithm.

The 3D radiation pattern at 26 GHz of the best obtained design by Jaya is depicted in Figure 4.28. The corresponding surface current distribution for the same frequency is illustrated in Figure 4.29. The radiation pattern is very close to uniform in space, while the surface current shows higher values in the feed line and in the area near the shorting bar. The radiation patterns for three plane cuts are depicted in Figures 4.30–4.32. The distribution of the radiated power is judged satisfactory as the Gain holds values near the maximum, inside wide areas in space, making the antenna capable of transmitting and receiving, effectively, signals toward and from a large number of various directions. This performance is desirable in the case of mobile communications.

The frequency response is plotted for both S_{11} and AR magnitudes in Figure 4.33. At the design frequency of 26 GHz the S_{11} value is -18.6 dB, while the AR magnitude is 2.9. The obtained design is ultra wide band and the ($S_{11} < -10$ dB) bandwidth ranges from 21.26 GHz to 35 GHz. However, the S_{11}–AR bandwidth is narrower, and it is about 3.97%.

The VSWR versus frequency plot is depicted in Figure 4.34. This is similar with the previous graph that shows a wideband behavior. The Gain versus frequency plot is illustrated in Figure 4.35. Between 26 GHz and 28 GHz the antenna Gain is close to 6 dBi. At the design frequency at 26 GHz the Gain has a value of 5.64 dBi, which is quite satisfactory. Overall, Jaya found an antenna design that covers the design requirements for 5G operation and CP.

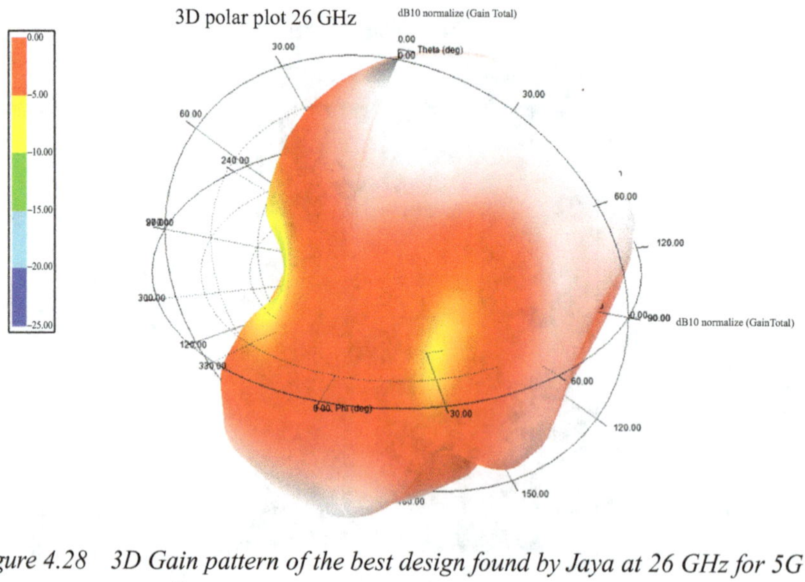

Figure 4.28 3D Gain pattern of the best design found by Jaya at 26 GHz for 5G operation

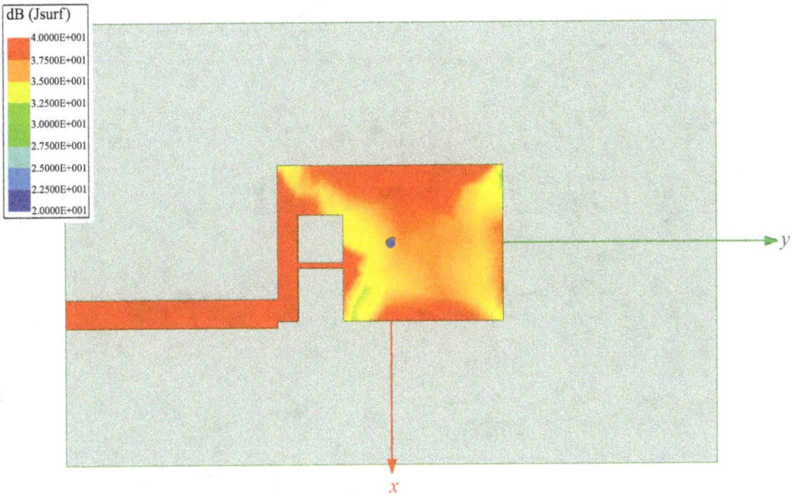

Figure 4.29 Simulated surface current distribution of the best design at 26 GHz for 5G operation

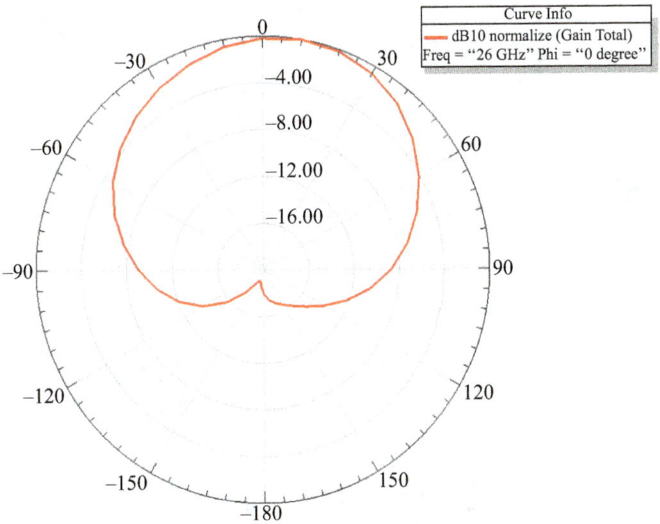

Figure 4.30 Radiation pattern of the best obtained design by Jaya at 26 GHz for 5G operation ($\phi = 0°$)

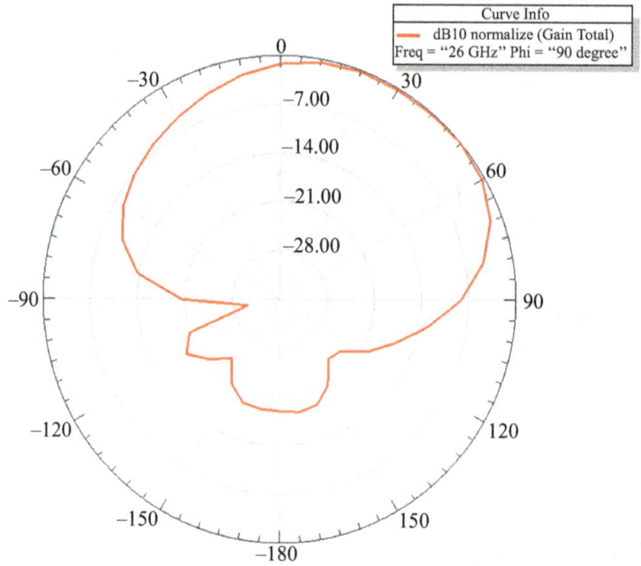

Figure 4.31 Radiation pattern of the best obtained design by Jaya at 26 GHz for 5G operation ($\phi = 90°$)

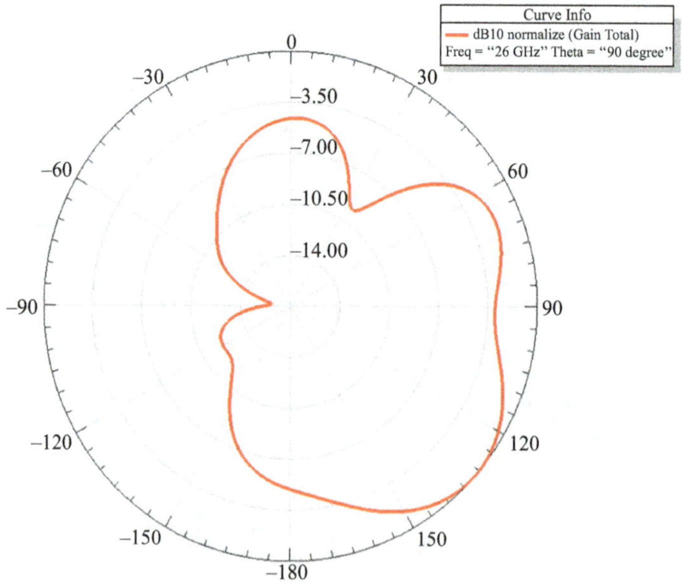

Figure 4.32 Radiation pattern of the best obtained design by Jaya at 26 GHz for 5G operation ($\theta = 90°$)

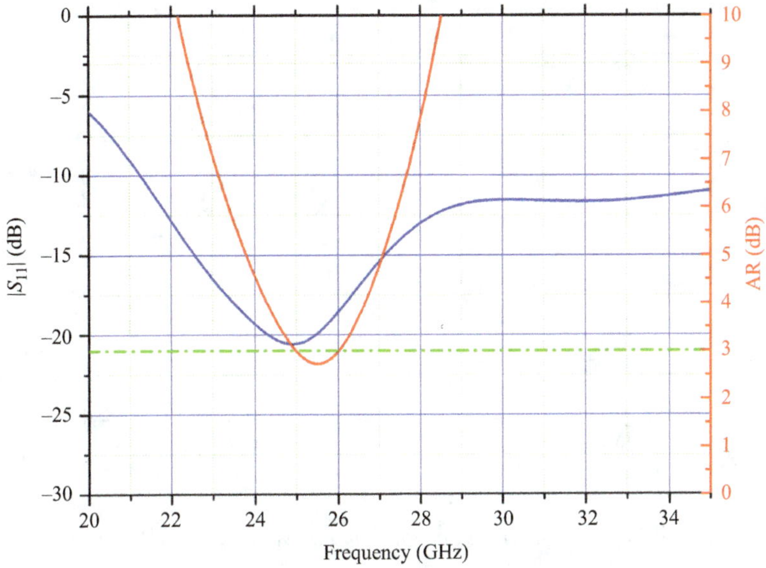

Figure 4.33 *S_{11}–AR plot of the best antenna best design found by Jaya at 26 GHz for 5G operation*

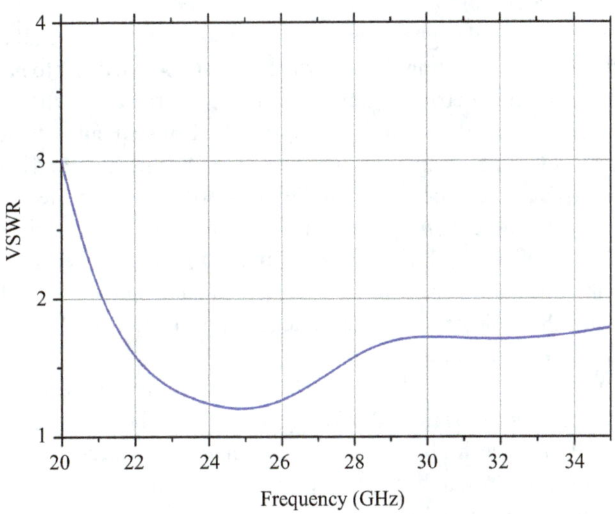

Figure 4.34 *VSWR plot of the best design found by Jaya at 26 GHz for 5G operation*

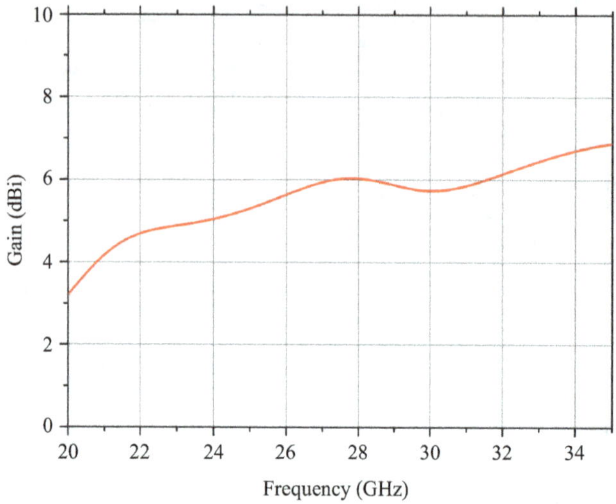

Figure 4.35 Gain plot of the best design found by Jaya at 26 GHz for 5G operation

4.3 Arbitrary-shaped patch antenna design

The design case of this section is that of arbitrary-shaped dual-band patch antenna. The basic idea here is to optimize not only the appropriate geometrical parameters but also to find a best shape for the antenna. An example of this type of arbitrary-shaped patch antenna is shown in Figure 4.36.

The patch antenna is designed on an Arlon substrate with $\varepsilon_r = 3.38$. The antenna utilizes aperture-coupled feeding. The aperture is H-shaped in order to have dual-band operation. Thus, the geometrical parameters of the aperture and the feed line have also to be determined by the optimization process. Consequently, to fully describe the design of a random patch antenna, a total number of nine geometrical parameters need to be specified as it can be seen from Figure 4.36. Moreover, the patch is divided into an $m \times n$ grid, where m, n are integers. Each element of the grid can be metal ("binary one") or dielectric ("binary zero"). Figure 4.37 shows a typical case of such a grid. Thus, the total number of decision variables equals $m \times n + 9$. This problem is classified into the category of mixed-integer optimization problems that contain both real and discrete variables.

A randomly shaped dual-band patch antenna can be designed using the biogeography-based optimization (BBO) algorithm [30]. The BBO algorithm is incorporated with a full-wave high-frequency EM solver to extract the optimal design parameters. The random patch antenna is dual tuned at f_1 GHz and f_2 GHz. The objective function is given by [31]

$$F(\bar{x}) = \max\left(S_{11}^{f_1\mathrm{GHz}}, S_{11}^{f_2\mathrm{GHz}}\right) \tag{4.6}$$

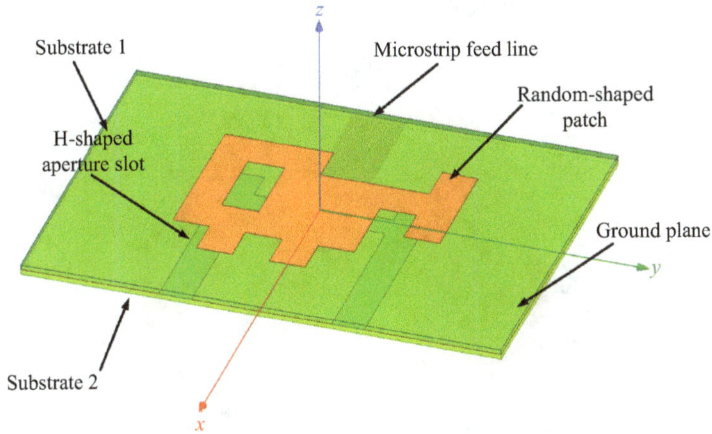

Figure 4.36 Geometry of an aperture-coupled random dual-band patch antenna

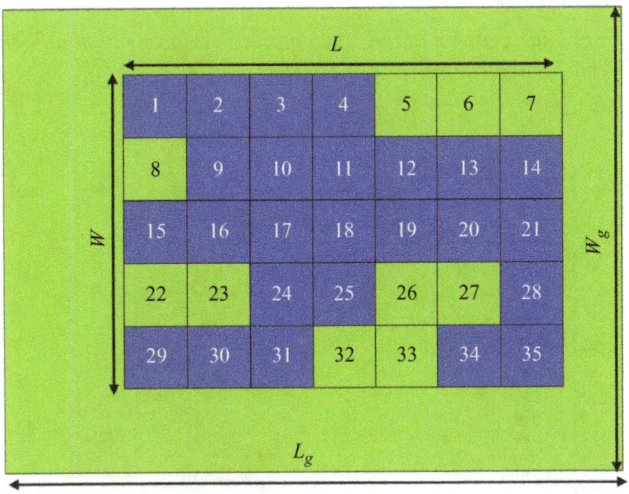

Figure 4.37 Geometry of 7 × 5 example random patch antenna

In this case, we design a random patch antenna for dual-band operation at 2.4 GHz (Wi-Fi) and at 2.6 GHz (4G/LTE (long-term evolution)). We apply the BBO algorithm for ten independent runs. We set the population size is set to 20 and the maximum number of iterations is set to 200. Figures 4.38 and 4.39 show the 3D radiation pattern at 2.4 GHz and at 2.6 GHz. Additionally, Figures 4.40 and 4.41 plot the surface current distribution at both frequencies. We notice that both radiation patterns are close to omnidirectional in space, while the surface current show higher values at particular areas different for every frequency. The radiation patterns for three plane cuts are depicted in Figures 4.42–4.44.

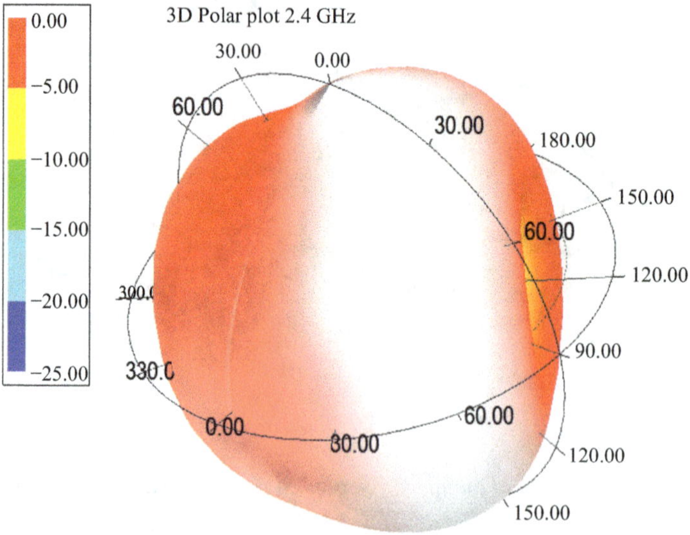

Figure 4.38 3D Gain pattern of the best obtained random patch antenna for dual-band operation at 2.4 GHz

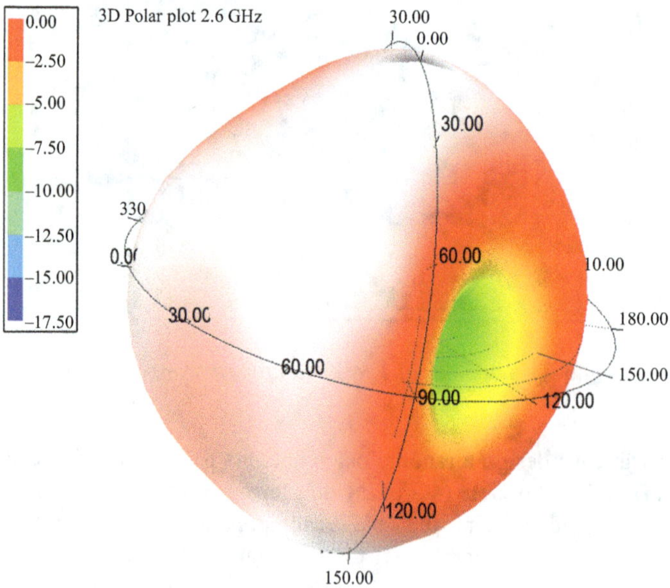

Figure 4.39 3D Gain pattern of the best obtained random patch antenna for dual band operation at 2.6 GHz

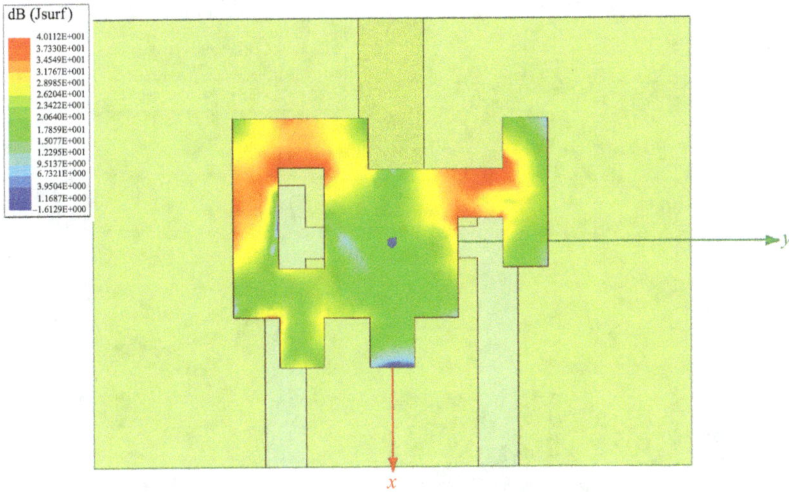

Figure 4.40 Simulated surface current distribution of the best obtained random patch antenna for dual-band operation at 2.4 GHz

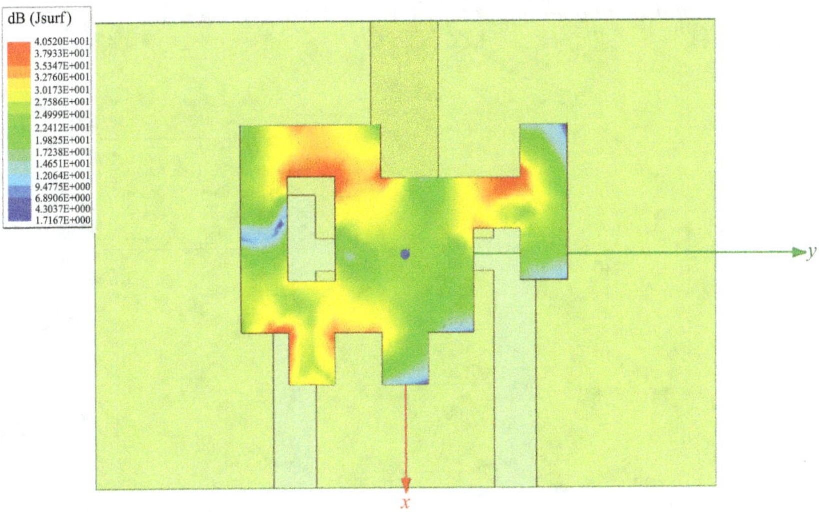

Figure 4.41 Simulated surface current distribution of the best obtained random patch antenna for dual-band operation at 2.6 GHz

Figure 4.45 illustrates the S_{11} plot versus frequency for the best random patch antenna design. We can easily notice that the proposed antenna is wide band and its operational bandwidth ($S_{11} < -10$ dB) is from 2.34 GHz to 2.65 GHz, thus covering both Wi-Fi and LTE bands. Moreover, the S_{11} values are -21.42 dB and -29.75 dB at 2.4 GHz and 2.6 GHz, respectively. The VSWR plot versus frequency is depicted

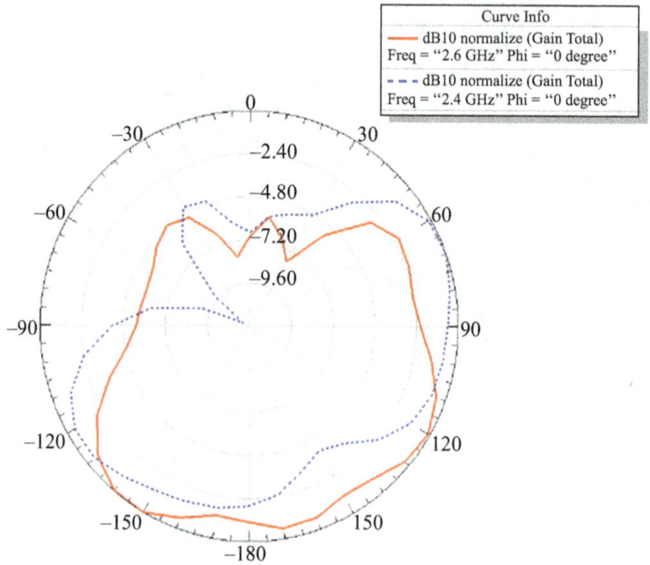

Figure 4.42 Radiation pattern of the best obtained random patch antenna for dual-band operation at 2.4 GHz and 2.6 GHz ($\phi = 0°$)

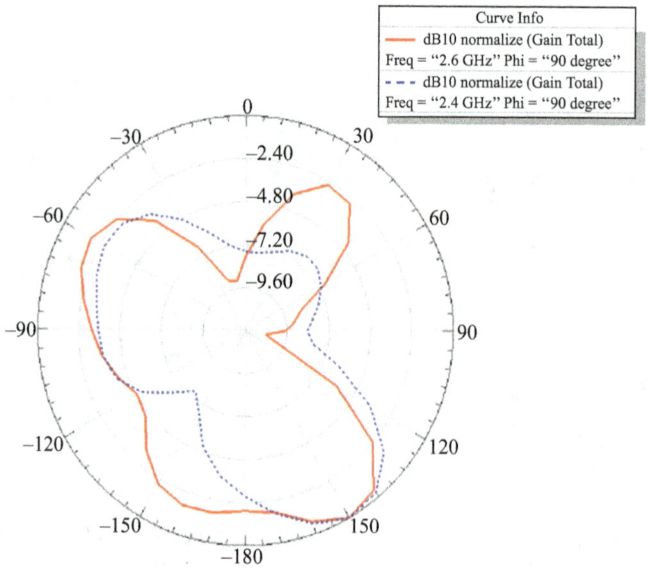

Figure 4.43 Radiation pattern of the best obtained random patch antenna for dual-band operation at 2.4 GHz and 2.6 GHz ($\phi = 90°$)

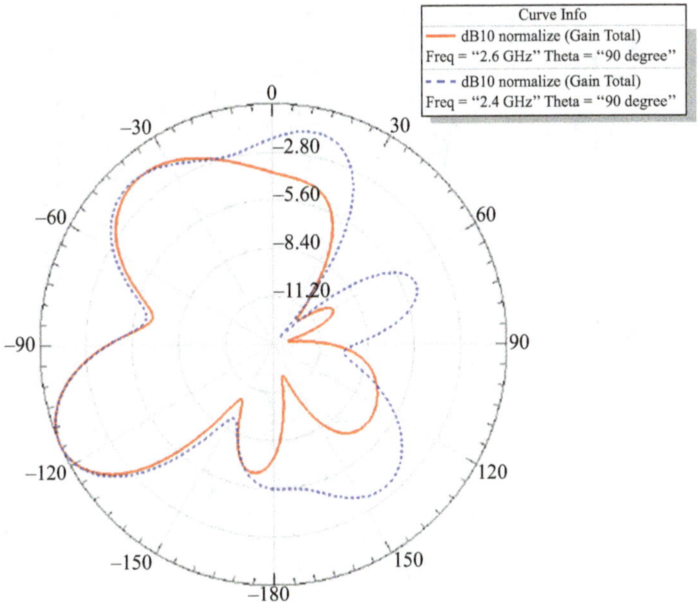

Figure 4.44 *Radiation pattern of the best obtained random patch antenna for dual-band operation at 2.4 GHz and 2.6 GHz ($\theta = 90°$)*

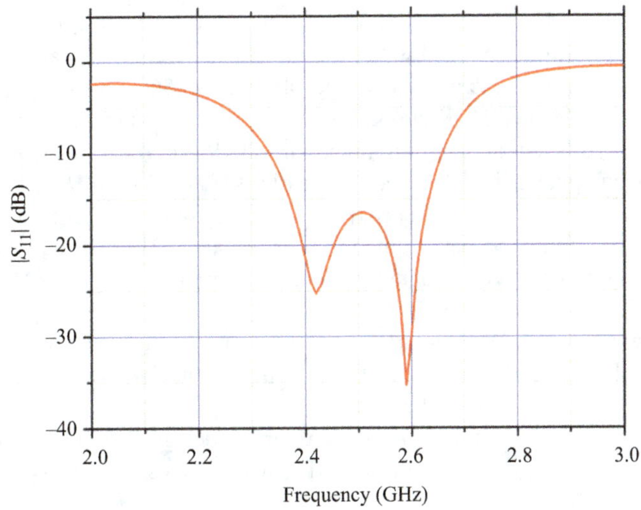

Figure 4.45 *S_{11} plot of the best obtained random patch antenna for dual-band operation at 2.4 GHz and 2.6 GHz*

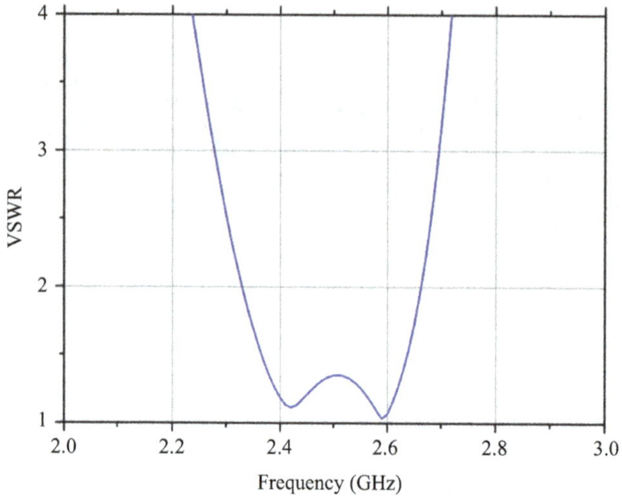

Figure 4.46 VSWR plot of the best obtained random patch antenna for dual-band operation at 2.4 GHz and 2.6 GHz

in Figure 4.46, the ($VSWR < 2$) bandwidth is about 310 MHz. Overall, the obtained result is satisfactory and fully complies with the desired antenna characteristics.

References

[1] Yeung SH, Man KF, Luk KM, *et al.* A trapeziform U-slot folded patch feed antenna design optimized with jumping genes evolutionary algorithm. IEEE Transactions on Antennas and Propagation. 2008;56(2):571–577.

[2] Bach TB, Manh LH, Khac KN, *et al.* Evolved design of microstrip patch antenna by genetic programming. In: 2019 International Conference on Electromagnetics in Advanced Applications (ICEAA); 2019. p. 1393–1397.

[3] Salucci M, Oliveri G, Rocca P, *et al.* A system-by-design approach for efficient multiband patch antennas design. In: 2017 International Applied Computational Electromagnetics Society Symposium – Italy (ACES); 2017. p. 1–2.

[4] Weng Z, Guo D, Wu Y, *et al.* A 2.45 GHz microstrip patch antenna evolved for WiFi application. In: 2015 IEEE Congress on Evolutionary Computation (CEC); 2015. p. 1191–1195.

[5] Jiao R, Sun Y, Sun J, *et al.* Antenna design using dynamic multi-objective evolutionary algorithm. IET Microwaves, Antennas Propagation. 2018;12(13):2065–2072.

[6] Gjokaj V, Doroshewitz J, Nanzer J, *et al.* A design study of 5G antennas optimized using genetic algorithms. In: 2017 IEEE 67th Electronic Components and Technology Conference (ECTC); 2017. p. 2086–2091.

[7] Jin N and Rahmat-Samii Y. Parallel particle swarm optimization and finite-difference time-domain (PSO/FDTD) algorithm for multiband and wide-band patch antenna designs. IEEE Transactions on Antennas and Propagation. 2005;53(11):3459–3468.

[8] Rahmat-Samii Y. Modern antenna designs using nature inspired optimization techniques: Let Darwin and the bees help designing your multi band MIMO antennas. In: 2007 IEEE Radio and Wireless Symposium; 2007. p. 463–466.

[9] Rahmat-Samii Y, Kovitz JM, and Rajagopalan H. Nature-inspired optimization techniques in communication antenna designs. Proceedings of the IEEE. 2012;100(7):2132–2144.

[10] Zhang L, Cui Z, Jiao YC, *et al.* Broadband patch antenna design using differential evolution algorithm. Microwave and Optical Technology Letters. 2009;51(7):1692–1695.

[11] Rocca P, Oliveri G, and Massa A. Differential evolution as applied to electromagnetics. IEEE Antennas and Propagation Magazine. 2011;53(1):38–49.

[12] Goudos SK, Siakavara K, Samaras T, *et al.* Self-adaptive differential evolution applied to real-valued antenna and microwave design problems. IEEE Transactions on Antennas and Propagation. 2011;59(4):1286–1298.

[13] Gangopadhyaya M, Mukherjee P, Sharma U, *et al.* Design optimization of microstrip fed rectangular microstrip antenna using differential evolution algorithm. In: 2015 IEEE 2nd International Conference on Recent Trends in Information Systems (ReTIS); 2015. p. 49–52.

[14] Goudos SK, Tsiflikiotis A, Babas D, *et al.* Evolutionary design of a dual band E-shaped patch antenna for 5G mobile communications. In: 2017 6th International Conference on Modern Circuits and Systems Technologies (MOCAST); 2017. p. 1–4.

[15] Goudos SK, Yioultsis TV, Dalidou K, *et al.* A low cost wide band and circularly polarized modified half e-shaped patch antenna for 5G mobile communications. In: IET Conference Publications, EuCAP 2018. vol. 2018; 2018.

[16] Goudos SK, Yioultsis TV, Boursianis AD, *et al.* Application of new hybrid Jaya grey wolf optimizer to antenna design for 5G communications systems. IEEE Access. 2019;7:71061–71071.

[17] ANSYS. Electromagnetics Suite, ANSYS: User's Guide, Version 16.1; 2015.

[18] Bekers DJ, Monni S, van den Berg SM, *et al.* Optimization of phased arrays integrated with FSS and feeding elements based on parametric models. In: Antennas and Propagation, 2007. EuCAP 2007. The Second European Conference on; 2007. p. 1–7.

[19] Ramasami V. A HFSS API to Control HFSS from Matlab; 2017.

[20] Zhu W, Xiao S, Yuan R, *et al.* Broadband and dual circularly polarized patch antenna with H-shaped aperture. In: International Symposium on Antennas and Propagation Conference Proceedings; Kaohsiung, Taiwan. 2014, p. 549–550. doi: 10.1109/ISANP.2014.7026769.

[21] Hoseini Izadi O and Mehrparvar M. A compact microstrip slot antenna with novel E-shaped coupling aperture. In: 5th International Symposium on

Telecommunications; Tehran, Iran. 2010, p. 110–114, doi: 10.1109/ISTEL.2010.5734008.

[22] Bo-yu X, Guang-qiu Z, and Zheng T. Design of reflect array antenna element based on Hour-Glass shaped coupling aperture. In: Proceedings of the 9th International Symposium on Antennas, Propagation and EM Theory; Guangzhou, China. 2010, p. 155–158. doi: 10.1109/ISAPE.2010.5696420.

[23] Jang TH, Kim HY, Song IS, *et al.* A wideband aperture efficient 60-GHz series-fed E-shaped patch antenna array with copolarized parasitic patches. IEEE Transactions on Antennas and Propagation. 2016;64(12):5518–5521.

[24] Aliakbari H, Abdipour A, Mirzavand R, *et al.* A single feed dual-band circularly polarized millimeter-wave antenna for 5G communication. In: 2016 10th European Conference on Antennas and Propagation, EuCAP 2016; 2016.

[25] Mak KM, Lai HW, Luk KM, *et al.* Circularly polarized patch antenna for future 5G mobile phones. IEEE Access. 2014;2:1521–1529.

[26] Pozar DM. Microstrip antenna aperture-coupled to a microstripline. Electronics Letters. 1985;21(2):49–50.

[27] Rao RV, Savsani VJ, and Vakharia DP. Teaching-learning-based optimization: A novel method for constrained mechanical design optimization problems. CAD Computer Aided Design. 2011;43(3):303–315.

[28] Kovitz JM, Rajagopalan H, and Rahmat-Samii Y. Circularly polarised half E-shaped patch antenna: A compact and fabrication-friendly design. IET Microwaves, Antennas Propagation. 2016;10(9):932–938.

[29] Brest J, Boškovć B, Greiner S, *et al.* Performance comparison of self-adaptive and adaptive differential evolution algorithms. Soft Computing. 2007;11(7):617–629.

[30] Simon D. Biogeography-based optimization. IEEE Transactions on Evolutionary Computation. 2008;12(6):702–713.

[31] Goudos SK, Siakavara K, and Kalialakis C. Application of opposition-based learning concepts for arbitrary patch antenna design for wireless communications. In: 2017 International Workshop on Antenna Technology: Small Antennas, Innovative Structures, and Applications (iWAT 2017); Athens. 2017, p. 171–174. doi: 10.1109/IWAT.2017.7915350.

Chapter 5
Microwave structures design using EAs

This chapter presents design cases from different microwave structure cases.

5.1 Design of microwave broadband absorbers

The problem of a planar microwave absorber design lies in the minimization of the reflection coefficient of an incident plane wave in a multilayer structure for a desired range of angles and frequencies. The reflection coefficient depends on the thickness and the electric and magnetic properties of each layer. Several studies, which address this problem, exist in the literature.

Evolutionary algorithms (EAs) like genetic algorithms (GAs) have been in several occasions applied in absorber design [1–7]. In [8], a microGA algorithm with a predefined materials database was applied. The major drawback of a GA approach is the difficulty in implementation due to the algorithm-inherited complexity and the required long computational time.

Moreover, the application of swarm intelligence optimizers in electromagnetic (EM) design problems has attracted several researchers [9–12]. Particle swarm optimization (PSO) has also been used successfully in absorber design problems [3,5,7,13–18]. Recently, artificial bee colony (ABC) [19] has been applied in several cases to microwave absorber design [20–22]. Differential evolution (DE) has also been applied to absorber design problem [23,24]. A comparative study between PSO and DE for the microwave absorber design problems is reported in [25]. Moreover, other emerging nature-inspired algorithms have been deployed in the literature for solving the abovementioned problem [26–28].

5.1.1 Problem formulation

For a multilayer planar structure given in Figure 5.1, the unknowns are the number and the characteristics of each layer. The characteristics sought are the thickness and the frequency-dependent (in general) complex permittivity and permeability, which are given by

$$\varepsilon_i(f) = \varepsilon_0[\varepsilon'_i(f) - j\varepsilon''_i(f)] \tag{5.1}$$

$$\mu_i(f) = \mu_0[\mu'_i(f) - j\mu''_i(f)] \tag{5.2}$$

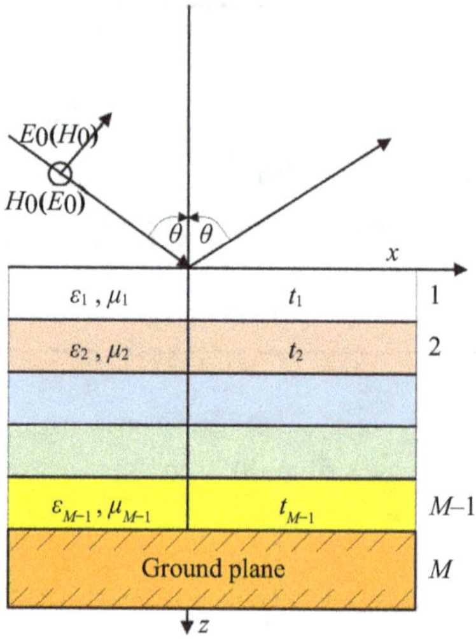

Figure 5.1 Multilayer absorber geometry

The terms ε_0 and μ_0 are the free space permittivity and permeability, respectively. The imaginary parts in (5.1) and (5.2) represent dielectric and magnetic losses, respectively. Lossless $\varepsilon'' = 0$, $\mu' = 1$, $\mu'' = 0$ and lossy dielectrics ($\varepsilon'' \neq 0$, $\mu' = 1$, $\mu'' = 0$) have frequency-independent permittivity in a relatively wide range. The imaginary part in (5.1) can be alternatively represented by the conductivity given by $\sigma = \varepsilon_0\varepsilon'' 2\pi f$. The last layer of Figure 5.1 is considered to be perfect electric conductor. All layers are assumed infinite. The incident plane wave may have transverse electric (TE) or transverse magnetic (TM) polarization. The plane of incidence is normal to the multilayer structure and the incident media is free space. For this type of multilayer structure, the general expression of the reflection coefficients at the ith layer R_i^{TE} or R_i^{TM} for an incident plane wave is found by using the following recursive formula [29]:

$$R_i^{TE/TM} = \frac{r_i^{TE/TM} + R_{i+1}^{TE/TM} e^{-2jk_{i+1}t_{i+1}}}{1 + r_i^{TE/TM} R_{i+1}^{TE/TM} e^{-2jk_{i+1}t_{i+1}}} \tag{5.3}$$

where

$$r_i^{TE} = \begin{cases} \dfrac{\mu_{i+1}k_i - \mu_i k_{i+1}}{\mu_{i+1}k_i + \mu_i k_{i+1}}, & i < M \\ -1, & i = M \end{cases} \tag{5.4}$$

and

$$r_i^{TM} = \begin{cases} \dfrac{\varepsilon_{i+1}k_i - \varepsilon_i k_{i+1}}{\varepsilon_{i+1}k_i + \varepsilon_i k_{i+1}}, & i < M \\ 1, & i = M \end{cases} \tag{5.5}$$

where M is the number of layers, k_i is the wavenumber of the ith layer given by $k_i = 2\pi f\sqrt{\varepsilon_i \mu_i}$. The absorber design is expressed as the minimization problem of the quantity $R_1{}^{TE/TM}$ (expressed in dB) given next:

$$R_1{}^{TE/TM} = 20\log\{\max|R(f,\theta)|f \in B, \theta \in A\} \tag{5.6}$$

where $\max|R(f,\theta)|$ is the maximum reflection coefficient over the desired frequency and angle range for a given polarization. B denotes the desired set of frequencies, while A is the desired set of angles. The absorber total thickness is the sum of the layer thicknesses:

$$T_{tot} = \sum_{i=1}^{M-1} t_i \tag{5.7}$$

Several design approaches exist for the earlier problem. An effective way to do so is to use an exact penalty method and to combine the two objective–functions in a single one. The objective–function can be expressed as [24]

$$F(f,\theta) = R(f,\theta) + \Xi \times \max\{0, T_{tot} - T_{des}\} \tag{5.8}$$

where Ξ is a very large number, T_{tot} is the total thickness calculated, and T_{des} is the maximum total thickness desired. By minimizing the previous formula with a global optimizer, a solution can be found. The same algorithm can run for different values of T_{des}, and different designs for different total thicknesses can be found. A multi-objective (MO) optimization method can be used instead for this problem. Such a method offers a whole set of solutions belonging to the Pareto front. The problem is then expressed as a bi-objective problem given by minimization of the objectives:

$$f_1(\bar{x}) = \min R_1{}^{TE/TM}(\bar{x}), \qquad f_2(\bar{x}) = \min T_{tot}(\bar{x}) \tag{5.9}$$

where $\bar{x}$ is the vector of the configuration of the absorber that includes both material types and layer thicknesses.

5.1.2 Single-objective absorber optimization

In this section, we compare the application of four different biogeography-based optimization (BBO) [30] models to single objective microwave absorber design problem. These models are the original BBO linear model [30], the sinusoidal model presented in [31], and the nonlinear models 7 and 8 reported in [32]. In all examples that follow the habitat modification probability, P_{mod}, is set to one, and the maximum mutation rate, $m_{\max}$, is set equal to 0.005. The maximum immigration rate I, and the maximum emigration rate E are both set both to one. The motivation for using BBO to this problem type is given next.

Similar to the other EAs such as GAs, ABC, and PSO, in the BBO approach, there is a way of sharing information between solutions [30]. This feature makes BBO suitable for the same types of problems that the other algorithms are used for, namely, high-dimensional data. Additionally, BBO has some unique features that are different from those found in the other EAs. For example, quite different from GAs, ant colony optimization, and PSO, from one generation to the next, the set of the

BBO's solutions is maintained and improved using the migration model, where the emigration and immigration rates are determined by the fitness of each solution. BBO differs from PSO in the fact that PSO solutions do not change directly; the velocities change. The BBO solutions share directly their attributes using the migration models. The migration operator provides BBO with a good exploitation ability.

These differences can make BBO outperform other algorithms [30,31,33]. It must be pointed out if PSO or ABC are constrained to discrete space then the next generation will not necessarily be discrete [33]. However, this is not true for BBO; if BBO is constrained to a discrete space then the next generation will also be discrete to the same space. As the authors in [33] suggest that this indicates that BBO could perform better than other EAs on combinatorial optimization problems, which makes BBO suitable for application to the microwave absorber problem. The main computational cost of EAs is in the evaluation of the objective–function. The BBO mechanism is simple, like that of PSO and ABC. Therefore, for most problems, the computational cost of BBO and other EAs will be the same since it will be dominated by objective–function evaluation [31]. The objective–function was that given in (5.8). For all examples, the population size and the number of generations were set to 100 and 1,000, respectively. Each algorithm runs for 20 independent trials.

The examples use the same predefined materials database as in [24]. This database consists of all types of materials: lossless dielectrics, lossy dielectrics, lossy magnetics, and relaxation-type magnetic materials. This is given in Table 5.1. Thus, the decision variables for this case are the material number and the layer thickness (which can be varied from 0 to maximum value). The maximum layer thickness was set (as in [24]) equal to 2 mm.

First, we present as in [24] two cases of a five-layer broadband absorber in the frequency range from 2GHz to 8GHz for normal incidence and TE polarization (designs HF1, HF2 [24]). For HF1, we set the maximum total thickness desired T_{des} to 5 mm, while for HF2 it is $T_{des} = 3$ mm. Figure 5.2 presents the boxplot of the algorithms' results for the HF1 case. We notice that BBO nonlinear model 7 has obtained the best results in terms of smaller dispersion of values and best results. The linear model obtained the results with the larger dispersion of values. Moreover, the convergence rate plot for this case is depicted in Figure 5.3. We notice all BBO algorithms converge at similar speed. Additionally, it seems that all require fewer than 100 iterations to converge near the final value. The frequency response of best obtained absorber from BBO nonlinear model 7 is plotted in Figure 5.4. The obtained design performs similarly with that from [24] designed with jDE. Moreover, the new found design is thinner than that of [24]. The maximum reflection coefficient in the desired frequency zone is −24.53 dB, which is close to the value from (−24.75 dB). Table 5.2 holds the layer parameters for this case.

The next case of HF2 is thinner, thus the maximum reflection coefficient in the desired frequency zone is higher. Figure 5.5 depicts the boxplot for the HF2 case. It is clear that all algorithms obtain results with large dispersion of values. For this case, the BBO linear model found the best result. Again as in the previous convergence plot, Figure 5.6 shows that the convergence speed is the same for all algorithms. The

Table 5.1 Predefined materials database

Lossless dielectric materials ($\mu' = 1, \mu'' = 0$)

No	ε'
1	10
2	50

Lossy magnetic materials ($\varepsilon' = 15, \varepsilon'' = 0$)

$$\mu = \mu' - j\mu'', \mu'(f) = \frac{\mu'(1GHz)}{f^a}, \mu''(f) = \frac{\mu''(1GHz)}{f^b}$$

No	μ'(1 GHz)	A	μ''(1 GHz)	b
3	5	0.974	10	0.961
4	3	1.000	15	0.957
5	7	1.000	12	1.000

Lossy dielectric materials ($\mu' = 1, \mu'' = 0$)

$$\varepsilon = \varepsilon' - j\varepsilon''\varepsilon'(f) = \frac{\varepsilon'(1GHz)}{f^a}, \varepsilon''(f) = \frac{\varepsilon''(1GHz)}{f^b}$$

No	ε'(1 GHz)	a	ε''(1 GHz)	b
6	5	0.861	8	0.569
7	8	0.778	10	0.682
8	10	0.778	6	0.861

Relaxation-type magnetic materials ($\varepsilon' = 15, \varepsilon'' = 0$)

$$\mu = \mu' - j\mu'', \mu'(f) = \frac{\mu_m f_m^2}{f^2 + f_m^2}, \mu''(f) = \frac{\mu_m f_m f}{f^2 + f_m^2}$$

No	μ_m	f_m
9	35	0.8
10	35	0.5
11	30	1.0
12	18	0.5
13	20	1.5
14	30	2.5
15	30	2.0
16	25	3.5

frequency response of the best obtained absorber by BBO linear model is plotted in Figure 5.7. We notice that the BBO obtained a design that is thicker than the one in [24] with thickness 2.801 mm (instead of 2.588 mm). However, the new design performs better in the desired frequency zone with maximum reflection coefficient −20.77 dB (instead of −20.03 dB in [24]). Table 5.3 shows the layer parameters for the best obtained results for the HF2 design case.

The next design case is that of a five-layer broadband absorber DES1 [24]. For this case, we set $T_{des} = 6$ mm, while the design requirements for optimization are from

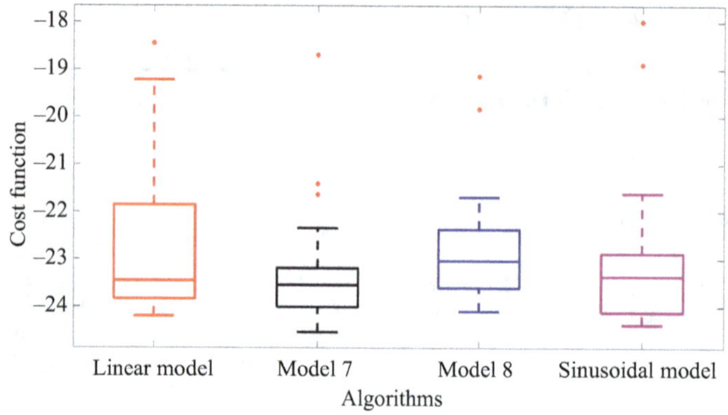

Figure 5.2 HF1 design five-layer broadband absorber from 2 GHz to 8 GHz for normal incidence and TE/TM polarization, T_{des} = 5 mm. Boxplot of all algorithms results

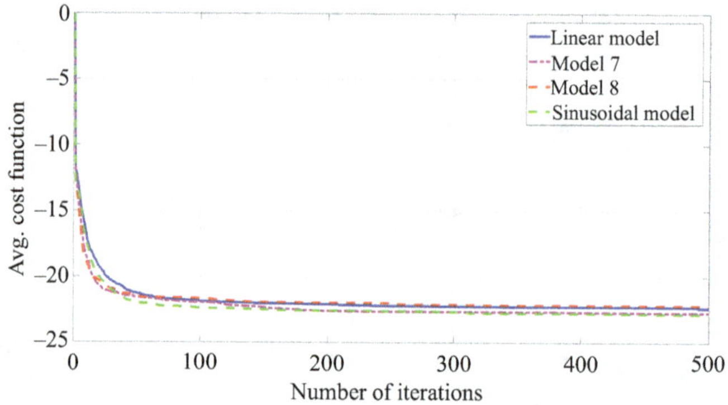

Figure 5.3 HF1 design five-layer broadband absorber 2 GHz to 8 GHz for normal incidence and TE/TM polarization, T_{des} = 5 mm. Convergence rate graph

3 GHz to 6 GHz, for angle of incidence $0° - 10°$ and for both polarizations. Thus, the problem is quite complex. The boxplot for this case is depicted in Figure 5.8. We notice that BBO model 7 obtained the smaller dispersion of values. However, BBO with sinusoidal model derived the best result. Figure 5.9 presents the convergence rate plot. All algorithms converge at similar speed. Figure 5.10 plots the frequency response of the best obtained result for normal incidence. We notice that BBO sinusoidal model derived a result with better frequency response for normal incidence than the one from [24]. The maximum reflection coefficient in the desired frequency zone is −31.70 dB (instead of −25.98 dB in [24]). Additionally, the new found design is thinner than in [24] with total thickness of 5.771 mm (6 mm in [24]). The frequency

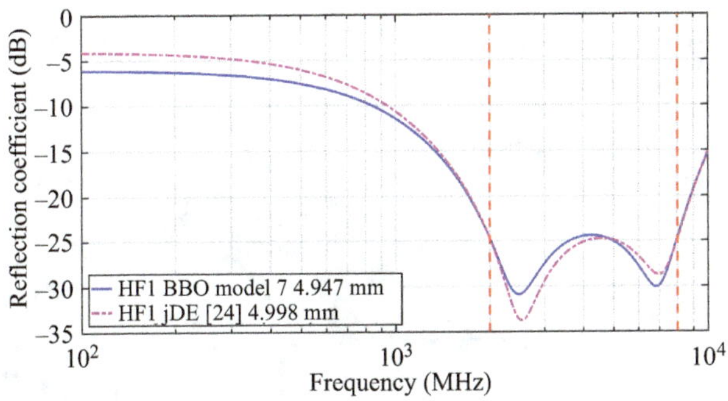

Figure 5.4 HF1 design five-layer broadband absorber 2 GHz–8 GHz for normal incidence and TE/TM polarization, T_{des} = 5 mm. Reflection coefficient graph

Table 5.2 Design HF1

Layer	Material	Thickness (mm)
1	14	0.459
2	6	1.804
3	6	0.772
4	5	1.318
5	15	0.594
6	Ground plane	

Layer parameters for five-layer broadband absorber optimized for normal incidence, TE/TM polarization for $f = 2 - 8$ GHz.

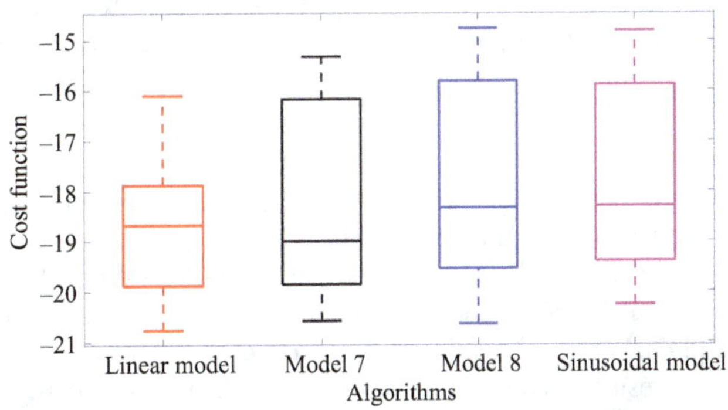

Figure 5.5 HF2 design five-layer broadband absorber from 2 GHz to 8 GHz for normal incidence and TE/TM polarization, T_{des} = 3 mm. Boxplot of all algorithms results

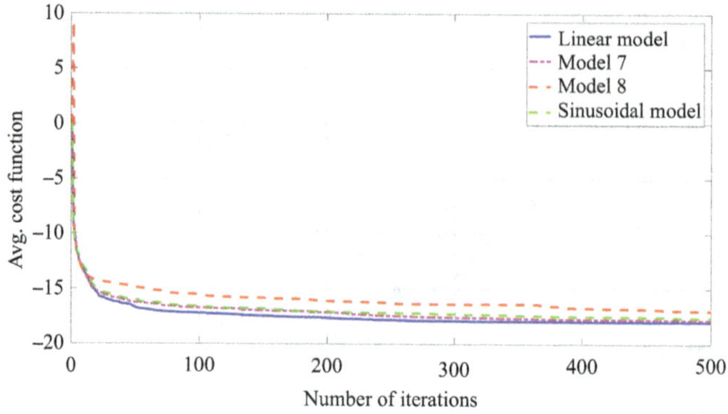

Figure 5.6 HF2 design five-layer broadband absorber 2–8 GHz for normal incidence and TE/TM polarization, T_{des} = 3 mm. Convergence rate graph

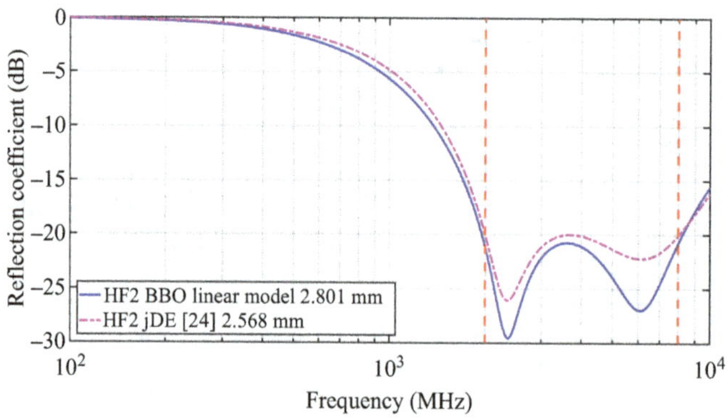

Figure 5.7 HF2 design five-layer broadband absorber from 2 GHz to 8 GHz for normal incidence and TE/TM polarization, T_{des} = 3 mm. Reflection coefficient graph

response for different angles of incidence and polarizations is presented in Figure 5.11. Table 5.4 reports the layer parameters for this case.

A five-layer broadband absorber in the frequency range from 500 MHz to 8 GHz optimized for normal incidence and TE/TM polarization is given in the next example (design DES2 [24]). We set the maximum desired thickness $T_{des} = 5$ mm. The boxplot of the algorithms results is shown in Figure 5.12. The sinusoidal model obtained the best result, while model 7 and the linear model found results with smaller dispersion of values. The convergence rate graph is presented in Figure 5.13. One may notice that BBO linear model and the model 7 converge slightly faster than the other algorithms. Figure 5.14 shows the frequency response of the best absorber design obtained by BBO

Table 5.3 Design HF2

Layer	Material	Thickness (mm)
1	16	0.573
2	6	0.811
3	2	0.398
4	13	0.813
5	14	0.206
6	Ground plane	

Layer parameters for five-layer broadband absorber optimized for normal incidence, TE polarization for $f = 2 - 8$ GHz, $T_{des} = 3$ mm.

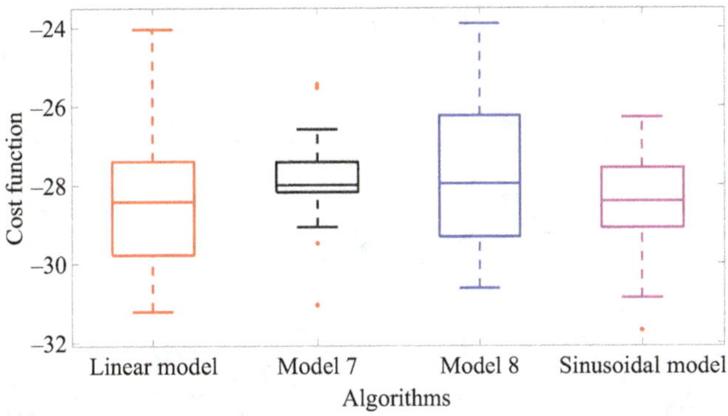

Figure 5.8 DES1 design five-layer broadband absorber for angle of incidence $0°-10°$, both polarizations from 3 GHz to 6 GHz, $T_{des} = 6$ mm. Boxplot of all algorithms results

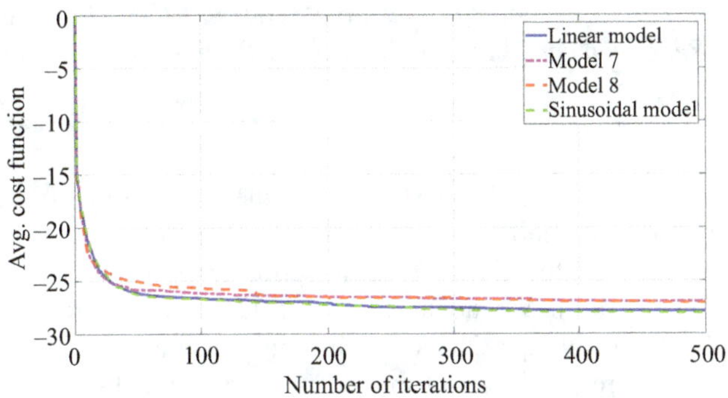

Figure 5.9 DES1 design five-layer broadband absorber for angle of incidence $0°-10°$, both polarizations from 3 GHz to 6 GHz, $T_{des} = 6$ mm. Convergence rate graph

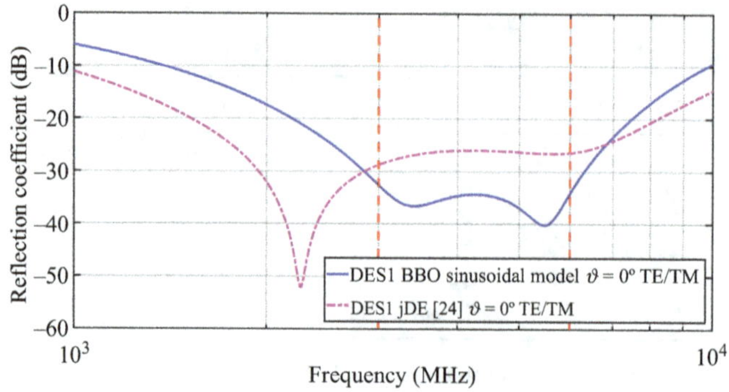

Figure 5.10 DES1 design five-layer broadband absorber for angle of incidence 0°−10°, both polarizations from 3 GHz to 6 GHz, $T_{des} = 6$ mm. Reflection coefficient graph for normal incidence

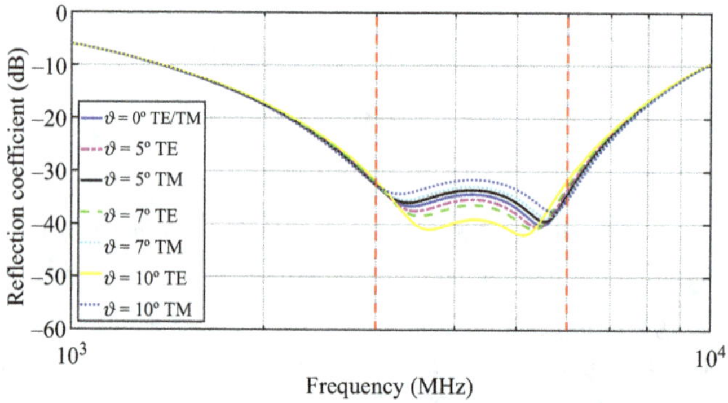

Figure 5.11 DES1 design five-layer broadband absorber for angle of incidence 0°−10°, both polarizations from 3 GHz to 6 GHz, $T_{des} = 6$ mm. Reflection coefficient graph for 0°−10° angle of incidence

sinusoidal model, which appears to be below −20 dB in the desired frequency range. The maximum reflection coefficient has a value of −20.70 dB (instead of −20.26 dB in [24]). Moreover, BBO found a slightly thinner design than the one in [24] (4.863 mm instead of 4.998 mm). Additionally, Figure 5.15 shows the frequency response for the best design for oblique incidence for angle of incidence 0° − 40° and TE polarization. The corresponding plot for TM polarization and oblique incidence is depicted in Figure 5.16. It is apparent that although the design was optimized only for normal incidence, it also shows a smooth behavior in the desired frequency zone for oblique incidence for both polarizations. Table 5.5 reports the layer parameters for the best obtained result.

Table 5.4 *DES1 design five-layer broadband absorber for angle of incidence 0°–10°, both polarizations from 3 GHz to 6 GHz, $T_{des} = 6$ mm*

Layer	Material	Thickness (mm)
1	14	0.518
2	6	1.677
3	6	1.453
4	1	1.44
5	14	0.683
6	Ground plane	

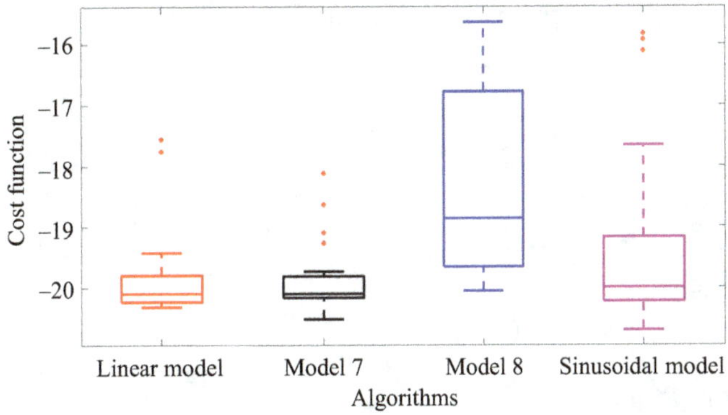

Figure 5.12 *DES2 design five-layer broadband absorber for normal incidence, TE/TM polarization, from 500 MHz to 8 GHz, $T_{des} = 5$ mm. Boxplot of all algorithms results*

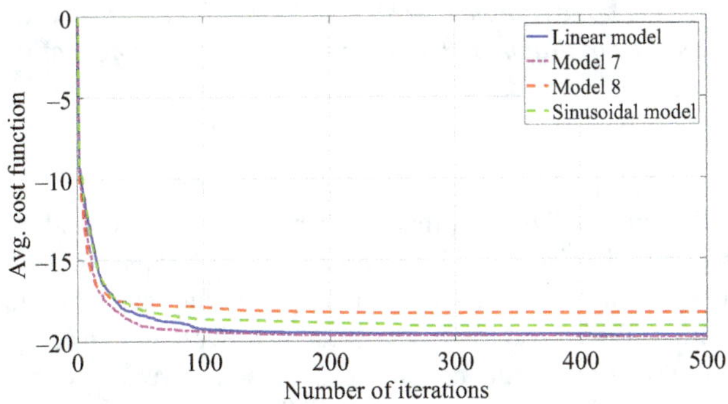

Figure 5.13 *DES2 design five-layer broadband absorber for normal incidence, TE/TM polarization, from 500 MHz to 8 GHz, $T_{des} = 5$ mm. Convergence rate graph*

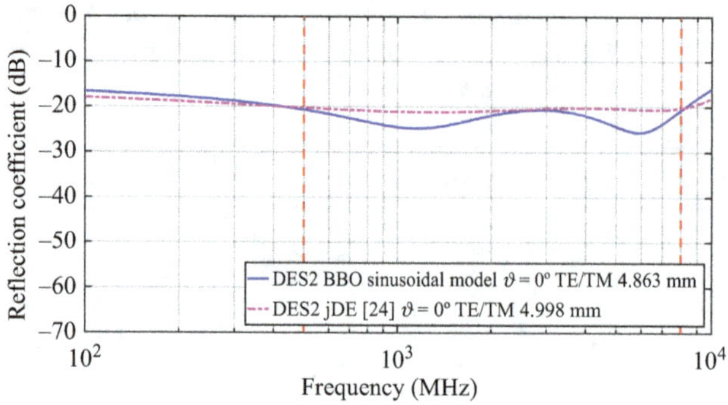

Figure 5.14 DES2 design five-layer broadband absorber for normal incidence, TE/TM polarization, from 500 MHz to 8 GHz, T_{des} = 5 mm. Reflection coefficient graph for normal incidence

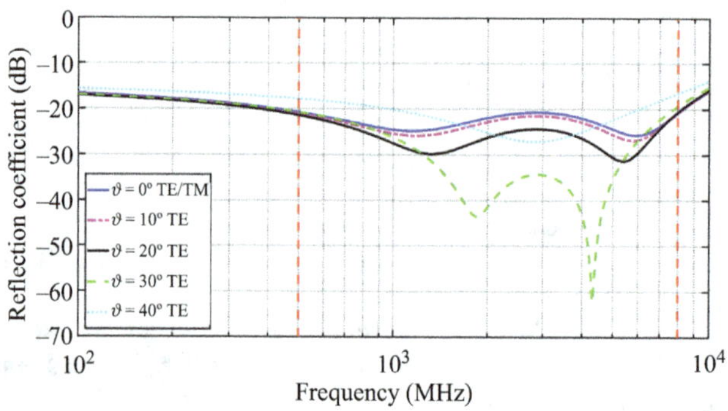

Figure 5.15 DES2 design five-layer broadband absorber for normal incidence, TE/TM polarization, 500 MHz to 8 GHz, T_{des} = 5 mm. Reflection coefficient graph for TE polarization, 0°−40° angle of incidence

The next design case DES3 is that of a seven-layer broadband absorber in the frequency range from 100 MHz to 20 GHz optimized for normal incidence and TE/TM polarization [34]. We notice that the problem requires finding a wideband design. We set the maximum desired thickness $T_{des} = 7$ mm in order to obtain comparable results with [34]. Figure 5.17 shows the boxplot of the results. We notice that BBO model 7 managed to obtain the results with the smaller dispersion of values. However, the sinusoidal model found the best absorber design. The convergence rate graph is depicted in Figure 5.18. All algorithms converge at similar speed. The frequency response of the best design found for normal incidence is presented in Figure 5.19.

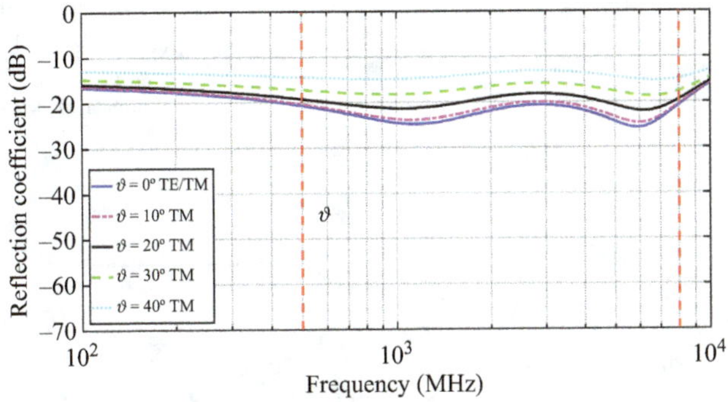

Figure 5.16 DES2 design five-layer broadband absorber for normal incidence, TE/TM polarization, 500 MHz to 8 GHz, $T_{des} = 5$ mm. Reflection coefficient graph for TM polarization, 0°−40° angle of incidence

Table 5.5 DES2 design five-layer broadband absorber for normal incidence, TE/TM polarization, from 500 MHz to 8 GHz, $T_{des} = 5$ mm

Layer	Material	Thickness (mm)
1	16	0.480
2	6	1.160
3	5	0.815
4	4	1.428
5	5	0.980
6	Ground plane	

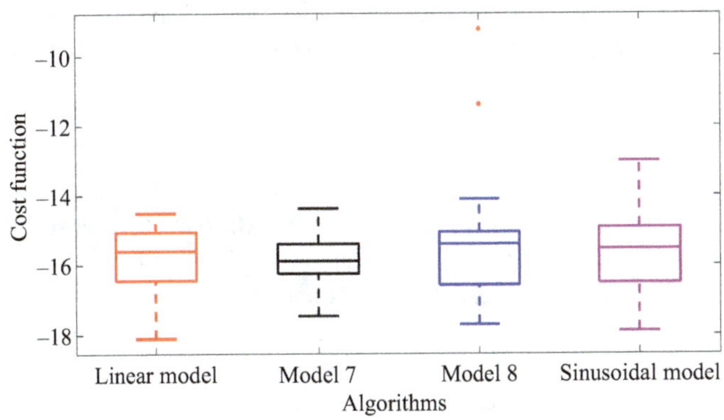

Figure 5.17 DES3 design seven-layer broadband absorber for normal incidence from 100 MHz to 20 GHz, $T_{des} = 7$ mm. Boxplot of all algorithms results

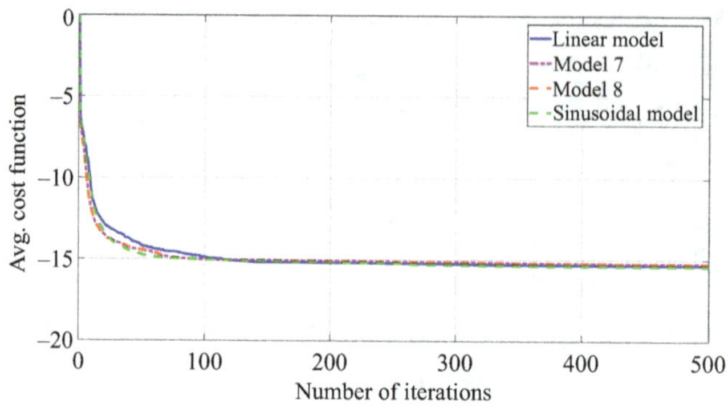

Figure 5.18 DES3 design seven-layer broadband absorber for normal incidence from 100 MHz to 20 GHz, T_{des} = 7 mm. Convergence rate graph

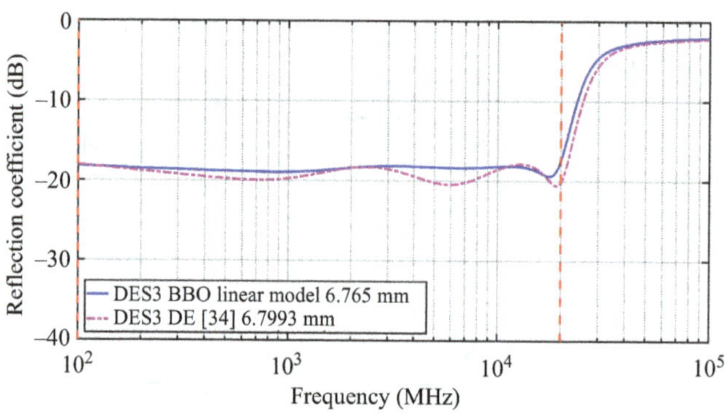

Figure 5.19 DES3 design seven-layer broadband absorber for normal incidence from 100 MHz to 20 GHz, T_{des} = 7 mm. Reflection coefficient graph for normal incidence

The maximum reflection coefficient is −18.11 dB, while the total absorber thickness is 6.765 mm. Both results outperform the design from [34] with corresponding values −18 dB and 6.7993 mm. The best absorber found shows also a smooth behavior for oblique incidence for both polarizations. Figure 5.20 depicts the frequency response of the best design for $0^\circ - 10^\circ$ angle of incidence. Table 5.6 reports the layer parameters of the best obtained absorber design for this case.

Finally, we redesign the same DES3 design case for 100 MHz to 20 GHz optimized for angle of incidence 50°. This design problem is more difficult to solve than in the previous case. The boxplot for this case is depicted in Figure 5.21. Again, as in previous case BBO model 7 obtained the results with the smaller dispersion of values.

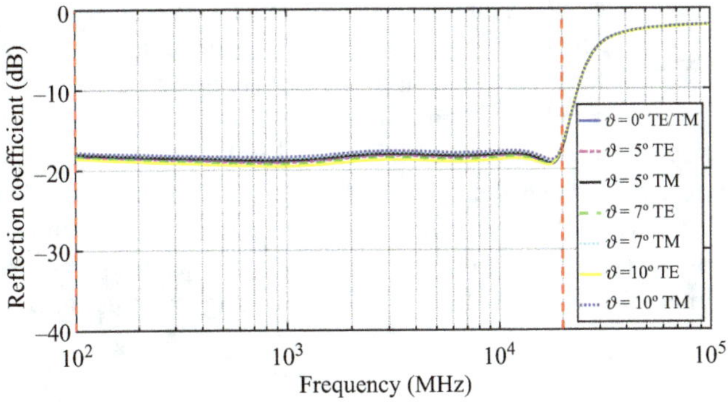

Figure 5.20 *DES3 design seven-layer broadband absorber for normal incidence from 100 MHz to 20 GHz, T_{des} = 7 mm. Reflection coefficient graph for TE/TM polarization, 0°−10° angle of incidence*

Table 5.6 *DES3 design seven-layer broadband absorber for normal incidence from 100 MHz to 20 GHz, T_{des} = 7 mm*

Layer	Material	Thickness (mm)
1	14	0.224
2	6	2.000
3	15	0.500
4	5	0.108
5	6	0.891
6	5	1.798
7	4	1.244
8	Ground plane	

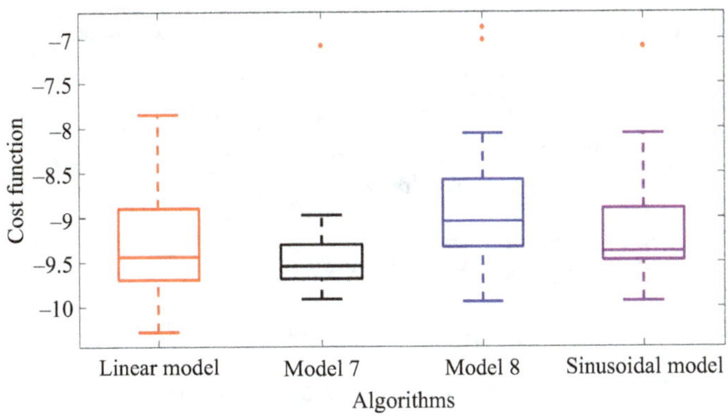

Figure 5.21 *DES3 design seven-layer broadband absorber for angle of incidence 50° from 100 MHz to 20 GHz. Boxplot of all algorithms results*

The original BBO linear model obtained the best value, however BBO model 7 derived the best mean and median results. Figure 5.22 plots the convergence rate graph. It appears that all algorithms converge similarly fast. Figure 5.23 shows the frequency response of the best absorber design found by BBO linear model, which presents a maximum reflection coefficient value of −10.3 dB close to that of [34] (−10.4 dB). Moreover, Figure 5.24 shows the frequency response of the best absorber design for normal incidence as well. We notice that although the design was optimized for oblique incidence, it presents a smooth behavior for normal incidence as well. Table 5.7 lists the layer parameters of the best obtained absorber design for this case.

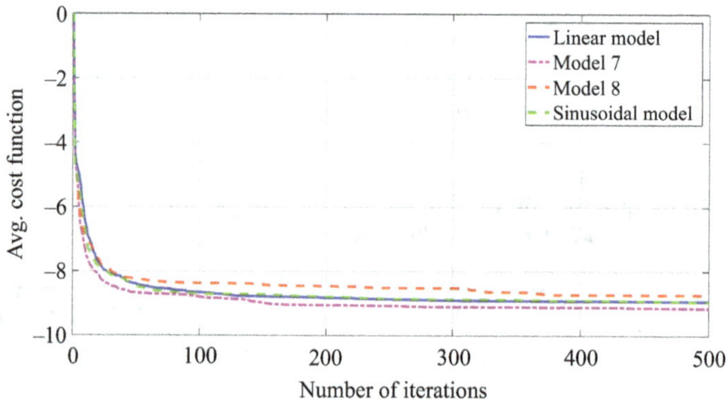

Figure 5.22 DES3 design seven-layer broadband absorber for angle of incidence 50° from 100 MHz to 20 GHz. Convergence rate graph

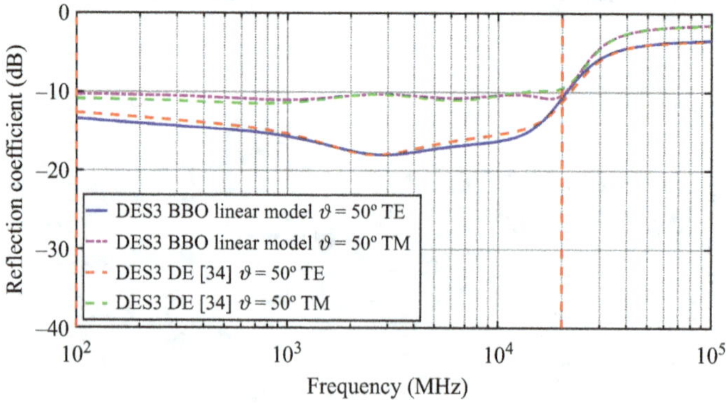

Figure 5.23 DES3 design seven-layer broadband absorber for angle of incidence 50° from 100 MHz to 20 GHz. Reflection coefficient graph for normal incidence

5.1.2.1 Statistical analysis of results

In order to evaluate the BBO models performance, we analyze the results obtained from all previous design cases. Table 5.8 lists the mean values obtained in all design cases. We notice that BBO model 7 obtained the best result in four out of the six cases. This is evident also from the Table 5.9 which shows the algorithms ranking first and second. Again, it is apparent that BBO model 7 outperforms the other algorithms. The BBO linear model comes second in most of the cases. This can be verified if we perform a Friedman ranking test [35,36]. Table 5.10 reports the algorithms' rankings

Table 5.7 DES3 design seven-layer broadband absorber for angle of incidence 50° from 100 MHz to 20 GHz

Layer	Material	Thickness (mm)
1	16	0.222
2	6	2.000
3	14	0.271
4	14	0.335
5	6	0.96
6	5	1.917
7	5	1.232
8	Ground plane	

Table 5.8 Algorithms mean results for all absorber design cases

BBO migration model	HF1	HF2	DES1	DES2	DES3	DES3 $\vartheta = 50°$
Linear model	−22.59	**−18.77**	−28.46	−19.80	−15.78	−9.32
Model 7	**−23.19**	−18.30	−27.84	**−19.84**	**−15.88**	**−9.39**
Model 8	−22.70	−17.95	−27.73	−18.37	**−15.88**	−8.86
Sinusoidal model	−22.93	−17.78	**−28.58**	−19.25	−15.68	−9.13

Smaller values are indicated in bold font.

Table 5.9 Algorithm ranking first and second for all cases

Algorithm	Ranking first	Ranking second
Linear model	1	3
Model 7	4	1
Model 8	1	0
Sinusoidal model	1	1

Table 5.10 Average rankings achieved by Friedman test

Method	Average rank	Normalized values	Rank
Linear model	2.33	1.47	2
Model 7	**1.58**	**1.00**	**1**
Model 8	3.25	2.05	4
Sinusoidal model	2.83	1.79	3

Smaller values are indicated in bold font.

according to Friedman ranking test. BBO model 7 obtained the best rank value of 1.58, while BBO linear model is second.

Overall, the best performance in absorber design problems has been obtained by BBO model 7, which outperforms the original linear model in most of the cases. Regarding convergence speed it seems that in general no BBO model has a clear advantage over the others. All algorithms converge at similar speed.

5.1.3 Multi-objective absorber optimization

In this subsection, we perform MO optimization to the absorber design case. This case is well suited for MO since both objectives are conflicting. The two objectives are the ones from (5.9). We apply three MO algorithms to the absorber design problem: the non-dominated sorting GA-II (NSGA-II) [37], the generalized DE (GDE3) [38], and the multi-objective BBO (MOBBO) from [39]. The absorber design case selected for MO is the seven-layer broadband absorber in the frequency range from 100 MHz to 20 GHz optimized for normal incidence and TE/TM polarization [34], which was solved in the previous section using single objective optimization.

Figure 5.25 shows the obtained Pareto front. We notice that GDE3 clearly outperforms the other algorithms for this case. This is also evident from the MO performance indicators listed in Table 5.11. It is apparent that the GDE3 results obtained the smaller indicator values. The main benefit of having the Pareto front is that now we have a set of solutions and we can choose the one that best fits the design requirements in every case. If we simply want to select the best compromised solution that its objective–function values are a compromise between the whole Pareto front, then a decision maker has to be selected.

In this case, we select the fuzzy decision maker described in previous section is obtained. The frequency response for normal incidence of the best compromised solution is presented in Figure 5.26. This is a solution found by MOBBO and is 5.776- mm thick. The frequency response is quite close to that of [34]. The frequency response for oblique incidence of $0° - 10°$ and both polarizations is depicted in Figure 5.27. The absorber behavior is smooth in the desired frequency band. The same absorber is also evaluated for angle of incidence 50° and the corresponding frequency response is plotted in Figure 5.28 . We notice that the absorber response is similar to that of the

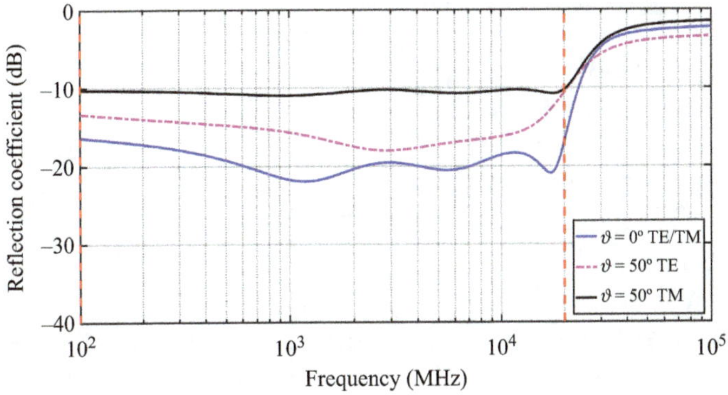

Figure 5.24 DES3 design seven-layer broadband absorber for angle of incidence 50° from 100 MHz to 20 GHz. Reflection coefficient graph for TE/TM polarization, 0°–50° angle of incidence

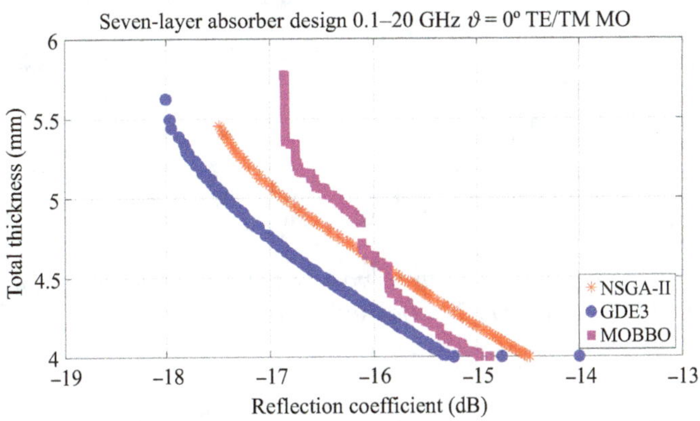

Figure 5.25 Comparative results 2D Pareto front of f_1 and f_2

Table 5.11 Performance indicators of the Pareto fronts

Performance indicator	NSGA-II	GDE3	MOBBO
QI_{R2}	3.90E−02	**1.14e−09**	7.43E−02
$QI_{\varepsilon}1$	1.29E−01	**2.82e−06**	2.84E−01
$QI_{\bar{H}}$	1.66E−01	**3.55e−07**	2.31E−01

Smaller values are indicated in bold font.

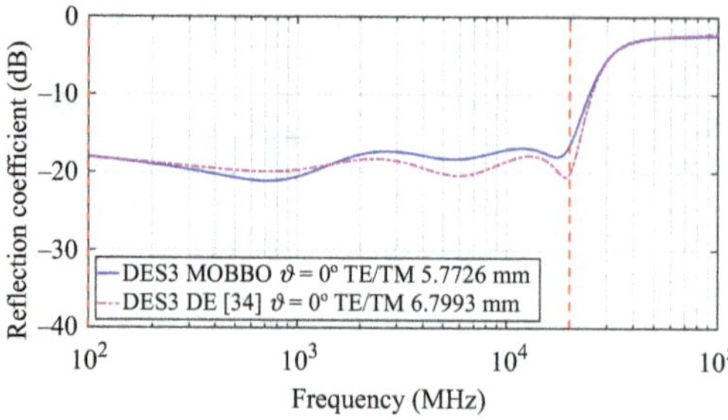

Figure 5.26 DES3 design seven-layer broadband absorber for normal incidence from 100 MHz to 20 GHz. Reflection coefficient graph for normal incidence

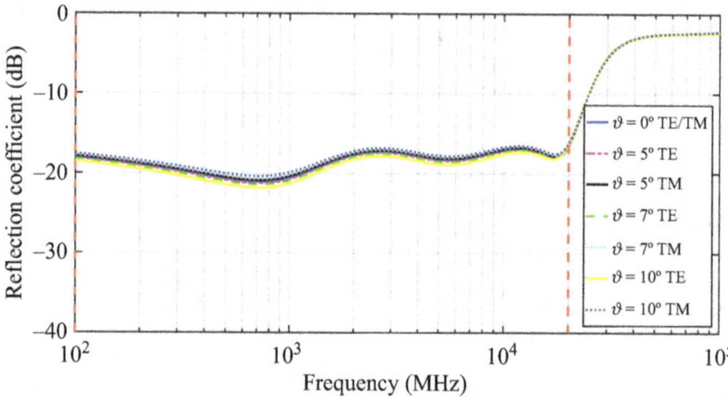

Figure 5.27 DES3 design seven-layer broadband absorber for normal incidence from 100 MHz to 20 GHz. Reflection coefficient graph for TE/TM polarization, 0°−10° angle of incidence

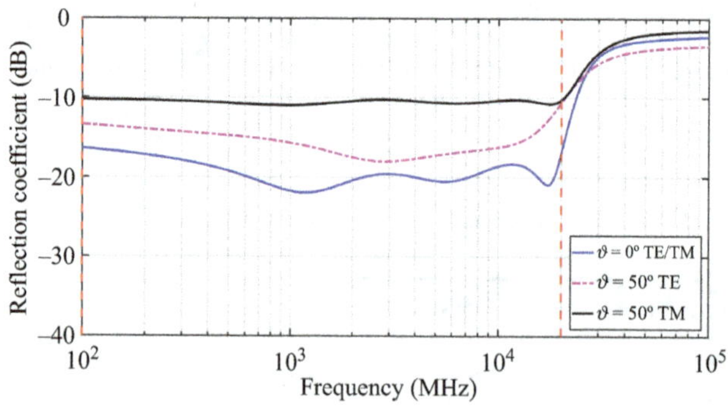

Figure 5.28 DES3 design seven-layer broadband absorber for normal incidence from 100 MHz to 20 GHz. Reflection coefficient graph for TE/TM polarization, 50° angle of incidence

Table 5.12 DES3 design seven-layer broadband absorber for normal incidence from 100 MHz to 20 GHz. Best compromised solution obtained by MOBBO

Layer	Material	Thickness (mm)
1	16	0.222
2	6	2.000
3	14	0.271
4	14	0.335
5	6	0.96
6	5	1.917
7	5	1.232
8	Ground plane	

previous design for angle of incidence 50°. Table 5.12 lists the layer parameters of the best compromised solution. Overall, the MO optimization succeeded in producing the Pareto front with several different solutions. GDE3 seems to be the most suitable algorithm for this case.

5.2 Dielectric filters design

Microwave filters are among the important components of a modern wireless communication system. Several papers exist in the literature that address the filter design problem [40–53]. Band-pass filter design is presented in [40,42,49,52,53] using different filter structures like hexagonal loop resonators, radial line stubs, parallel-coupled lines, spiral shaped resonators, and half mode substrate-integrated, folded waveguide. Microstrip ultra-wideband filter design is given in [44,50], while in [46–48] dual-band filters are designed.

The filter design problem is in general MO. In the case of multilayer dielectric filters, there are two basic design objectives. The first is the minimization of the reflection coefficient of the multilayer structure for an incident plane wave in the passband frequency zone. The second is the minimization of the transmission coefficient in the stopband frequency zone. Both reflection and transmission coefficients depend on the thickness and the electric properties of each layer. The selection of the optimal permittivity for each layer from a predefined database of commonly available materials is also an important requirement.

Additional constraints can also be imposed in the previously described problem. Such design constraints require that the reflection coefficient value in the passband and stopband zones should not lie above or below a predefined level, respectively. Moreover, as in [54] we consider an additional constraint of desired total layer thickness. Similar design objectives and constraints exist also for other filter types like microstrip filters [55].

EAs like GAs have been applied to a variety of microwave component design problems [2,8,56–60]. Multilayer dielectric filters have been studied in the literature using different EAs. In [56], a binary-coded GA is used for multilayer dielectric filter synthesis. The unknown variables are the layer thickness and permittivity, which are allowed to vary.

The same technique is also used in [59] for frequency selective surfaces filter design. In [61], a single objective approach is proposed using a self-adaptive EA. This is produced from the aggregation of the objective–functions and a penalty term. The layer materials are chosen from two predefined materials databases consisting of 15 and 44 materials. In [62] an MO EA is used for the generation of the Pareto front for the constraint dielectric filter design problem.

5.2.1 Problem formulation

Figure 5.29 shows the structure of the multilayer dielectric filter. The optimization variables for this case are the thickness and the EM characteristics of each layer. These characteristics are the frequency-dependent (in general) complex permittivity and permeability given by

$$\varepsilon_m(f) = \varepsilon_0[\varepsilon'_m(f) - j\varepsilon''_m(f)] \tag{5.10}$$

$$\mu_m(f) = \mu_0[\mu'_m(f) - j\mu''_m(f)] \tag{5.11}$$

The terms ε_0 and μ_0 are, respectively, the free space permittivity and permeability. We assume that the filter is composed of low-loss dielectric nonmagnetic materials ($\varepsilon'' \ll 1, \mu' = 1, \mu'' = 0$), which have frequency-independent permittivity in the desired frequency range. Thus, in the case of a dielectric filter with M layers, the number of the unknown variables is $2M$. For this type of multilayer structure, the

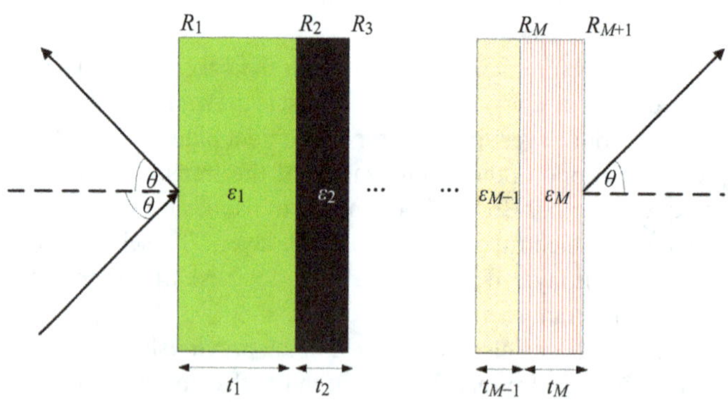

Figure 5.29 Multilayer dielectric filter structure

general expression of the reflection coefficients $R_m{}^{TE}$ and $R_m{}^{TM}$ at the mth layer for the TE and the TM modes is derived from the recursive formula [54,63]:

$$R_m{}^{TE/TM} = \frac{r_m^{TE/TM} + R_{m+1}^{TE/TM} e^{-2jk_{m+1}t_{m+1}}}{1 + r_m^{TE/TM} R_{m+1}^{TE/TM} e^{-2jk_{m+1}t_{m+1}}}, \quad m = M, M-1, \ldots, 1 \tag{5.12}$$

$$R_{M+1}^{TE/TM} = r_{M+1}^{TE/TM} = \frac{n_M - n_0}{n_M - n_0} \tag{5.13}$$

$$r_m^{TE/TM} = \frac{n_{m-1} - n_m}{n_{m-1} + n_m}, \quad m = 1, 2, \ldots, M \tag{5.14}$$

$$n_m = \begin{cases} \sqrt{\varepsilon'_m - \sin^2\theta} & \text{TE mode} \\ \dfrac{\varepsilon'_m}{\sqrt{\varepsilon'_m - \sin^2\theta}} & \text{TM mode} \end{cases} \tag{5.15}$$

$$k_m = \frac{2\pi f}{c\sqrt{\varepsilon'_m}} \tag{5.16}$$

where M denotes the number of layers, t_m and ε'_m are the mth layer thickness and the dielectric constant, respectively. The dielectric filter design problem can be expressed by the minimization of the two objective–functions

$$F_1(\bar{x}) = \sum_p \left[\left| R_{TE}(\bar{x}, f_p) \right|^2 + \left| R_{TM}(\bar{x}, f_p) \right|^2 \right] \tag{5.17}$$

$$F_2(\bar{x}) = \sum_s \left[\left(1 - |R_{TE}(\bar{x}, f_s)|^2\right) + \left(1 - |R_{TM}(\bar{x}, f_s)|^2\right) \right] \tag{5.18}$$

Additionally, the design problem is subject to the following constraints:

$$g_1(\bar{x}) = 20\log\left(R_{TE}(\bar{x}, f_p^c)\right) < C_{TE}^p \tag{5.19}$$

$$g_2(\bar{x}) = 20\log\left(R_{TM}(\bar{x}, f_p^c)\right) < C_{TM}^p \tag{5.20}$$

$$g_3(\bar{x}) = 20\log\left(R_{TE}(\bar{x}, f_s^c)\right) > C_{TE}^s \tag{5.21}$$

$$g_4(\bar{x}) = 20\log\left(R_{TM}(\bar{x}, f_s^c)\right) > C_{TM}^s \tag{5.22}$$

$$g_5(\bar{x}) = T_{tot}(\bar{x}) \leq T_{des} \tag{5.23}$$

where $R_{TE} = R_1{}^{TE}$ and $R_{TM} = R_1{}^{TM}$ are the overall reflection coefficients (or the reflection coefficients at the first layer) of the filter structure, respectively, for the TE and the TM modes, $\bar{x} = (t_1, \varepsilon_1, t_2, \varepsilon_2, \ldots, t_M, \varepsilon_M)$ the vector of the unknown variables (layer thickness and dielectric constant), T_{tot} is the total layer thickness of the design found, and T_{des} is the desired total layer thickness.

Moreover, f_p and f_s define correspondingly the passband and the stopband frequency ranges, while f_p^c and f_s^c define, respectively, the passband and the stopband

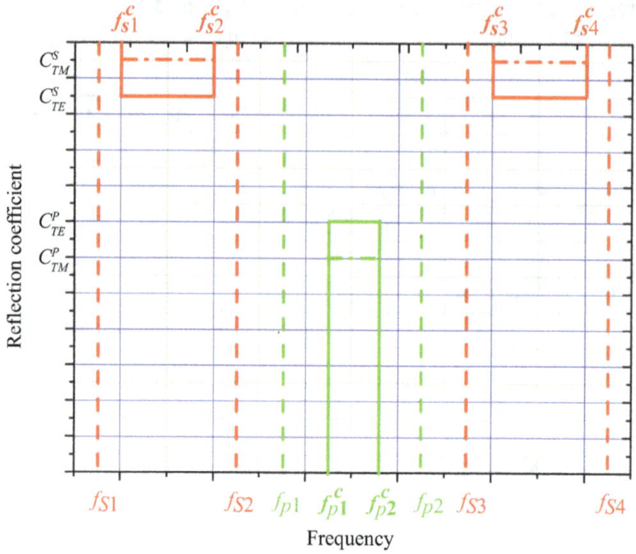

Figure 5.30 Dielectric filter design specifications

frequencies where constrains must be satisfied. C_{TE}^{p} and C_{TM}^{p} are the maximum allowable reflection coefficient values in the passband frequency ranges for TE and TM modes, respectively, while C_{TE}^{s} and C_{TM}^{s} accordingly are the minimum allowable reflection coefficient values in the stopband frequency ranges. The frequency ranges where constraints apply are in general different than the frequency ranges of the objective–functions. They are defined narrower than those of the objective–functions as in [62]. Therefore, the constraint functions make the problem more strict and difficult. Figure 5.30 shows graphically the constraint functions.

5.2.2 Single-objective optimization of dielectric filters

In the case of single objective optimization, the dielectric filter design problem is formulated as

$$\begin{aligned} F(\bar{x}) = {} & \alpha_1 F_1(\bar{x}) + \alpha_2 F_1(\bar{x}) + \Xi(\max(0, g_1(\bar{x}) - C_{TE}^{p}) + \max(0, g_2(\bar{x}) - C_{TM}^{p}) \\ & + \max(0, C_{TE}^{s} - g_3(\bar{x})) + \max(0, C_{TM}^{s} - g_4(\bar{x})) \\ & + \max(0, g_5(\bar{x}) - T_{des})) \end{aligned} \tag{5.24}$$

where α_1 and α_2 are suitable weight factors, Ξ is a penalty factor usually set to very large value like 10^{20}.

Next, we present several different filter design cases. In all design cases, the angle of incidence is set to $\vartheta = 45°$ as in the literature [54,61,62,64]. In all cases, we apply the original BBO algorithm and BBO with different oppositional-based learning (OBL) schemes. Namely, the OBL algorithms are opposition-based BBO (OBBO), quasi OBBO (QOBBO), quasi reflection OBBO (QROBBO), current optimum OBBO (COOBBO), generalized OBBO (GOBBO), centroid OBBO (COBBO), extended OBBO (EOBBO), and reflected EOBBO (REOBBO). Thus, there are a total of nine different BBO variants. The corresponding OBL schemes were described in Chapter 2.

In all cases, we set the layer thickness to be within [1, 10] mm. In all cases, the population size is set to 100 and the maximum number of independent trias is 20. In all examples that follow the habitat modification probability, P_{mod}, is set to 1, and the maximum mutation rate, m_{max}, is set equal to 0.005. The maximum immigration rate I, and the maximum emigration rate E are both set both to one. The maximum number of iterations depends on the design case. As in previous papers, we assume that the resolution of layer thickness is 0.001 mm (1 μm). These dielectric multilayer filters could be fabricated using thin glue layers among different dielectrics [59]. We assume that these glue layers are very thin and therefore not taken into account as in [54,56,59,61,62].

The first three examples compare the results with those found in [64]. These are a low-pass, a band-pass, and a band-stop filter as in [64], where MO algorithms (MOPSO-fs and NSGA-II) were applied. In these cases, we consider seven-layer filters. The design frequency range is set from 24 GHz to 36 GHz. In these cases, we use the same predefined material database as in [61,62,64]. The first database consists of 15 commercially available dielectric materials with real permittivity values. Table 5.13 lists the database 1 real permittivity values. We assume that these values remain unchanged in the desired frequency range.

We must also point out that in convergence rate graphs that follow the y-axis depict the $\log_{10}(F(\bar{x}))$ values instead of objective–function values. The idea is that this helps one to better visualize the graph and the number of iterations on x-axis where the objective–function becomes feasible. This is depicted in the convergence rate figures by a vertical line dropping close to origin of axis.

Table 5.13 Database 1. Predefined materials database of 15 commercially available low-loss dielectric materials

Material no.	ε_r	**Material no.**	ε_r
1	1.01	9	3.38
2	2.2	10	4.48
3	2.33	11	4.5
4	2.5	12	6
5	2.94	13	6.15
6	3	14	9.2
7	3.02	15	10.2
8	3.27		

First, we consider a low-pass filter design case. We select the passband and stopband frequencies to 24 GHz $\leq f_p < 30$ GHz and 30 GHz $\leq f_s < 36$ GHz, respectively. The ranges of frequencies where constraints apply are defined 24 GHz $\leq f_p^c <$ 28 GHz and 32 GHz $\leq f_s^c < 36$ GHz. As in [64], we set the total desired thickness to 17 mm. Table 5.14 shows all the design specifications for this case. The maximum number of iterations is set to 2,000 for this case.

Table 5.15 reports the algorithms' comparative results. We notice that GOBBO obtained the best result. However, the best mean value is derived by REOBBO. The SRate column in Table 5.15 denotes the success rate percentage. This means that if the value is 100, the algorithm in all runs obtained a solution that satisfies all constraints. If the success rate is less than 100, then the algorithm did not succeed in all runs to find a solution. That way we can measure the effectiveness of an algorithm in a complex problem. We notice that QROBBO and COBBO have a success rate of 95%, while all the other algorithms managed to obtain a solution in all runs.

Table 5.14 Design specifications for a seven-layer low-pass filter case

Frequency band	**Limits**	**Level (dB) reflection coefficient**
Stopband	30 GHz $\leq f_s < 36$ GHz	
Passband	24 GHz $\leq f_p < 30$ GHz	
First constraint stopband	32 GHz $\leq f_s^c < 36$ GHz	>-5
Constraint passband	24 GHz $\leq f_p^c < 28$ GHz	<-10
Second constraint stopband	34 GHz $\leq f_s^c < 36$ GHz	>-5
Maximum total thickness (mm)	**17 mm**	

Table 5.15 Seven-layer low-pass dielectric filter using database 1 materials (comparative results)

Algorithm	**Best**	**Worst**	**Mean**	**Median**	**St. Dev.**	**SRate**
BBO	7.869	11.936	9.114	8.837	1.066	100
OBBO	7.521	12.594	8.783	**8.531**	1.053	100
QOBBO	7.159	12.449	8.977	8.720	1.269	100
QROBBO	7.087	11.108	8.879	8.910	0.965	95
GOBBO	**6.684**	**10.546**	8.782	8.856	**0.986**	100
COOBBO	7.494	12.757	9.051	8.637	1.567	100
COBBO	7.469	12.546	8.970	8.856	1.228	95
EOBBO	7.449	12.082	9.017	8.727	1.255	100
REOBBO	6.984	11.515	**8.713**	8.656	1.090	100

Smaller values are indicated in bold font.

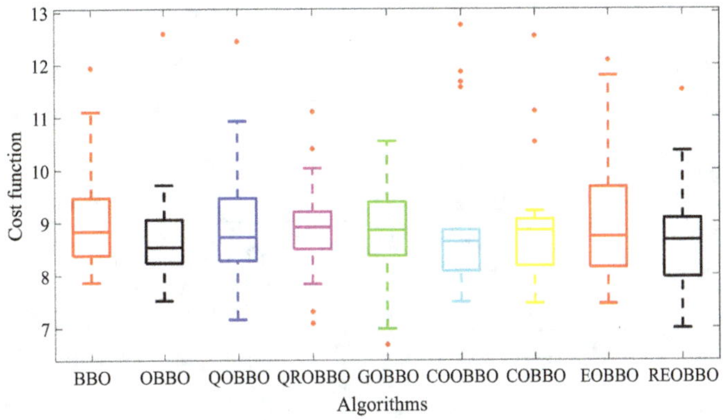

Figure 5.31 Seven-layer low-pass dielectric filter using database 1 materials ($\vartheta = 45°$). Boxplot of all algorithms results

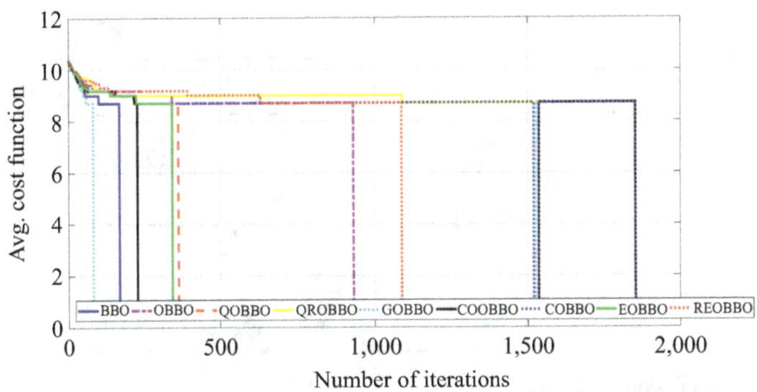

Figure 5.32 Seven-layer low-pass dielectric filter using database 1 materials ($\vartheta = 45°$). Convergence rate graph

The boxplot of Figure 5.31 shows the distribution of values among the solutions. In these plots, only the feasible solutions are displayed. We notice that there is a high dispersion of values in all algorithms, COOBBO and QOBBO obtained the larger dispersion of all algorithms. The corresponding convergence rate plot for this case is depicted in Figure 5.32. We notice that some algorithms require in average fewer iterations than the maximum number. In general, the convergence speeds are quite different. We notice that GOBBO converges faster than the other algorithms and requires less than 100 iterations for each final value. BBO and COOBBO converge fast as well and require less than 250 iterations for convergence. Figure 5.33 presents the frequency response of the best filter found by GOBBO. The total thickness of the design is 15.74 mm compared with 16.24 mm from [64]. The frequency response of

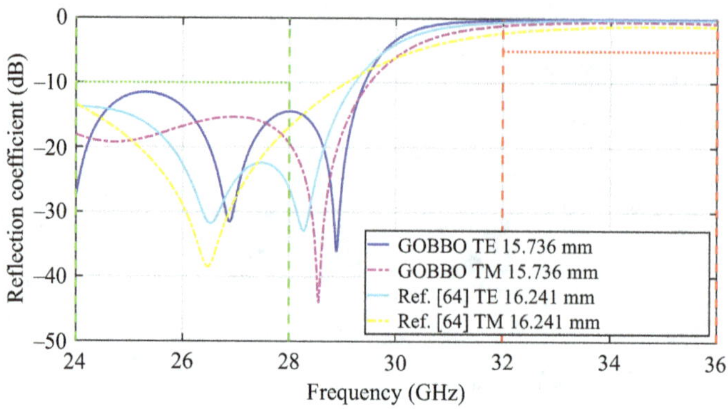

Figure 5.33 Seven-layer low-pass dielectric filter using database 1 materials ($\vartheta = 45°$). Frequency response graph

Table 5.16 Design specifications for the seven-layer low-pass filter case

Layer	Material	Thickness (mm)
1	5	2.422
2	14	2.331
3	2	1.588
4	13	2.771
5	2	1.691
6	14	2.262
7	4	2.671

Best solution obtained by GOBBO.

the new design is very close to that of [64] for the desired frequency range. Table 5.16 lists the layer parameters of the best solution found by GOBBBO.

Next, we present a constrained band-pass filter design. We select the passband and stopband frequencies to 28 GHz $\leq f_p < 32$ GHz and 24 GHz $\leq f_s < 28$ GHz, 32 GHz $\leq f_s < 36$ GHz, respectively. The range of constraints are set to 29 GHz $\leq f_p^c < 31$ GHz and 24 GHz $\leq f_s^c < 26$ GHz, 34 GHz $\leq f_s^c < 36$ GHz. The total desired thickness is set to 30 mm. This design problem is more difficult than the previous one and requires the satisfaction of more constraints. Table 5.17 lists all the design specifications for this case. The maximum number of iterations is set as in the previous case to 2,000. Table 5.18 holds the algorithms' comparative results. It is evident that the original BBO obtained the best result; however, OBBO derived the best mean and median results. Additionally, for all algorithms the success rate was less than 100 with EOBBO to have the largest value of 95%.

Figure 5.34 shows the boxplot of the results. COBBO obtained the results with the smallest dispersion of values. Figure 5.35 depicts the convergence rate plot. All

Table 5.17 Design specifications for a seven-layer band-pass filter case

Frequency band	**Limits**	**Level (dB) reflection coefficient**
First stopband	24 GHz $\leq f_s <$ 28 GHz	
Passband	28 GHz $\leq f_p <$ 32 GHz	
Second stopband	32 GHz $\leq f_s <$ 36 GHz	
First constraint stopband	24 GHz $\leq f_s^c <$ 26 GHz	>-5
Constraint passband	29 GHz $\leq f_p^c <$ 31 GHz	<-10
Second constraint stopband	34 GHz $\leq f_s^c <$ 36 GHz	>-5
Maximum total thickness (mm)	**30 mm**	

Table 5.18 Seven-layer band-pass dielectric filter using database 1 materials (comparative results)

Algorithm	**Best**	**Worst**	**Mean**	**Median**	**St. Dev.**	**SRate**
BBO	**8.950**	15.591	10.809	10.647	1.389	85
OBBO	9.562	12.621	**10.701**	**10.305**	0.865	70
QOBBO	9.511	15.013	11.247	10.869	1.516	85
QROBBO	9.132	12.149	11.078	11.183	0.813	80
GOBBO	9.444	16.098	11.003	10.745	1.525	75
COOBBO	10.176	13.641	11.085	10.906	0.934	70
COBBO	9.865	**11.302**	10.729	10.787	**0.421**	70
EOBBO	9.098	12.189	10.732	10.846	0.777	**95**
REOBBO	9.652	13.639	11.015	10.652	1.055	80

Smaller values are indicated in bold font.

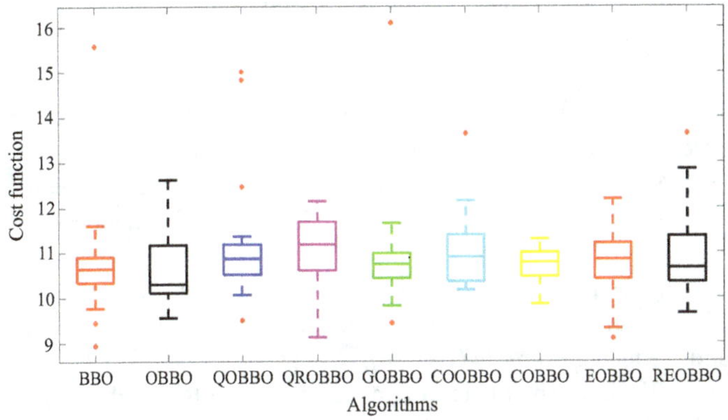

Figure 5.34 Seven-layer band-pass dielectric filter using database 1 materials ($\vartheta = 45°$). Boxplot of all algorithms results

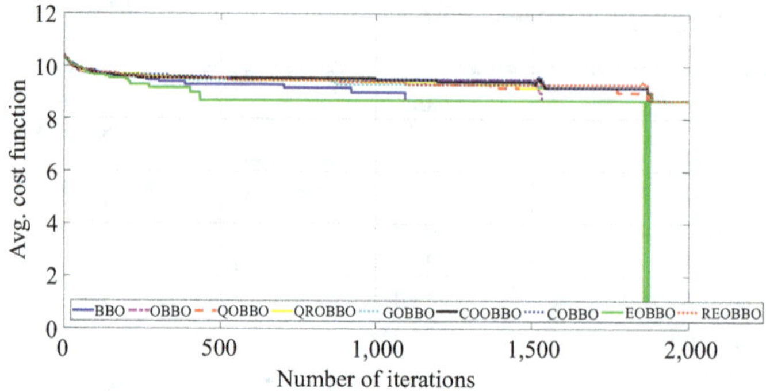

Figure 5.35 Seven-layer band-pass dielectric filter using database 1 materials ($\vartheta = 45°$). Convergence rate graph

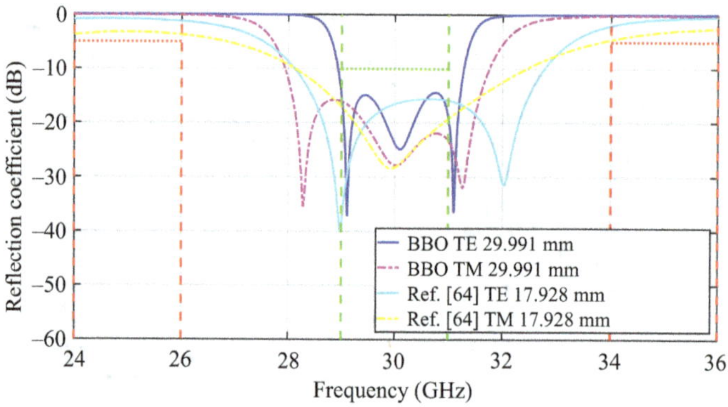

Figure 5.36 Seven-layer band-pass dielectric filter using database 1 materials ($\vartheta = 45°$). Frequency response graph

algorithms converge at similar speeds. The frequency response of the best design found is plotted in Figure 5.36. The obtained best design has 29.99 mm total thickness (instead of 33.44 mm found in [62]). The frequency response seems to be slightly better than the one from [64]. Table 5.19 reports the layer data for this best case obtained by BBO.

Next, we present an example of a band-stop filter design. The lower and upper cutoffs are set to 28 GHz and 32 GHz, respectively. Therefore, the passband and stopband frequencies are 24 GHz $\leq f_p <$ 28 GHz, 32 GHz $\leq f_p <$ 36 GHz, and 28 GHz $\leq f_s <$ 32 GHz. The range of constraints are set to 24 GHz $\leq f_p^c <$ 26 GHz, 34 GHz $\leq f_p^c <$ 36 GHz, and 29 GHz $\leq f_s^c <$ 31 GHz. We place the total desired thickness to 36 mm. Table 5.20 lists the design requirements for this case. The algorithms' results are reported in Table 5.21. We notice that the original BBO algorithm obtained the best and mean value results. Only two out of the nine algorithms found a feasible solution in all runs, EOBBO and GOBBO. The box-plot graph is shown in Figure 5.37. We notice that as in previous case, the COBBO derived the smaller dispersion of values. Figure 5.38 plots the convergence rate

Table 5.19 Design specifications for the seven-layer band-pass filter case

Layer	Material	Thickness (mm)
1	14	3.339
2	1	4.193
3	14	4.967
4	1	4.604
5	14	4.990
6	1	3.724
7	13	4.174

Best solution obtained by BBO.

Table 5.20 Design specifications for a seven-layer band-stop filter case

Frequency band	Limits	Level (dB) reflection coefficient
First passband	24 GHz $\leq f_p <$ 28 GHz	
Stopband	28 GHz $\leq f_s <$ 32 GHz	
Second passband	32 GHz $\leq f_p <$ 36 GHz	
First constraint passband	24 GHz $\leq f_p^c <$ 26 GHz	<-10
Constraint stopband	29 GHz $\leq f_s^c <$ 31 GHz	>-5
Second constraint passband	34 GHz $\leq f_p^c <$ 36 GHz	<-10
Maximum total thickness (mm)	**36 mm**	

Table 5.21 Seven-layer band-stop dielectric filter using database 1 materials (comparative results)

Algorithm	Best	Worst	Mean	Median	St. Dev.	SRate
BBO	**9.109**	15.216	**13.146**	13.563	1.548	80
OBBO	11.081	14.494	13.183	**13.246**	**0.955**	90
QOBBO	10.899	14.714	13.216	13.379	0.991	90
QROBBO	11.522	**14.290**	13.189	13.334	0.958	85
GOBBO	10.805	16.641	13.836	14.124	1.242	**100**
COOBBO	11.722	14.827	13.428	13.361	0.691	80
COBBO	12.345	14.384	13.389	13.386	0.482	70
EOBBO	10.790	15.939	13.536	13.571	1.148	**100**
REOBBO	11.292	15.080	13.541	13.590	0.781	90

Smaller values are indicated in bold font.

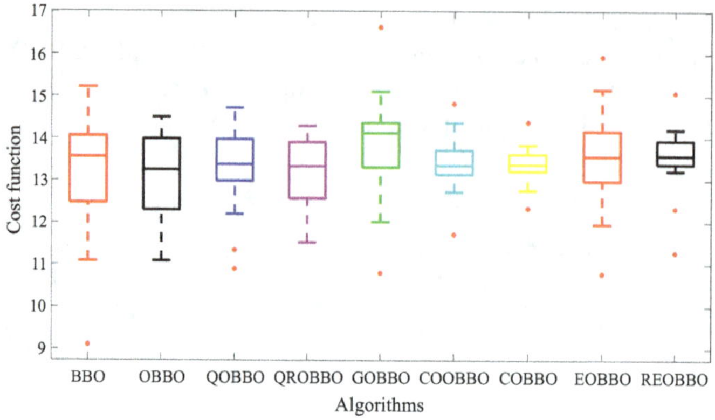

Figure 5.37 Seven-layer band-stop dielectric filter using database 1 materials ($\vartheta = 45°$). Boxplot of all algorithms results

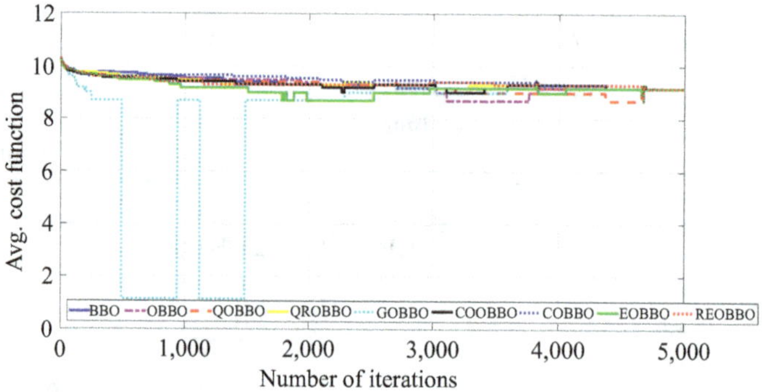

Figure 5.38 Seven-layer band-stop dielectric filter using database 1 materials ($\vartheta = 45°$). Convergence rate graph

graph. All algorithms converge similarly fast, except GOBBO that is faster. The frequency response of the best design found by BBO is depicted in Figure 5.39. The best design found has a total thickness of 35.87 mm, and its frequency response is smoother than the one from [64]. The layer parameters for this case are listed in Table 5.22.

In the next examples, we will use the same predefined materials database as in [54]. This database, which we call database 2, consists of 44 commercially available low-loss dielectric materials. As with the previous database, we assume that permittivity values remain constant over the desired frequency range. Table 5.23 lists the material characteristics of database 2. The motivation for using this database is

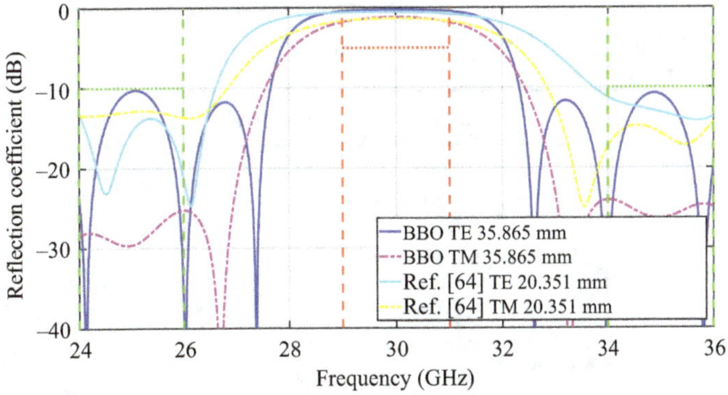

Figure 5.39 Seven-layer band-stop dielectric filter using database 1 materials ($\vartheta = 45°$). Frequency response graph

Table 5.22 Design specifications for the seven-layer band-stop filter case

Layer	**Material**	**Thickness (mm)**
1	3	5.557
2	2	5.737
3	14	4.230
4	5	4.818
5	14	4.230
6	2	5.739
7	3	5.554

Best solution obtained by BBO.

that as in [54] filters with fewer layers can be obtained. Using this database, we will design more complex filters, namely, a bandpass, a dual-band, and a tri-band filter. We select the population size to be 100 and the number of iterations to 2,000 for all algorithms.

First, we design a three-layer band-pass filter. We may notice that instead of seven-layers in the previous example, here we design a three-layer filter. The corresponding example from [54] uses four layers for the same specifications. The design requirements are presented in Table 5.24. We notice that the specifications are more strict than the ones in Table 5.17. For this case, the maximum total filter thickness is set to 10 mm.

Table 5.25 holds the comparative results. We notice that for this case QROBBO obtains the best result as well the smaller mean value and the smaller standard deviation value. The success rate of the algorithm is 100. We notice that only three out of the

Table 5.23 Database 2. Predefined materials database of 44 commercially available low-loss dielectric materials

Material no.	ε_r	tanδ	Material no.	ε_r	tanδ
1	1.03	0.0001	23	4.5	0.002
2	1.06	0.0001	24	5	0.002
3	1.2	0.0100	25	6	0.002
4	1.25	0.005	26	6.15	0.0027
5	1.5	0.01	27	6.5	0.002
6	1.7	0.004	28	7	0.002
7	2.1	0.0001	29	8	0.002
8	2.28	0.003	30	9	0.002
9	2.2	0.0005	31	9.2	0.002
10	2.33	0.0007	32	9.6	0.002
11	2.4	0.0022	33	9.8	0.0001
12	2.5	0.0005	34	10	0.002
13	2.54	0.0004	35	10.2	0.0023
14	2.75	0.0025	36	11	0.002
15	2.94	0.0012	37	12	0.002
16	3	0.002	38	13	0.002
17	3.02	0.0016	39	14	0.002
18	3.27	0.002	40	15	0.002
19	3.38	0.0027	41	16	0.002
20	3.5	0.002	42	20	0.002
21	3.8	0.002	43	25	0.002
22	4	0.002	44	30	0.002

Table 5.24 Design specifications for a three-layer band-pass filter case

Frequency band	Limits	Level (dB) reflection coefficient
First stopband	24 GHz $\leq f_s <$ 28 GHz	
Passband	28 GHz $\leq f_p <$ 32 GHz	
Second stopband	32 GHz $\leq f_s <$ 36 GHz	
First constraint stopband	24 GHz $\leq f_s^c <$ 26 GHz	>-5
Constraint passband	29 GHz $\leq f_p^c <$ 31 GHz	<-10
Second constraint stopband	34 GHz $\leq f_s^c <$ 36 GHz	>-5
Maximum total thickness (mm)	**10 mm**	

nine algorithms have 100% success rate. The boxplot of the algorithm's results is plotted in Figure 5.40. The interquartile range of all boxplots seems quite similar; however, most of the algorithms' results have obtained a large number of outliers. The convergence rate graph for this case is shown in Figure 5.41. We notice that the

Table 5.25 Three-layer band-pass dielectric filter using database 2 materials (comparative results)

Algorithm	Best	Worst	Mean	Median	St. Dev.	SRate
BBO	11.435	14.725	12.340	12.318	0.690	**100**
OBBO	11.443	13.365	12.184	**12.002**	0.485	95
QOBBO	11.662	21.593	13.201	12.388	2.373	95
QROBBO	**11.278**	**12.869**	**12.143**	12.126	**0.476**	**100**
GOBBO	11.580	14.724	12.330	12.114	0.923	90
COOBBO	11.704	16.991	12.777	12.253	1.394	95
COBBO	11.322	21.299	12.954	12.160	2.437	90
EOBBO	11.396	19.232	12.601	12.056	1.861	**100**
REOBBO	11.432	13.775	12.373	12.227	0.655	90

Smaller values are indicated in bold font.

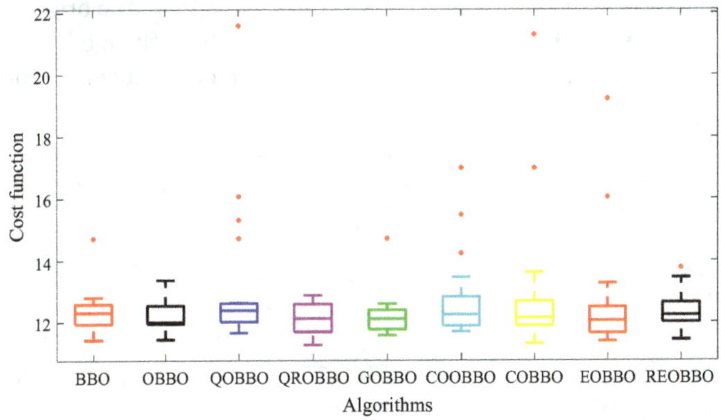

Figure 5.40 Three-layer band-pass dielectric filter using database 2 materials ($\vartheta = 45°$). Boxplot of all algorithms results

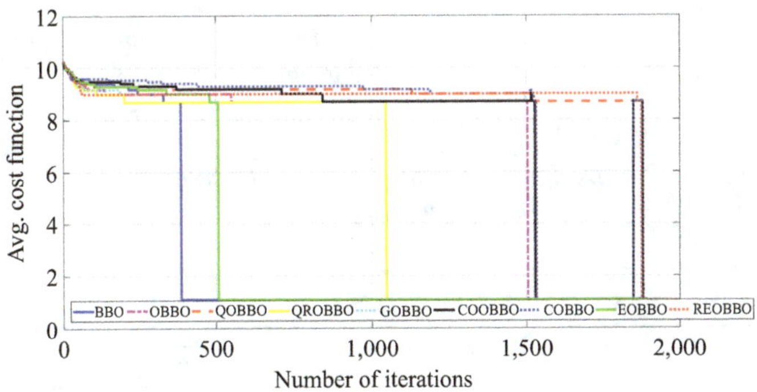

Figure 5.41 Three-layer band-pass dielectric filter using database 2 materials ($\vartheta = 45°$). Convergence rate graph

original BBO and EOBBO converge faster than the other algorithms at about 500 iterations. The frequency response of the best obtained filter by QROBBO is plotted in Figure 5.42. We notice that the newly found design satisfies the design requirements and, however, is slightly worse than the result from [54]. Table 5.26 reports the layer data of the best solution by QROBBO.

The next example is a four-layer dual-band filter design case. Table 5.27 holds the desired design specifications. We set the desired reflection coefficient values in the passband and stopband frequency ranges to $C^p_{TE} = C^p_{TM} = -10$ dB and $C^s_{TE} = C^s_{TM} = -5$ dB, respectively. Moreover, the total desired thickness is set to 30 mm. For this design case, more constraints are added. Thus, the problem becomes more difficult to solve than the previous case. The maximum number of iterations is set to 5,000.

Table 5.28 holds the comparative results. It is interesting to notice that not all algorithms obtained feasible solutions. OBBO and QROBBO could not find a feasible solution in all runs. Moreover, the other algorithms obtained a feasible solution in only few of the runs and their success rate ranges from 5% to 15%. The original BBO and COOBBO obtained the best success rate, while REOBBO obtained the best result. Due to the small number of successful runs, the boxplot is not shown for this case.

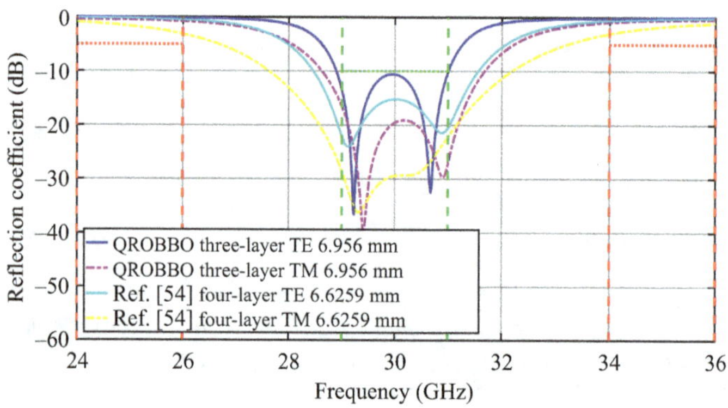

Figure 5.42 Three-layer band-pass dielectric filter using database 2 materials ($\vartheta = 45°$). Frequency response graph

Table 5.26 Design specifications for the three-layer band-pass filter case

Layer	Material	Thickness (mm)
1	40	2.675
2	4	1.980
3	42	2.301

Best solution obtained by QROBBO.

Table 5.27 Design specifications for a four-layer dual-band filter case

Frequency band	**Limits**	**Level (dB) reflection coefficient**
First stopband	11 GHz $\leq f_s <$ 15 GHz	
First passband	15 GHz $\leq f_p <$ 19 GHz	
Second stopband	19 GHz $\leq f_s <$ 23 GHz	
Second passband	23 GHz $\leq f_p <$ 27 GHz	
Third stopband	27 GHz $\leq f_s <$ 31 GHz	
First constraint stopband	11 GHz $\leq f_s^c <$ 12 GHz	>-5
First constraint passband	16 GHz $\leq f_p^c <$ 18 GHz	<-10
Second constraint stopband	21 GHz $\leq f_s^c <$ 22 GHz	>-5
Second constraint passband	24 GHz $\leq f_p^c <$ 26 GHz	<-10
Third constraint stopband	30 GHz $\leq f_s^c <$ 31 GHz	>-5
Maximum total thickness (mm)	**30 mm**	

Table 5.28 Four-layer dual-band dielectric filter using database 2 materials (comparative results)

Algorithm	**Best**	**Worst**	**Mean**	**Median**	**St. Dev.**	**SRate**
BBO	28.055	31.307	30.177	31.169	1.839	**15**
OBBO	N/A	N/A	N/A	N/A	N/A	0
QOBBO	31.289	31.289	31.289	31.289	0.000	5
QROBBO	N/A	N/A	N/A	N/A	N/A	0
GOBBO	27.770	28.059	27.914	27.914	0.204	10
COOBBO	27.367	29.529	28.313	28.042	1.106	**15**
COBBO	29.827	30.337	30.082	30.082	0.360	10
EOBBO	28.372	28.372	28.372	28.372	0.000	5
REOBBO	**27.226**	**27.226**	**27.226**	**27.226**	0.000	5

Smaller values are indicated in bold font.

The convergence rate graph is depicted in Figure 5.43. We notice that all algorithms converge at similar speed, and all require almost the maximum number of iterations for obtaining a feasible solution. The frequency response of the best obtained design is shown in Figure 5.44. The new design is thinner than the one from [54]. The frequency response of the best design by REOBBO is worse than the one from [54] but it fully satisfies all the constraints. The layer parameters for this case are reported in Table 5.29.

An additional design case of a tri-band five-layer design case is presented next. This case has a total of eight constraint functions [54]. Table 5.30 shows the design specifications. As previously, we set the desired reflection coefficient

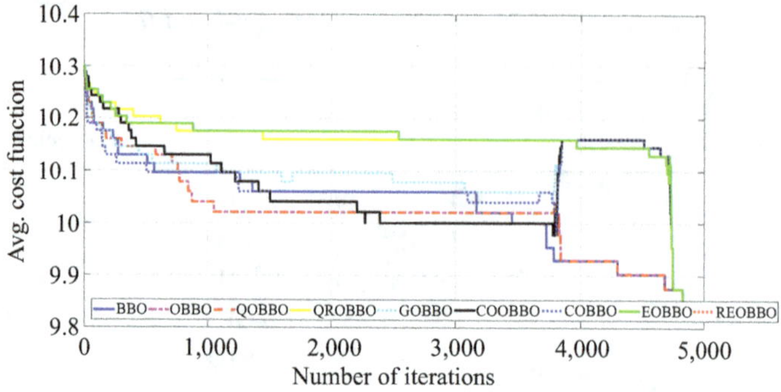

Figure 5.43 Four-layer dual-band dielectric filter using database 2 materials ($\vartheta = 45°$). Convergence rate graph

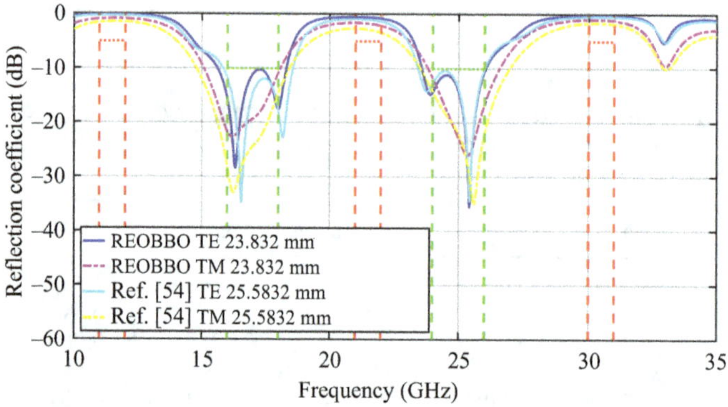

Figure 5.44 Four-layer dual-band dielectric filter using database 2 materials ($\vartheta = 45°$). Frequency response graph

Table 5.29 Design parameters for the dual-band filter case

Layer	Material	Thickness (mm)
1	23	3.534
2	6	6.675
3	43	3.623
4	13	10.000

Best solution obtained by REOBBO.

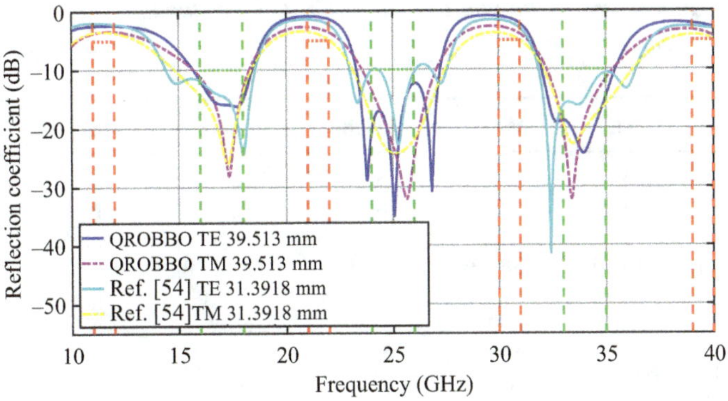

Figure 5.45 Five-layer tri-band dielectric filter using database 2 materials ($\vartheta = 45°$). Frequency response graph

Table 5.30 Design specifications for a five-layer tri-band filter case

Frequency band	**Limits**	**Level (dB) reflection coefficient**
First stopband	11 GHz $\leq f_s <$ 15 GHz	
First passband	15 GHz $\leq f_p <$ 19 GHz	
Second stopband	19 GHz $\leq f_s <$ 23 GHz	
Second passband	23 GHz $\leq f_p <$ 27 GHz	
Third stopband	27 GHz $\leq f_s <$ 32 GHz	
Third passband	32 GHz $\leq f_p <$ 36 GHz	
Fourth stopband	36 GHz $\leq f_s <$ 40 GHz	
First constraint stopband	11 GHz $\leq f_s^c <$ 12 GHz	>−5
First constraint passband	16 GHz $\leq f_p^c <$ 18 GHz	<−10
Second constraint stopband	21 GHz $\leq f_s^c <$ 22 GHz	>−5
Second constraint passband	24 GHz $\leq f_p^c <$ 26 GHz	<−10
Third constraint stopband	30 GHz $\leq f_s^c <$ 31 GHz	>−5
Third constraint passband	33 GHz $\leq f_p^c <$ 35 GHz	<−10
Fourth constraint stopband	39 GHz $\leq f_s^c <$ 40 GHz	>−5
Maximum total thickness (mm)	**40 mm**	

values in the passband and stopband frequency ranges to $C_{TE}^p = C_{TM}^p = -10$ dB and $C_{TE}^s = C_{TM}^s = -5$ dB, respectively. Moreover, the total desired thickness is set to 40 mm. The comparative results are reported in Table 5.31. We notice that only two of the algorithms QROBBO and GOBBO obtained a feasible result. All the other algorithms failed to obtain feasible solutions. The total thickness of the best result

Table 5.31 Five-layer tri-band dielectric filter using database 2 materials (comparative results)

Algorithm	Best	Worst	Mean	Median	St. Dev.	SRate
BBO	N/A	N/A	N/A	N/A	N/A	0
OBBO	N/A	N/A	N/A	N/A	N/A	0
QOBBO	N/A	N/A	N/A	N/A	N/A	0
QROBBO	**38.130**	**38.130**	**38.130**	**38.130**	0	**10**
GOBBO	39.103	39.103	39.103	39.103	0	10
COOBBO	N/A	N/A	N/A	N/A	N/A	0
COBBO	N/A	N/A	N/A	N/A	N/A	0
EOBBO	N/A	N/A	N/A	N/A	N/A	0
REOBBO	N/A	N/A	N/A	N/A	N/A	0

Smaller values are indicated in bold font.

Table 5.32 Design parameters for the tri-band filter case

Layer	Material	Thickness (mm)
1	3	6.858
2	8	11.243
3	39	4.827
4	11	10.848
5	5	5.737

Best solution obtained by QROBBO.

is 39.51 mm (instead of 31.39 mm in [54]). The frequency response of the design is quite similar to that of [54] and slightly better in some frequencies (see Figure 5.45). Table 5.32 holds the layer parameters of the best result.

5.2.2.1 Statistical analysis of results

The optimization procedure in the previous subsection shows that the multilayer dielectric filter is a complex problem. The single objective optimization procedure using an exact penalty method did not manage to obtain feasible solutions in all cases. Table 5.33 lists the algorithms' mean results for the feasible solutions. In the design cases of the dual-band and the tri-band filters some algorithms could not find feasible solutions. Table 5.34 shows the algorithms ranking first and second. We notice that QROBBO and REOBBO ranked first at two out of the six design cases. The ranks according to Friedman test are reported in Table 5.35. We notice that OBBO obtained the lower rank; however, very close to GOBBO and QROBBO. Table 5.35 does not give a clear indication of the best BBO algorithm for filter design. Differences are very small.

Table 5.33 Algorithms mean results for all filter design cases

Algorithm	Lowpass	Bandpass 1	Stopband	Bandpass 2	Dualband	Triband
BBO	9.114	10.809	**13.146**	12.340	30.177	N/A
OBBO	8.783	**10.701**	13.183	12.184	N/A	N/A
QOBBO	8.977	11.247	13.216	13.201	31.289	N/A
QROBBO	8.879	11.078	13.189	**12.143**	N/A	**38.130**
GOBBO	8.782	11.003	13.836	12.330	27.914	39.103
COOBBO	9.051	11.085	13.428	12.777	28.313	N/A
COBBO	8.970	10.729	13.389	12.954	30.082	N/A
EOBBO	9.017	10.732	13.536	12.601	28.372	N/A
REOBBO	**8.713**	11.015	13.541	12.373	**27.226**	N/A

Smaller values are indicated in bold font.

Table 5.34 Algorithm ranking first and second for all cases

Algorithm	Ranking first	Ranking second
BBO	1	0
OBBO	1	2
QOBBO	0	0
QROBBO	2	0
GOBBO	0	3
COOBBO	0	0
COBBO	0	1
EOBBO	0	0
REOBBO	2	0

Table 5.35 Average rankings achieved by Friedman test

Method	Average rank	Normalized values	Rank
BBO	5.00	1.33	5
OBBO	**3.75**	**1.00**	**1**
QOBBO	6.83	1.82	8
QROBBO	4.08	1.09	3
GOBBO	3.83	1.02	2
COOBBO	6.33	1.69	9
COBBO	5.17	1.38	6
EOBBO	5.50	1.47	7
REOBBO	4.50	1.20	4

Smaller values are indicated in bold font.

However, if we take into account the algorithms' success rates then we get a quite different view of the results. Table 5.36 shows the success rates for the design cases. We notice that EOBBO obtained the best result in four out of the six design cases. The Friedman test again shows the ranks according to success rate. This is presented

Table 5.36 Algorithms' success rate results for all filter design cases

Algorithm	Lowpass	Bandpass 1	Stopband	Bandpass 2	Dualband	Triband
BBO	100	85	80	100	15	0
OBBO	100	70	90	95	0	0
QOBBO	100	85	90	95	5	0
QROBBO	95	80	85	100	0	10
GOBBO	100	75	100	90	10	10
COOBBO	100	70	80	95	15	0
COBBO	95	70	70	90	10	0
EOBBO	100	95	100	100	5	0
REOBBO	100	80	90	90	5	0

Table 5.37 Average rankings achieved by Friedman test for success rate

Method	Average rank	Normalized values	Rank
BBO	3.92	1.15	2
OBBO	5.92	1.73	8
QOBBO	4.58	1.34	4
QROBBO	5.17	1.51	5
GOBBO	4.08	1.20	3
COOBBO	5.33	1.56	6
COBBO	7.17	2.10	9
EOBBO	**3.42**	**1.00**	**1**
REOBBO	5.42	1.59	7

Smaller values are indicated in bold font.

in Table 5.37. We notice that the differences between the algorithms are more distinct in this case. EOBBO obtained the best rank, while the original BBO and GOBBO follow. It is interesting to notice that GOBBO is the only algorithm that derived a feasible solution in all six design cases.

Overall, we may conclude that GOBBO is the most well-suited BBO algorithm for multilayer dielectric filter problem that finds a solution in all cases.

5.2.3 Multi-objective optimization

As with the microwave absorbers described previously, MO EAs can be used successfully for microwave multi-band filter design. In [54], a filter design method based on the GDE3 algorithm has been reported. In this subsection, we apply MOEAs to the similar filter design problems as those described in the previous section. Namely, we use NSGA-II [37], the GDE3 [38], and the MOBBO from [39].

The first three design problems use the database of 15 predefined materials (see Table 5.13), while the next three the other database of 44 predefined materials (see Table 5.23). The design specifications are the same as those reported in Section 5.2.2.

The first design case is that of a five-layer low-pass filter. The population size is set to 100 and the maximum number of iterations is set to 1,000 for all algorithms. Figure 5.46 shows the Pareto front obtained. It is apparent that GDE3 solutions dominate clearly the other algorithms. This is also evident from the performance indicators that are reported in Table 5.38. All indicators for GDE3 are zero, which shows that GDE3 clearly outperforms the other algorithms. We select the best compromised solution obtained by GDE3. This is a design case with total thickness 6.10 mm (instead of 16.24 mm in [64]). The frequency response is depicted in Figure 5.47. We notice that although the design is quite thinner than the one in [64] the response is quite smooth and satisfactory. The layer parameters for this case are reported in Table 5.39.

Next, we design a five-layer band-pass filter. As in the previous example, the population size is set to 100 and the maximum number of iterations is set to 1,000 for all algorithms. Figure 5.48 plots the Pareto front. We notice that the NSGA-II dominates in most cases the other algorithms. Moreover, the Pareto front obtained

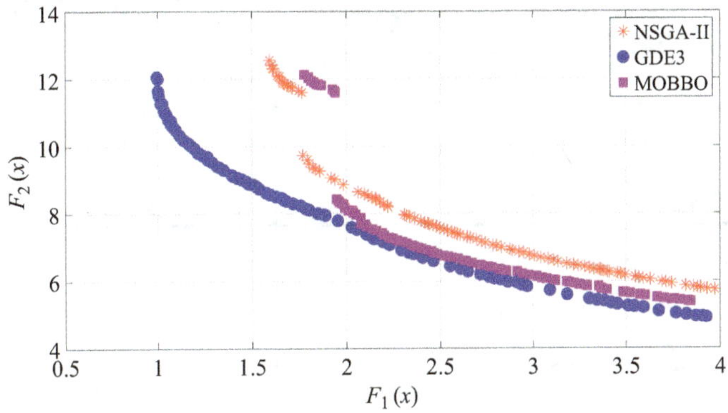

Figure 5.46 Five-layer low-pass dielectric filter using database 1 materials ($\vartheta = 45°$). Pareto front obtained by all algorithms

Table 5.38 Five-layer low-pass dielectric filter using database 1 materials

Performance indicator	**NSGA-II**	**GDE3**	**MOBBO**
QI_{R2}	5.89E−02	**0.00e+00**	5.74E−02
$QI_{\varepsilon}1$	2.02E−01	**0.00e+00**	2.62E−01
$QI_{\bar{H}}$	2.29E−01	**0.00e+00**	1.97E−01

Performance indicators of the Pareto fronts. Bold font indicates the lower values.

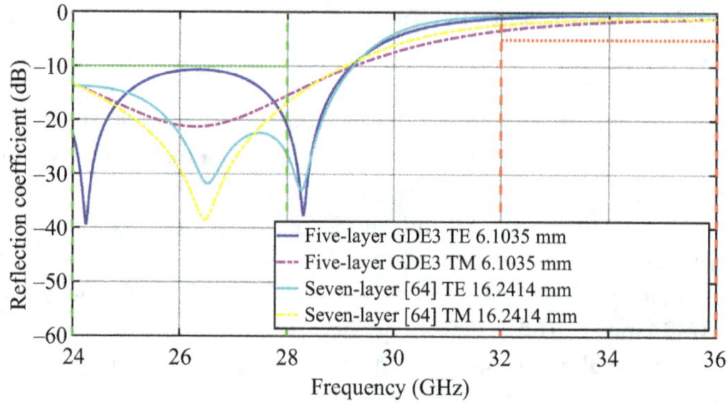

Figure 5.47 Five-layer low-pass dielectric filter using database 1 materials ($\vartheta = 45°$). Frequency response

Table 5.39 Design parameters for the five-layer low-pass filter case

Layer	Material	Thickness (mm)
1	12	1.591
2	14	0.623
3	1	1.644
4	14	0.576
5	11	1.670

Best compromised solution obtained by GDE3.

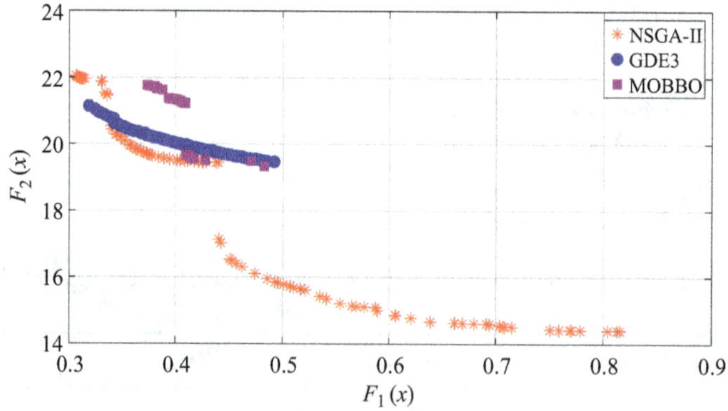

Figure 5.48 Five-layer band-pass dielectric filter using database 1 materials ($\vartheta = 45°$). Pareto front obtained by all algorithms

by NSGA-II shows a wider spread. This fact is also reported in Table 5.40 where the smaller values for all indicators are derived for the Pareto front obtained by NSGA-II. Figure 5.49 presents the frequency response of the best compromised solution obtained by MOBBO (Table 5.41). The design is thinner than the one from [64]. However, its response is slightly worse than the one from [64].

Table 5.40 Five-layer band-pass dielectric filter using database 1 materials

Performance indicator	NSGA-II	GDE3	MOBBO
QI_{R2}	**2.10e−04**	2.25E−01	2.28E−01
$QI_{\varepsilon}1$	**4.13e−02**	6.62E−01	6.44E−01
$QI_{\bar{H}}$	**4.59e−03**	5.06E−01	5.32E−01

Performance indicators of the Pareto fronts. Bold font indicates the lower values.

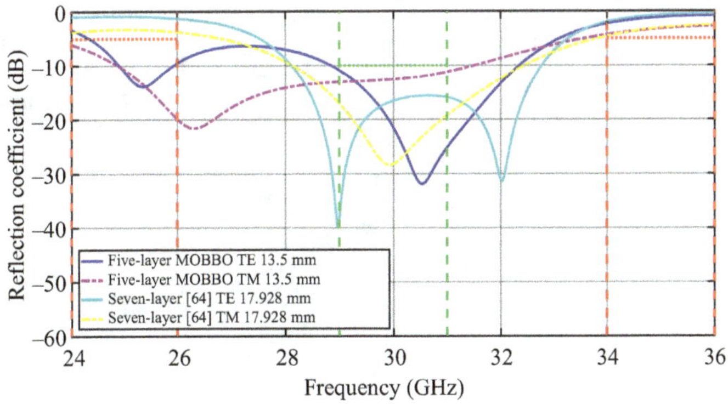

Figure 5.49 Five-layer band-pass dielectric filter using database 1 materials ($\vartheta = 45°$). Frequency response

Table 5.41 Design parameters for the five-layer band-pass dielectric filter

Layer	Material	Thickness (mm)
1	10	2.66
2	1	1.54
3	14	1.76
4	14	1.76
5	1	5.78

Best compromised solution obtained by MOBBO.

A five-layer stopband filter is the next design case. For this case because the problem complexity increases, we set the population size to 200 and the maximum number of iterations to 5,000 for all algorithms. The Pareto front for this case is shown in Figure 5.50. It is not clear from the Pareto front which algorithm performs better. This can be deducted from Table 5.42. We notice that NSGA-II emerges as the best algorithm; however, GDE3 indicators are very close. The best compromised solution obtained by NSGA-II is 25.60 mm thick. Figure 5.51 presents its frequency response, which seems smooth in the desired frequency bands. The layer data for this case are listed in Table 5.43.

The design cases that follow use the database 2 of 44 predefined materials. We present a three-layer band-pass filter. For this case, the maximum number of iterations is set to 2,000 and the population size is equal to 100. The Pareto front is drawn in Figure 5.52. NSGA-II seems to dominate in most of the cases. This is also clear from the performance indicators listed in Table 5.44. NSGA-II is clearly the best algorithm. One may notice that the number of layer is three instead of four in [54].

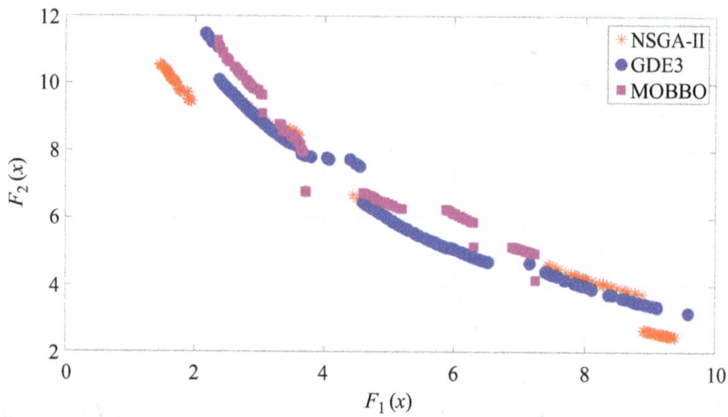

Figure 5.50 Five-layer stopband dielectric filter using database 1 materials ($\vartheta = 45°$). Pareto front obtained by all algorithms

Table 5.42 Five-layer stopband dielectric filter using database 1 materials

Performance indicator	NSGA-II	GDE3	MOBBO
QI_{R2}	**2.11e−03**	1.23E−02	3.16E−02
$QI_{\varepsilon}1$	**3.96e−02**	9.71E−02	1.86E−01
$QI_{\bar{H}}$	**1.16e−02**	6.07E−02	1.06E−01

Performance indicators of the Pareto fronts. Bold font indicates the lower values.

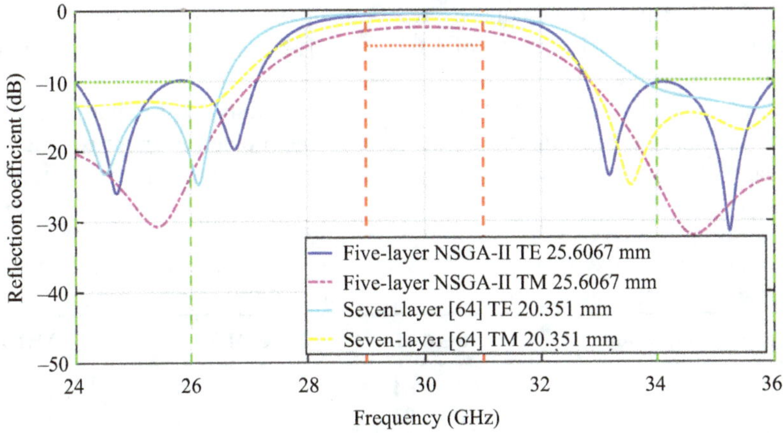

Figure 5.51 Five-layer stopband dielectric filter using database 1 materials ($\vartheta = 45°$). Frequency response

Table 5.43 Design parameters for the five-layer stopband dielectric filter

Layer	**Material**	**Thickness (mm)**
1	3	5.363
2	1	5.672
3	14	4.025
4	1	5.615
5	4	4.932

Best compromised solution obtained by NSGA-II.

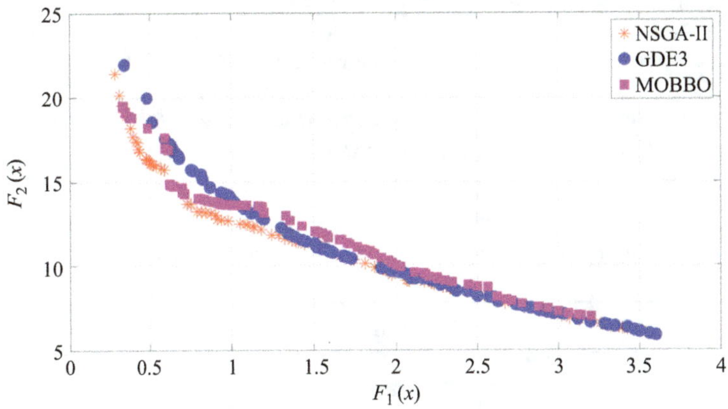

Figure 5.52 Three-layer band-pass dielectric filter using database 2 materials ($\vartheta = 45°$). Pareto front obtained by all algorithms

Figure 5.53 illustrates the frequency response of the best compromised solution, which is 4.38-mm thick (instead of 6.63 mm in [54]). The layer data for this case are reported in Table 5.45.

A four-layer dual-band filter is selected next. This design case requires additional constraints than the previous band-pass case, so the problem is more complex. For this reason, the population size was set to 200 and the maximum number of iterations

Table 5.44 Three-layer band-pass dielectric filter using database 2 materials

Performance indicator	NSGA-II	GDE3	MOBBO
QI_{R2}	**8.04e−04**	8.03E−03	1.26E−02
$QI_{\varepsilon}1$	**3.28e−02**	6.02E−02	6.78E−02
$QI_{\bar{H}}$	**5.48e−03**	4.51E−02	5.36E−02

Performance indicators of the Pareto fronts. Bold font indicates the lower values.

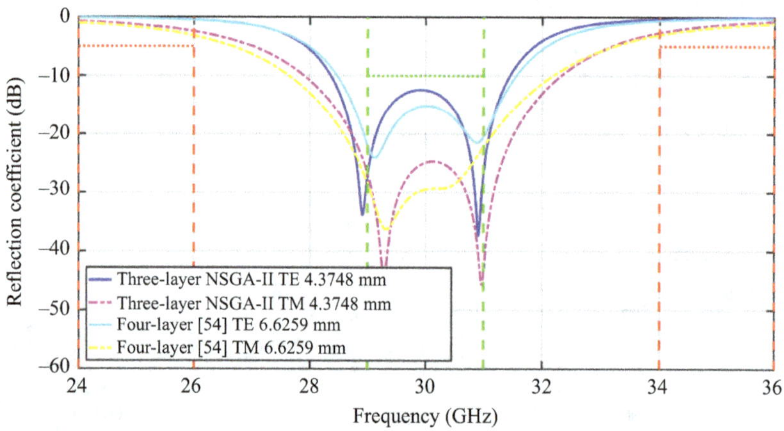

Figure 5.53 Three-layer band-pass dielectric filter using database 2 materials ($\vartheta = 45°$). Frequency response

Table 5.45 Design parameters for the three-layer band-pass dielectric filter

Layer	Material	Thickness (mm)
1	42	1.029
2	3	2.317
3	42	1.029

Best compromised solution obtained by NSGA-II.

was set to 5,000. Figure 5.54 depicts the Pareto front for this case. We notice that both NSGA-II and GDE3 results seem similar, while NSGA-II presents a wider spread of points. Table 5.46 shows the performance indicators for this case. NSGA-II obtains clearly the best performance. The frequency response of the best compromised solution obtained by MOBBO is illustrated in Figure 5.55. We notice that the new design is thinner than the one from [54] (23.87 mm instead of 25.58 mm in [54]). The layer data for this case are listed in Table 5.47.

The final example is that of a five-layer tri-band filter. This case is the most complex design problem of all examples presented earlier. As previously, the population size is selected to be 200 and the maximum number of iterations is 5,000. Figure 5.56 shows the Pareto front for this case. There is no clear indication about the algorithm that dominates. This is also evident in Table 5.48. NSGA-II is better than the other algorithms in two indicators, while GDE3 outperforms the other algorithms is terms of the binary ε-indicator. The frequency response of the best compromised solution is shown in Figure 5.57. The proposed solution is thinner than the

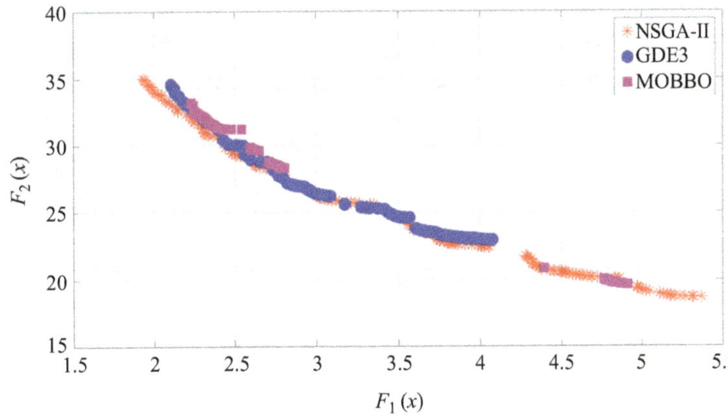

Figure 5.54 Four-layer dual-band dielectric filter using database 2 materials ($\vartheta = 45°$). Pareto front obtained by all algorithms

Table 5.46 Four-layer dual-band dielectric filter using database 2 materials

Performance indicator	NSGA-II	GDE3	MOBBO
QI_{R2}	**1.04e−03**	3.97E−02	3.70E−02
$QI_{\varepsilon}1$	**3.30e−02**	2.63E−01	2.42E−01
$QI_{\bar{H}}$	**4.45e−03**	1.09E−01	1.60E−01

Performance indicators of the Pareto fronts. Bold font indicates the lower values.

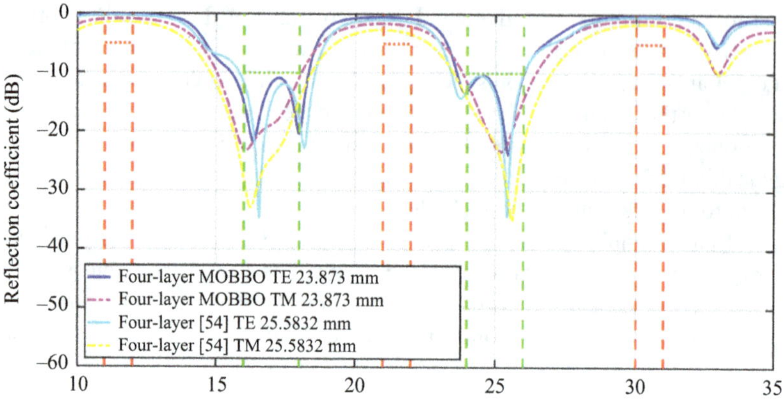

Figure 5.55 Four-layer dual-band dielectric filter using database 2 materials ($\vartheta = 45°$). Frequency response

Table 5.47 Design parameters for the four-layer dual-band dielectric filter

Layer	Material	Thickness (mm)
1	23	3.28
2	5	6.74
3	42	3.62
4	11	3.78

Best compromised solution obtained by NSGA-II.

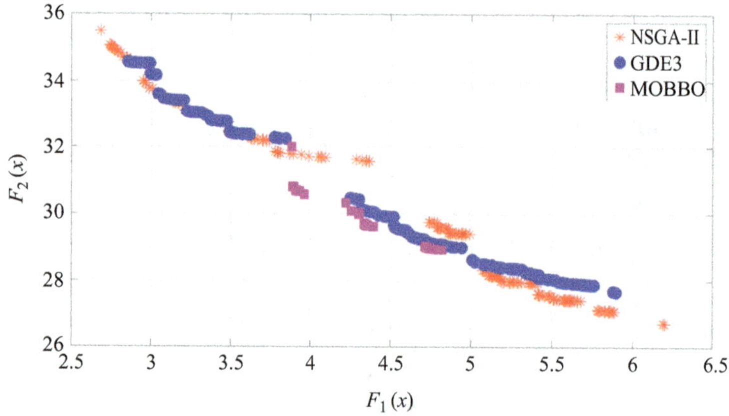

Figure 5.56 Five-layer tri-band dielectric filter using database 2 materials ($\vartheta = 45°$). Pareto front obtained by all algorithms

one in [54] (27.25 mm instead of 31.39 mm). It seems that the response of the new filter is slightly worse than the one in [54]. Table 5.49 lists the layer characteristics for this case.

Overall, the MO optimization obtained better solutions than the ones from single objective optimization. In most cases, designs with fewer layers were also found. The

Table 5.48 Five-layer tri-band dielectric filter using database 2 materials

Performance indicator	NSGA-II	GDE3	MOBBO
QI_{R2}	**1.36e−02**	1.57E−02	8.99E−02
$QI_{\varepsilon}1$	1.42E−01	**1.08e−01**	3.40E−01
$QI_{\bar{H}}$	**5.20e−02**	6.44E−02	2.01E−01

Performance indicators of the Pareto fronts. Bold font indicates the lower values.

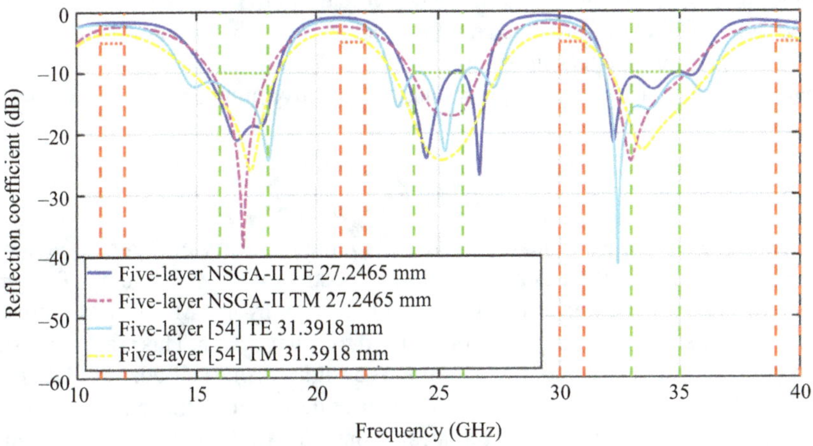

Figure 5.57 Five-layer tri-band dielectric filter using database 2 materials ($\vartheta = 45°$). Frequency response

Table 5.49 Design parameters for the five-layer tri-band dielectric filter

Layer	Material	Thickness (mm)
1	5	5.305
2	17	8.932
3	42	3.600
4	19	6.409
5	6	3.000

Best compromised solution obtained by NSGA-II.

main benefit by applying an MO algorithm is that we get a set of possible solutions and not a single one. NSGA-II outperformed the other algorithms in most of the cases.

5.3 Microstrip filters design

In the examples that follow, we will apply two emerging algorithms in different microstrip filter design cases, the harmony search (HS) algorithm [65], and the Jaya algorithm [66].

These are applied in conjunction with the commercially available EM solver FEKO. In order to integrate the in-house source code of the algorithms with FEKO, a wrapper program was created. FEKO, except of using a graphical user interface, offers the option to run the EM solver engine from command line. It requires an input file that defines the model geometry. This input file uses a script language that allows users to define variables and control options like the frequency range, the number of frequency points, and the required data in the output file. The wrapper creates a FEKO input file for each random vector created by the algorithms and runs FEKO. The output file that in our case is defined to contain the frequency and the S-parameters is read by the wrapper, and the objective and constraint functions are evaluated.

5.3.1 Microstrip band-pass filter

The design of a band-pass microstrip filter with the geometry of Figure 5.58 is presented here [67]. The filter is fabricated in a substrate with $\varepsilon_r = 9$ and is 0.66-mm thick. We model the filter in FEKO [68]. FEKO is a hybrid MoM/FEM software. In [69,70], the space-mapping technique is used for filter design. This is accomplished in conjunction with FEKO. In [67], the filter is designed using self-adaptive DE, more specifically the jDE [71] algorithm. Such a filter design problem can be defined by two objectives subject to two constraint functions. The first objective is to maximize

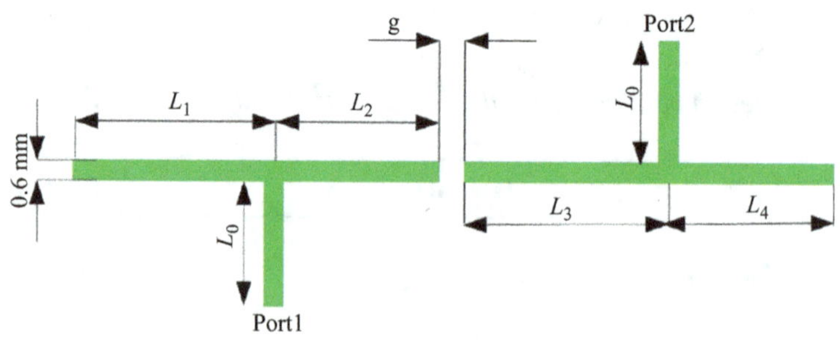

Figure 5.58 Band-pass microstrip filter geometry

the $|S_{21}|$ in the passband frequency range. The second objective is to minimize the $|S_{21}|$ in the stopband frequency range. Additionally, constraints can be set for $|S_{21}|$ levels in both the passband and the stopband frequency ranges. An effective way to combine the previous objectives in one objective–function is to use an exact penalty method [72]. Therefore, this design problem is defined by the minimization of the objective–function:

$$\begin{aligned} F(\bar{x}) &= -w_1 S_{21}^p + w_2 S_{21}^s \\ &\quad + \Xi \times \left| \left(\max\left\{0, S_{21}^s - C_{dB}^S\right\} + \min\left\{0, S_{21}^p - C_{dB}^p\right\} \right) \right| \end{aligned}$$

where (5.25)

$$\begin{aligned} S_{21}^p &= 20 \log\left\{\min |S_{21}(\bar{x},f)|, f \in f_p\right\} \\ S_{21}^s &= 20 \log\left\{\max |S_{21}(\bar{x},f)|, f \in f_s\right\} \end{aligned}$$

where $\bar{x} = (L_0, L_1, L_2, L_3, L_4, g)$ is the vector of filter geometry, f_p and f_s define the corresponding passband and the stopband frequency ranges, C_{dB}^P and C_{dB}^S define the minimum and maximum allowable values in the passband and stopband frequency ranges, respectively, and Ξ is a very large number. In a penalty method, the feasible region is expanded, but a large cost or "penalty" is added to the original objective–function for solutions that lie outside the original feasible region. Therefore, Ξ is chosen large enough to ensure that solutions that do not fulfill constrains result in large fitness values.

In this case, we apply the HS algorithm to the band-pass filter design problem. The harmony memory size or population size is set to 20, the harmony memory consideration rate is set to 0.90, the pitch adjusting rate is equal to 0.5, and pitch bandwidth is 0.2. The maximum number of iterations is set to 1,000.

Table 5.50 lists the design specifications for this filter. The convergence rate graph for this case is depicted in Figure 5.59. We notice that HS shows a fast convergence with less than 20 iterations, further iterations do not improve significantly the objective–function value. Figure 5.60 presents the frequency response of the best design found by HS. It appears that in the stopband frequency bands between 4.3 GHz and 4.6 GHz and between 5.2 GHz and 5.5 GHz, the $|S_{21}|$ value lies below −25 dB. The surface current distribution at 4.45 GHz (first stopband), 4.45 GHz (passband),

Table 5.50 Design specifications for the microstrip band-pass filter case

Frequency band	**Limits**	**Level $|S_{21}|$(dB)**
First stopband	$4.3\text{ GHz} \le f_s < 4.6\text{ GHz}$	<-20
Passband	$4.8\text{ GHz} \le f_p < 5\text{ GHz}$	>-5
Second stopband	$5.2\text{ GHz} \le f_s < 5.5\text{ GHz}$	<-20

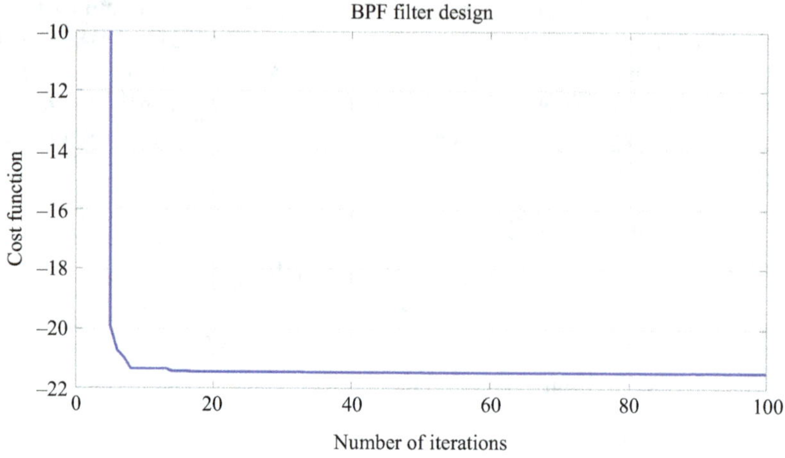

Figure 5.59 Average convergence rate plot for the microstrip band-pass filter case

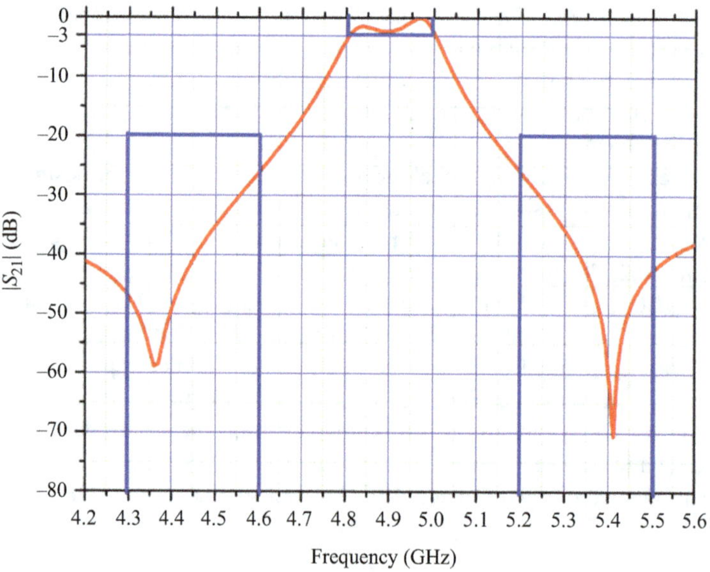

Figure 5.60 Best filter design case frequency response found by HS

5.35 GHz (second stopband) is presented in Figures 5.61–5.63. Table 5.51 shows the geometrical parameters of the best filter found by HS.

5.3.2 *Single band open-loop ring resonator filter*

Another popular microstrip filter is the open-loop ring resonator (OLRR) filter. OLRR filters, which consist of two uniform microstrip lines and pairs of open loops between

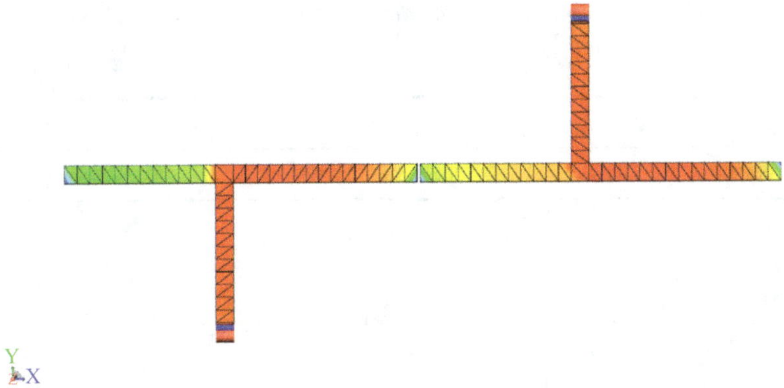

Figure 5.61 *Current distribution simulations for microstrip band-pass filter at 4.45 GHz (first stopband)*

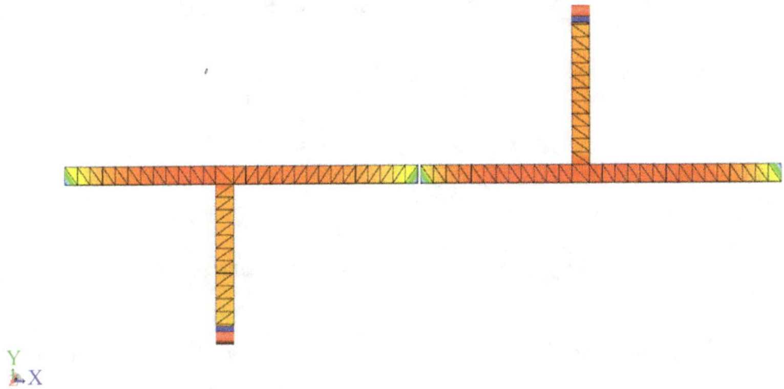

Figure 5.62 *Current distribution simulations for microstrip band-pass filter at 4.9 GHz (passband)*

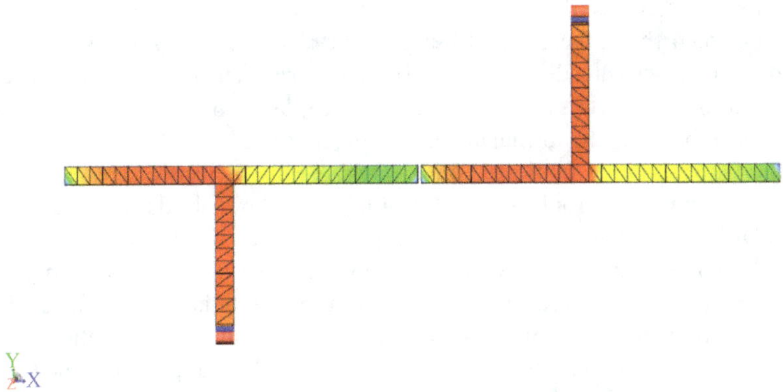

Figure 5.63 *Current distribution simulations for microstrip band-pass filter at 5.35 GHz (second stopband)*

Table 5.51 Design parameters for the microstrip band-pass filter case obtained by HS

L_0	L_1	L_2	L_3	L_4	g
5.000	5.418	6.500	5.329	6.800	0.110

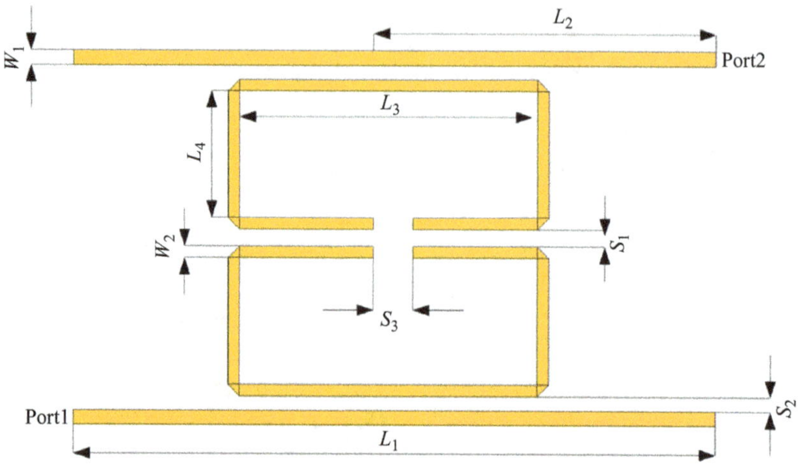

Figure 5.64 Open-loop ring resonator filter geometry

them, are widely used as the building block in several multiband band-pass filter design cases [73–75]. In [73], two pairs of folded OLRRs operating at two passbands are proposed to produce dual-band response. In [76], a tri-band band-pass filter is constructed by using three pairs of stub-loaded OLRRs. In [77], the space-mapping technique is used for OLRR filter design. In [54], OLRR filter is designed using MO DE, more precisely the GDE3 algorithm.

Figure 5.64 shows the OLRR filter geometry. The frequency response of such a filter depends on the filter dimensions and spacings between microstrip lines. The geometrical design parameters are nine ($W_1, W_2, L_1, L_2, L_3, L_4, S_1, S_2, S_3$), all expressed in mm. This problem can be formulated in one objective as in (5.25).

The OLRR filter is fabricated on a substrate with dielectric constant 10.2 and height 0.635 mm (RT/duroid 6,010.2 LM). Table 5.52 shows the design specifications for the OLRR filter [54]. The filter design procedure can be realized in conjunction with FEKO. In this case, we use the Jaya algorithm for filter design. The main benefit of Jaya is that it is a low-complexity algorithm that requires only the setting of population size and number of iterations. The maximum number of iterations is set to 300, while the population size is set to 30. For each FEKO run, 15 frequency sweeps are sampled in the frequency range 1.5–4.5 GHz.

Table 5.52 Design specifications for the OLRR band-pass filter case

Frequency band	**Limits**	**Level $\|S_{21}\|$(dB)**
First stopband	$1.5\,\text{GHz} \le f_s < 2.6\,\text{GHz}$	
Passband	$2.7\,\text{GHz} \le f_p < 3.3\,\text{GHz}$	
Second stopband	$3.4\,\text{GHz} \le f_s < 4.5\,\text{GHz}$	
First constraint stopband	$1.5\,\text{GHz} \le f_s^c < 2.5\,\text{GHz}$	<-20
Constraint passband	$2.8\,\text{GHz} \le f_p^c < 3.2\,\text{GHz}$	>-3
Second constraint stopband	$3.5\,\text{GHz} \le f_s^c < 4.5\,\text{GHz}$	<-20

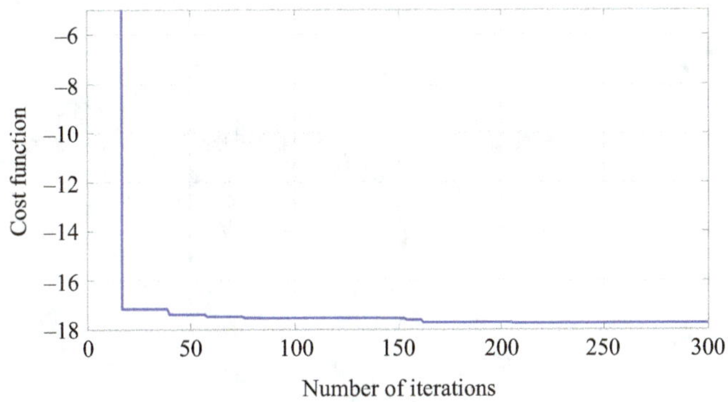

Figure 5.65 Average convergence rate plot for the OLRR filter case

Figure 5.65 shows the convergence rate graph for this case. We notice that Jaya obtained a feasible solution in average less than 50 iterations. The objective–function value is improving up to 200 iterations. Thus, Jaya obtained a fast convergence in this case. The frequency response of the best obtained OLRR filter is illustrated in Figure 5.66. We may notice that the solution found presents low S_{21} magnitude values also out of the desired stopband frequency bands. It appears that in the frequency bands between 1.0 GHz and 2.0 GHz and 4.0 GHz and 5.0 GHz, the S_{21} value lies below −40 dB. Figures 5.67–5.69 show the surface current distribution at 2.14 GHz (first stopband), 3 GHz (passband), 4.07 GHz (second stopband). We notice that the current distribution magnitude at the rings becomes high at the passband frequency, while it is low in the stopband. Table 5.53 shows the geometrical parameters of the best filter found.

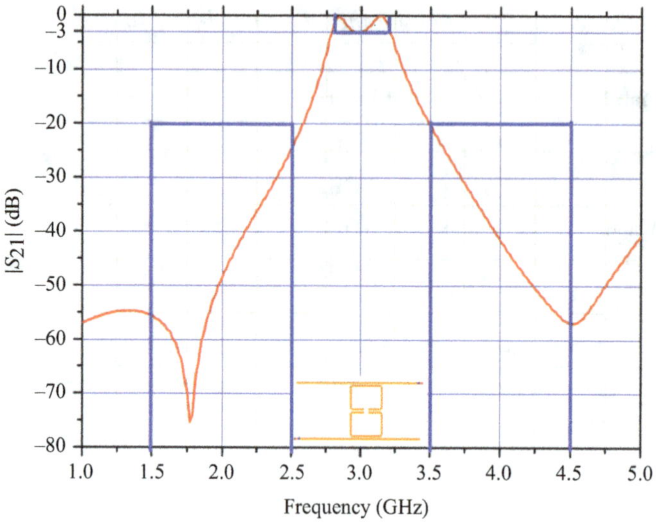

Figure 5.66 Best OLRR filter design case frequency response found by Jaya

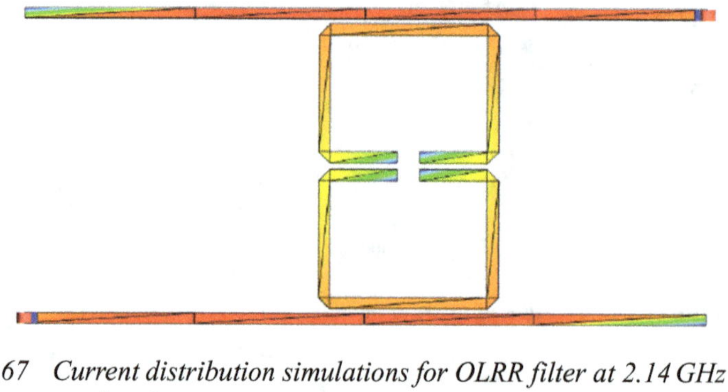

Figure 5.67 Current distribution simulations for OLRR filter at 2.14 GHz (first stopband)

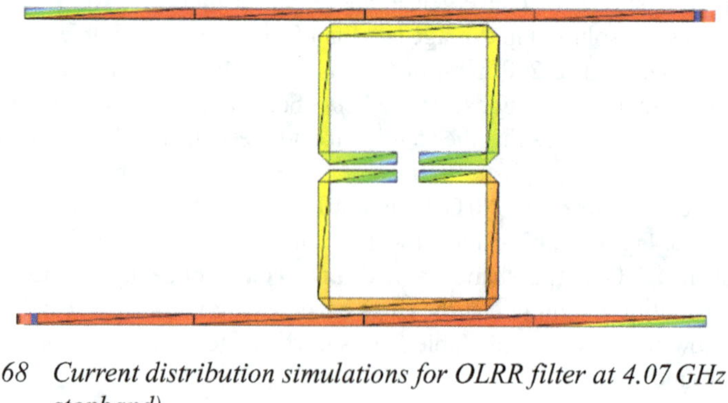

Figure 5.68 Current distribution simulations for OLRR filter at 4.07 GHz (second stopband)

5.3.3 Dual-band OLRR filter

Next, we design a dual-band OLRR filter. The geometry of such a filter is depicted in Figure 5.70. We notice that this filter has two pairs of rings one for each passband. The design parameters for this case are 11 (Figure 5.70) and they are given by the vector $\bar{x} = (W_1, W_2, L_1, L_2, L_3, L_4, L_5, S_1, S_2, S_3, G)$. This problem can be formulated in one objective as in (5.25). We design a filter for dual-band operation in both 5G networks (3.5 GHz) and vehicle-to-everything communication (5.8 GHz). The substrate is the

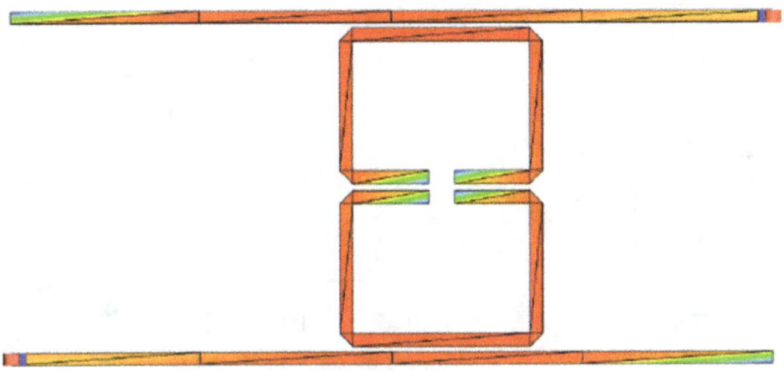

Figure 5.69 Current distribution simulations for OLRR filter at 3 GHz (passband)

Table 5.53 Design parameters for the OLRR band-pass filter case obtained by Jaya

Parameter	**Value (mm)**	**Parameter**	**Value (mm)**
W_1	0.38	L_1	25.00
W_2	0.42	L_2	11.3
S_1	0.22	L_3	5.8
S_2	0.17	L_4	4.13
S_3	0.85		

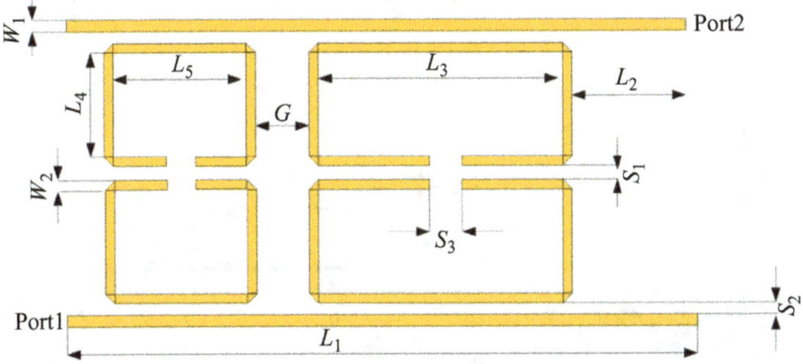

Figure 5.70 Dual-band open-loop ring resonator filter geometry

same as in previous OLRR design. We apply Jaya again to the DOLRR design problem. The maximum number of iterations is set to 500, while the population size is set to 30. For each FEKO run four frequency sweeps are sampled in the passband frequency ranges. For this case, we set $S_p = \{3.55, 3.6, 5.75, 5.8\}$.

The convergence rate graph is drawn in Figure 5.71. We notice that the algorithm requires 300 iterations to reach its final value. Figure 5.72 depicts the simulated frequency response of the best filter design. In the first passband between 3.43 GHz

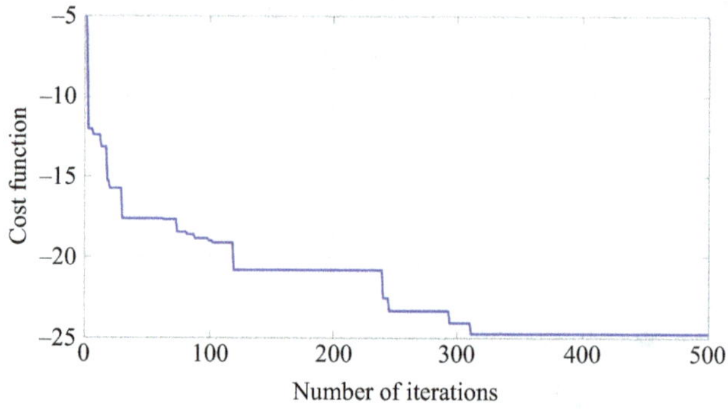

Figure 5.71 Average convergence rate plot for the dual-band OLRR filter case

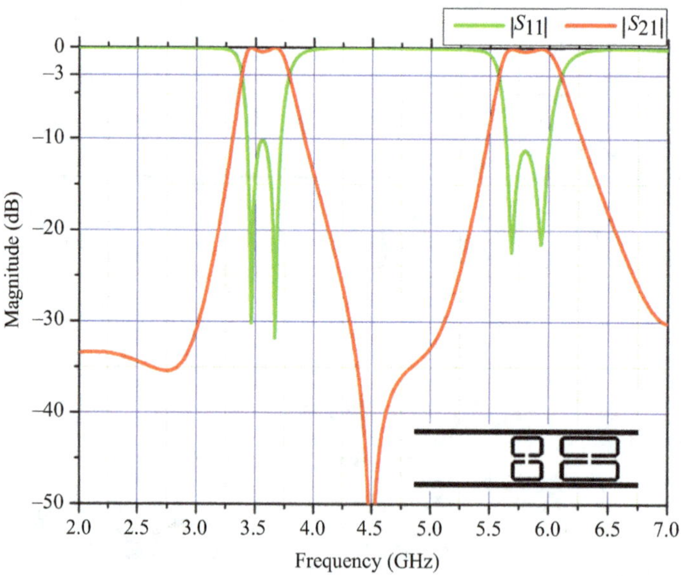

Figure 5.72 Best dual-band OLRR filter design case frequency response found by Jaya

and 3.72 GHz, the filter has a return loss greater than 10 dB and insertion loss less than 0.5 dB. In the second passband between 5.65 GHz and 6.0 GHz, the results also show a return loss greater than 10 dB and insertion loss less than 0.5 dB. The rejection band (between 4.13 GHz and 5.32 GHz) has an insertion loss less than 20 dB. In the first passband, the return loss is greater than 30 dB at both 3.47 GHz and 3.67 GHz. In the second passband, the return loss is greater than 20 dB at 5.68 GHz and 5.93 GHz.

Figures 5.73 and 5.74 illustrate the surface current distribution at 3.67 GHz (first passband), 5.68 GHz (second passband), and 4.5 GHz (stopband) (Figure 5.75).

Figure 5.73 Current distribution simulations for dual-band OLRR filter at 3.67 GHz (first passband)

Figure 5.74 Current distribution simulations for dual-band OLRR filter at 5.68 GHz (second passband)

Figure 5.75 Current distribution simulations for dual-band OLRR filter at 4.5 GHz (stopband)

Table 5.54 Design parameters for the best dual-band OLRR band-pass filter case obtained by Jaya

Parameter	Value (mm)	Parameter	Value (mm)
W_1	0.575	L_1	27.82
W_2	0.42	L_2	2.32
S_1	0.41	L_3	6.43
S_2	0.20	L_4	1.71
S_3	0.54	L_5	3.21
G	2.115		

We notice that we can clearly see the resonating pair of rings in each case. Table 5.54 lists the geometrical parameters of the best design found.

References

[1] Cui S, Mohan A, and Weile DS. Pareto optimal design of absorbers using a parallel elitist nondominated sorting genetic algorithm and the finite element-boundary integral method. IEEE Transactions on Antennas and Propagation. 2005;53(6):2099–2107.

[2] Cui S and Weile DS. Robust design of absorbers using genetic algorithms and the finite element-boundary integral method. IEEE Transactions on Antennas and Propagation. 2003;51(12):3249–3258.

[3] Cui S and Weile DS. Application of a parallel particle swarm optimization scheme to the design of electromagnetic absorbers. IEEE Transactions on Antennas and Propagation. 2005;53(11):3616–3624.

[4] Cui S, Weile DS, and Volakis JL. Novel planar electromagnetic absorber designs using genetic algorithms. IEEE Transactions on Antennas and Propagation. 2006;54(6):1811–1817.

[5] Michielssen E and Mittra R. Design of lightweight, broadband microwave absorbers using genetic algorithms. IEEE Transactions on Microwave Theory and Techniques. 1993;41(6):1024–1031.

[6] Weile DS, Michielssen E, and Goldberg DE. Genetic algorithm design of Pareto optimal broadband microwave absorbers. IEEE Transactions on Electromagnetic Compatibility. 1996;38(3):518–525.

[7] Goudos SK. A versatile software tool for microwave planar radar absorbing materials design using global optimization algorithms. Materials and Design. 2007;28(10):2585–2595.

[8] Chakravarty S, Mittra R, and Williams NR. Application of a microgenetic algorithm (MGA) to the design of broad-band microwave absorbers using multiple frequency selective surface screens buried in dielectrics. IEEE Transactions on Antennas and Propagation. 2002;50(3):284–296.

[9] Ababneh J, Khodier M, and Dib N. Synthesis of interdigital capacitors based on particle swarm optimization and artificial neural networks. International Journal of RF and Microwave Computer-Aided Engineering. 2006;16(4):322–330.
[10] Donelli M and Massa A. Computational approach based on a particle swarm optimizer for microwave imaging of two-dimensional dielectric scatterers. IEEE Transactions on Microwave Theory and Techniques. 2005;53(5):1761–1776.
[11] Khodier MM and Christodoulou CG. Linear array geometry synthesis with minimum sidelobe level and null control using particle swarm optimization. IEEE Transactions on Antennas and Propagation. 2005;53(8 II):2674–2679.
[12] Robinson J and Rahmat-Samii Y. Particle swarm optimization in electromagnetics. IEEE Transactions on Antennas and Propagation. 2004;52(2):397–407.
[13] Mouna H, Mekaladevi V, and Nirmala Devi M. Design of microwave absorbers using improvised particle swarm optimization algorithm. Journal of Microwaves, Optoelectronics and Electromagnetic Applications. 2018;17(2):188–200.
[14] Roy S, Roy SD, Tewary J, *et al.* Particle swarm optimization for optimal design of broadband multilayer microwave absorber for wide angle of incidence. Progress in Electromagnetics Research B. 2015;62(1):121–135.
[15] Chamaani S, Mirtaheri SA, and Shooredeli MA. Design of very thin wide band absorbers using modified local best particle swarm optimization. AEU – International Journal of Electronics and Communications. 2008;62(7):549–556.
[16] Chamaani S, Mirtaheri SA, Teshnehlab M, *et al.* Modified multi-objective particle swarm optimization for electromagnetic absorber design. Progress in Electromagnetics Research. 2008;79:353–366.
[17] Goudos SK. Application of a particle swarm optimizer to the design of microwave broadband absorbers. WSEAS Transactions on Communications. 2007;6(2):312–317.
[18] Goudos SK and Sahalos JN. Microwave absorber optimal design using multi-objective particle swarm optimization. Microwave and Optical Technology Letters. 2006;48(8):1553–1558.
[19] Karaboga D and Basturk B. A powerful and efficient algorithm for numerical function optimization: Artificial bee colony (ABC) algorithm. Journal of Global Optimization. 2007;39(3):459–471.
[20] Toktas A and Ustun D. A triple-objective optimization scheme using butterfly-integrated ABC algorithm for design of multilayer RAM. IEEE Transactions on Antennas and Propagation. 2020;68(7):5602–5612.
[21] Toktas A, Ustun D, Tekbas M. Global optimisation scheme based on triple-objective ABC algorithm for designing fully optimised multi-layer radar absorbing material. IET Microwaves, Antennas and Propagation. 2020;14(8):800–811.
[22] Yigit E and Duysak H. Determination of optimal layer sequence and thickness for broadband multilayer absorber design using double-stage artificial bee colony algorithm. IEEE Transactions on Microwave Theory and Techniques. 2019;67(8):3306–3317.

[23] Qing A. Design of thin wideband planar absorber using dynamic differential evolution and real electromagnetic composite materials. In: IEEE International Symposium on Antennas and Propagation (APSURSI 2011); Spokane, WA, USA. 2011, p. 2912–2915, doi: 10.1109/APS.2011.5997137.
[24] Goudos SK. Design of microwave broadband absorbers using a self-adaptive differential evolution algorithm. International Journal of RF and Microwave Computer-Aided Engineering. 2009;19(3):364–372.
[25] Goudos SK, Baltzis KB, Antoniadis K, *et al.* A comparative study of common and self-adaptive differential evolution strategies on numerical benchmark problems. Procedia Computer Science. 2011;3:83–88.
[26] Ranjan P, Choubey A, and Mahto SK. A novel approach for optimal design of multilayer wideband microwave absorber using wind driven optimization technique. AEU – International Journal of Electronics and Communications. 2018;83:81–87.
[27] Lu Y and Zhou Y. Design of multilayer microwave absorbers using hybrid binary lightning search algorithm and simulated annealing. Progress In Electromagnetics Research B. 2017;78:75–90.
[28] Ranjan P, Choubey A, and Mahto SK. Wide-angle polarization independent multilayer microwave absorber using wind driven optimization technique. International Journal of Applied Engineering Research. 2017;12(19): 8016–8025.
[29] Chew WC. Waves and Fields in Inhomogeneous Media. New York: Wiley-IEEE Press; 1995.
[30] Simon D. Biogeography-based optimization. IEEE Transactions on Evolutionary Computation. 2008;12(6):702–713.
[31] Ma H. An analysis of the equilibrium of migration models for biogeography-based optimization. Information Sciences. 2010;180(18):3444–3464.
[32] Guo W, Wang L, and Wu Q. An analysis of the migration rates for biogeography-based optimization. Information Sciences. 2014;254:111–140.
[33] Ma H, Simon D, Fei M, *et al.* On the equivalences and differences of evolutionary algorithms. Engineering Applications of Artificial Intelligence. 2013;26(10):2397–2407.
[34] Dib N, Asi M, and Sabbah A. On the optimal design of multilayer microwave absorbers. Progress in Electromagnetics Research C. 2010;13:171–185.
[35] Derrac J, García S, Molina D, *et al.* A practical tutorial on the use of nonparametric statistical tests as a methodology for comparing evolutionary and swarm intelligence algorithms. Swarm and Evolutionary Computation. 2011;1(1):3–18.
[36] García S, Molina D, Lozano M, *et al.* A study on the use of non-parametric tests for analyzing the evolutionary algorithms' behaviour: A case study on the CEC'2005 Special Session on Real Parameter Optimization. Journal of Heuristics. 2008;15(6):617–644.
[37] Deb K, Pratap A, Agarwal S, *et al.* A fast and elitist multiobjective genetic algorithm: NSGA-II. IEEE Transactions on Evolutionary Computation. 2002;6(2):182–197.

[38] Kukkonen S and Lampinen J. GDE3: The third evolution step of generalized differential evolution. In: Proceedings of the 2005 IEEE Congress on Evolutionary Computation, (CEC 2005). vol. 1; 2005. p. 443–450.
[39] Goudos SK, Plets D, Liu N, *et al.* A multi-objective approach to indoor wireless heterogeneous networks planning based on biogeography-based optimization. Computer Networks. 2015;91:564–576.
[40] Dai GL and Xia MY. Novel miniaturized bandpass filters using spiral-shaped resonators and window feed structures. Progress in Electromagnetics Research. 2010;100:235–243.
[41] Liu CC, Chang YH, Yang TJ, *et al.* Narrowband filter in a heterostructured multilayer containing ultrathin metallic films. Progress in Electromagnetics Research. 2009;96:329–346.
[42] Mo SG, Yu ZY, and Zhang L. Design of triple-mode bandpass filter using improved hexagonal loop resonator. Progress in Electromagnetics Research. 2009;96:117–125.
[43] Oraizi H and Esfahlan MS. Optimum design of lumped filters incorporating impedance matching by the method of least squares. Progress in Electromagnetics Research. 2010;100:83–103.
[44] Razalli MS, Ismail A, Mahdi MA, *et al.* Novel compact "via-less" ultra-wide band filter utilizing capacitive microstrip patch. Progress in Electromagnetics Research. 2009;91:213–227.
[45] Siakavara K and Damianidis C. Microwave filtering in waveguides loaded with artificial single or double negative materials realized with dielectric spherical particles in resonance. Progress in Electromagnetics Research. 2009;95:103–120.
[46] Vegesna S and Saed M. Novel compact dual-band bandpass microstrip filter. Progress in Electromagnetics Research B. 2010;(20):245–262.
[47] Velazquez-Ahumada MDC, Martel J, Medina F, *et al.* Application of stub loaded folded stepped impedance resonators to dual band filter design. Progress in Electromagnetics Research. 2010;102:107–124.
[48] Wang XH, Wang BZ, and Chen KJ. Compact broadband dual-band bandpass filters using slotted ground structures. Progress in Electromagnetics Research. 2008;82:151–166.
[49] Wang ZG, Li XQ, Zhou SP, *et al.* Half mode substrate integrated folded waveguide (HMSIFW) and partial H-plane bandpass filter. Progress in Electromagnetics Research. 2010;101:203–216.
[50] Wang Z, Lai Q, Xu R, *et al.* A millimeter-wave ultra-wideband four-way switch filter module based on novel three-line microstrip structure band-pass filters. Progress in Electromagnetics Research. 2009;94:297–309.
[51] Yang MH, Xu J, Zhao Q, *et al.* Compact, broad-stopband lowpass filters using SIRs-loaded circular hairpin resonators. Progress in Electromagnetics Research. 2010;102:95–106.
[52] Ye CS, Su YK, Weng MH, *et al.* Design of the compact parallel-coupled lines wideband bandpass filters using image parameter method. Progress in Electromagnetics Research. 2010;100:153–173.

[53] Zhang L, Yu Z, and Mo S. Novel planar multimode bandpass filters with radial-line stubs. Progress in Electromagnetics Research. 2010;101: 33–42.

[54] Goudos SK and Sahalos JN. Pareto optimal microwave filter design using multiobjective differential evolution. IEEE Transactions on Antennas and Propagation. 2010;58(1):132–144.

[55] Hong JS and Lancaster MJ. Microstrip Filters for RF/Microwave Applications. New York, NY: John Wiley and Sons; 2001.

[56] Chahravarty S and Mittra R. Design of microwave filters using a binary coded genetic algorithm. In: Antennas and Propagation Society International Symposium, 2000. IEEE. vol. 1; 2000. p. 144–147.

[57] Cui SM, Mohan A, and Weile DS. Pareto optimal design of absorbers using a parallel elitist nondominate sorting genetic algorithm and the finite element-boundary integral method. IEEE Transactions on Antennas and Propagation. 2005;53(6):2099–2107.

[58] Michielssen E, Sajer JM, Ranjithant S, *et al.* Design of lightweight, broad-band microwave absorbers using genetic algorithms. IEEE Transactions on Microwave Theory and Techniques. 1993;41(6–7):1024–1030.

[59] Yuan Y, Chan CH, Man KF, *et al.* A genetic algorithm approach to FSS filter design. In: IEEE Antennas and Propagation Society International Symposium. 2001;4:688–691.

[60] Zuffada C, Cwik T, and Ditchman C. Synthesis of novel all-dielectric grating filters using genetic algorithms. IEEE Transactions on Antennas and Propagation. 1998;46(5):657–663.

[61] Hoorfar A, Zhu J, and Nelatury S. Electromagnetic optimization using a mixed-parameter self-adaptive evolutionary algorithm. Microwave and Optical Technology Letters. 2003;39(4):267–271.

[62] Venkatarayalu NV, Ray T, and Gan YB. Multilayer dielectric filter design using a multiobjective evolutionary algorithm. IEEE Transactions on Antennas and Propagation. 2005;53(11):3625–3632.

[63] Chew WC. Waves and Fields in Inhomogeneous Media. New York, NY: IEEE Press; 1995.

[64] Goudos SK, Zaharis ZD, Salazar-Lechuga M, *et al.* Dielectric filter optimal design suitable for microwave communications by using multiobjective evolutionary algorithms. Microwave and Optical Technology Letters. 2007;49(10):2324–2329.

[65] Geem ZW, Kim JH, and Loganathan GV. A new heuristic optimization algorithm: Harmony search. Simulation. 2001;76(2):60–68.

[66] Rao RV. Jaya: A simple and new optimization algorithm for solving constrained and unconstrained optimization problems. International Journal of Industrial Engineering Computations. 2016;7(1):19–34.

[67] Goudos SK, Siakavara K, Samaras T, *et al.* Self-adaptive differential evolution applied to real-valued antenna and microwave design problems. IEEE Transactions on Antennas and Propagation. 2011;59(4):1286–1298.

[68] User's Manunal, FEKO Suite 5.2. 2003.

[69] Koziel S and Bandler JW. SMF: A user-friendly software engine for space-mapping-based engineering design optimization. In: Signals, Systems and Electronics, 2007. ISSSE '07. International Symposium on; 2007. p. 157–160.

[70] Koziel S, Bandler JW, and Madsen K. A space-mapping framework for engineering optimization; theory and implementation. IEEE Transactions on Microwave Theory and Techniques. 2006;54(10):3721–3730.

[71] Brest J, Greiner S, Boskovic B, *et al.* Self-adapting control parameters in differential evolution: A comparative study on numerical benchmark problems. IEEE Transactions on Evolutionary Computation. 2006;10(6):646–657.

[72] Pillo GD. Exact penalty methods. Algorithms for Continuous Optimization: The State of the Art. vol. 434. Netherlands: Kluwer Academic Publishers; 1994, p. 209–253.

[73] Chen CY and Hsu CY. A simple and effective method for microstrip dual-band filters design. IEEE Microwave and Wireless Components Letters. 2006;16(5):246–248.

[74] Mondal P and Mandal MK. Design of dual-band bandpass filters using stub-loaded open-loop resonators. IEEE Transactions on Microwave Theory and Techniques. 2008;56(1):150–155.

[75] Zhang XY, Chen JX, Xue Q, *et al.* Dual-band bandpass filters using stub-loaded resonators. IEEE Microwave and Wireless Components Letters. 2007;17(8):583–585.

[76] Lai X, Wu B, Su T, *et al.* A novel tri-band filter using stub-loaded open loop ring resonators. Microwave and Optical Technology Letters. 2010;52(3):523–526.

[77] Koziel S and Bandler JW. Space mapping with multiple coarse models for optimization of microwave components. IEEE Microwave and Wireless Components Letters. 2008;18(1):1–3.

Chapter 6
Design problems in wireless communications

This chapter presents representative examples of design problems in wireless communications using evolutionary algorithms (EAs).

6.1 Peak-to-average power ratio reduction in OFDM systems

Orthogonal frequency division multiplexing (OFDM) is a very popular technique that is utilized in a broad range of applications like digital audio broadcasting, digital video broadcasting, and wireless local area networks [1,2]. However, the OFDM systems require the solution of several research challenging issues. One of the major disadvantages of the OFDM-generated signals is the high value of peak-to-average power ratio (PAPR). The OFDM receiver detection efficiency is sensitive to nonlinear devices like the high power amplifier [3]. Therefore, the reduction of PAPR of an OFDM signal plays an important role for the full utilization of the OFDM technical characteristics.

A popular PAPR reduction method is partial transmit sequences (PTS) [4]. PTS obtains good PAPR reduction performance. The major drawback of PTS is that it requires an exhaustive search for finding optimal phase factors. Therefore, the PTS search complexity is high.

There are several efforts in literature that propose different methods in order to use PTS for PAPR reduction with low search complexity [4–33]. In [34], the author proposed a new technique based on biogeography-based optimization (BBO) for PTS search complexity reduction. More precisely, in [34] a new generalized oppositional BBO (GOBBO) algorithm is introduced. This GOBBO algorithm uses oppositional-based learning concepts. This algorithm was described in Chapter 2.

6.1.1 System model

An OFDM system splits the high-rate data steam into N low-rate data streams that are simultaneously transmitted using N subcarriers. Then, we can express the discrete-time signal of such a system as [1,2]

$$s_k = \frac{1}{\sqrt{N}} \sum_{n=0}^{N-1} S_n e^{j2\pi nk/LN} \quad k = 0, 1, \ldots, LN - 1 \tag{6.1}$$

where L denotes the oversampling factor, $\mathbf{S} = [S_0, S_1, \ldots, S_{N-1}]^T$ is the input signal block. For each symbol, we use phase-shift keying (PSK) or quadrature amplitude modulation (QAM). The signal PAPR of (6.1) can be defined as the ratio of the maximum to average power and in dB is formulated as [1,2]

$$PAPR(s) = 10\log_{10}\frac{\max\limits_{0\leq k\leq LN-1} |s_k|^2}{E\left[|s_k|^2\right]} \tag{6.2}$$

where $E\,[.]$ denotes the expected value operation. The PTS technique partitions the input data OFDM block into M-disjointed subblocks. These are denoted by the vector $\mathbf{S}_m$ $m = 1, 2, \ldots, M$ and oversampled by inserting $(L-1)\,N$ zeros. Thus, the PTS process is formulated mathematically as [4]

$$\mathbf{S} = \sum_{m=1}^{M} \mathbf{S}_m \tag{6.3}$$

Moreover, then a conversion process to time domain takes place. This is accomplished using LN point inverse fast Fourier transform (IFFT). The OFDM block formulation in time domain is then given by

$$\mathbf{s} = IFFT\left\{\sum_{m=1}^{M} \mathbf{S}_m\right\} = \sum_{m=1}^{M} IFFT\,\{\mathbf{S}_m\} = \sum_{m=1}^{M} \mathbf{s}_m \tag{6.4}$$

The PTS goal is to obtain a weighted combination of the M subblocks using $\mathbf{b} = [b_1, b_2, \ldots, b_M]^T$ complex phase factors in order to minimize PAPR. The transmitted signal in time domain after this combination is expressed as [4]

$$\mathbf{s}'(\mathbf{b}) = \sum_{m=1}^{M} b_m \mathbf{s}_m \tag{6.5}$$

Figure 6.1 depicts the block diagram of the PTS process.

We limit the phase factor possible values to a finite set in order to facilitate the search complexity reduction. Then, the set of allowable phase factors becomes [22]

$$A = \left\{e^{j2\pi n/W}\ |n = 0, 1, \ldots, W-1\right\} \tag{6.6}$$

where W denotes the number of allowed phase factors. Thus, if we have M subblocks and W phase factors, the total number of possible combinations is W^M. So as to achieve further search complexity reduction, we usually set fixed one phase factor. The optimization objective of the PTS process is defined as obtaining the optimal phase combination for minimum PAPR. Therefore, we may formulate the objective–function as [22]

minimize

$$F(\mathbf{b}) = 10\log_{10}\frac{\max\limits_{0\leq k\leq LN-1} |\mathbf{s}'(\mathbf{b})|^2}{E\left[|\mathbf{s}'(\mathbf{b})|^2\right]} \tag{6.7}$$

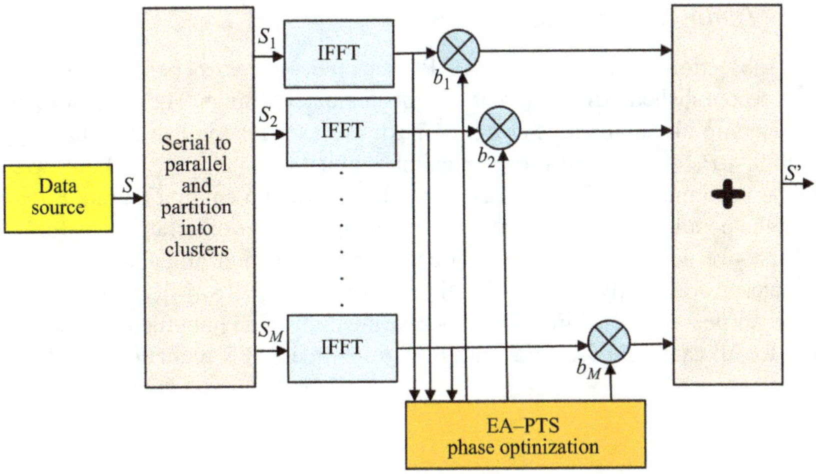

Figure 6.1 Block diagram of an evolutionary algorithm—PTS technique

subject to

$$\mathbf{b} \in \left\{e^{j\phi_m}\right\}^M \text{where } \phi_m \in \left\{\frac{2\pi n}{W} \,|n = 0, 1, \ldots, W - 1\right\} \tag{6.8}$$

For $W = 2$ $\mathbf{b} \in \{-1, 1\}^M$. We can choose a phase factor to be fixed without any performance loss. Consequently, there are $M - 1$ unknown variables to be found with an optimization procedure. Thus, the search space size is 2^{M-1}. We notice that the search complexity increases exponentially as M increases.

6.1.2 Simulation settings

In this section, we evaluate the GOBBO algorithm performance in the PTS problem. We perform several simulations in every optimization case. There are two main measurement criteria: the complementary cumulative distribution function (CCDF) and the computational complexity. We generate 10^5 random OFDM blocks in all simulations. The transmitted signal is oversampled by a factor $L = 4$. We set the symbol modulation to 16-QAM with $N = 512$ subcarriers that are divided into $M = 16$ random subblocks. We select phase factors for $W = 2$. We assume that the first phase factor is fixed. Thus, the total number of unknown phase factors is $M - 1$.

First, we give an example of tuning control parameters (opposition probability P_o and jumping rate j_r) by performing several simulations using GOBBO–PTS. Next, we evaluate the algorithms' performance using the fine-tuned control parameters by comparing results with different algorithms.

6.1.3 Tuning control parameters

In this subsection, we evaluate the GOBBO sensitivity to control parameter selection. This is accomplished using different control parameter settings. The GOBBO control parameters in all simulations are those next. We use, for the habitat modification probability, $P_{\text{mod}} = 1$, and for the maximum mutation rate, $m_{\max} = 0.005$. We set both the maximum immigration rate I, and the maximum emigration rate E to one. We set the population size to 30 and the maximum number of iterations to 30.

First, we need to evaluate the effect of the opposition probability P_o control parameter. Additionally, we use also the random search (RS) [4] method for comparison purposes by selecting 900 random evaluations. The jumping rate is set fixed to 0.3 for all cases. Figures 6.2 and 6.3 present the PAPR reduction performance

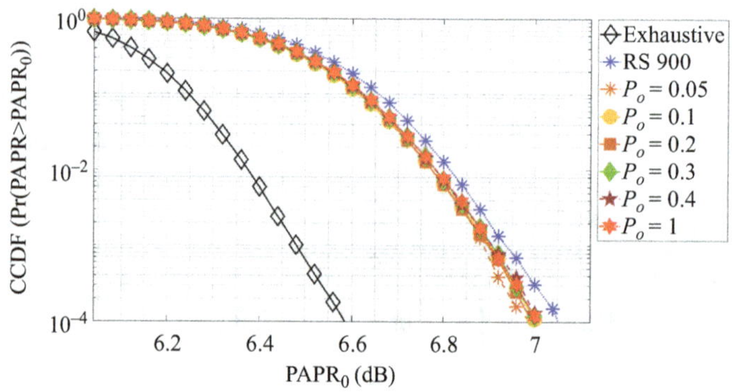

Figure 6.2 PAPR reduction performance comparison of GOBBO–PTS algorithm with different opposition probability values (total view)

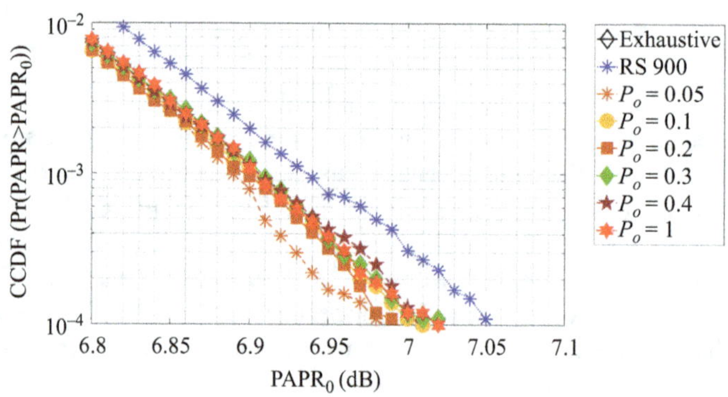

Figure 6.3 PAPR reduction performance comparison of GOBBO–PTS algorithm with different opposition probability values (detailed view)

comparison among different opposition probability values. Although the results are very close, it is apparent that for $P_o = 0.05$ the results are better than the others. Second best results are for $P_o = 0.1$. This is also evident from Table 6.1 that lists the comparative results for this case for different CCDF values. One may notice that for $P_o = 0.05$, we obtain the best results in general, while for $P_o = 0.1$ results are also very satisfactory. Thus, we may conclude that the GOBBO performance is relatively robust regarding this parameter for values less or equal to 0.1. For $P_o = 1$, the GOBBO transforms to the original OBBO algorithm. We observe that for this case, the results seem to become worse.

Moreover, we evaluate the GOBBO performance for different jumping rate values. The suggested value for jumping rate found in the literature is 0.3 [35]. We will evaluate if this is the proper jumping rate value for GOBBO when applied to the PTS problem. As it was found previously, the more suitable value for the opposition probability is 0.05. Figures 6.4 and 6.5 depict the PAPR reduction performance for different jumping rate values. One may notice that all CCDF curves are very close,

Table 6.1 PAPR reduction performance comparison of GOBBO–PTS algorithm with different opposition probability values for different CCDF values

Method	**1.00e−01**	**5.00e−02**	**1.00e−02**	**5.00e−03**	**1.00e−03**	**5.00e−04**
$P_o = 0.05$	6.62	6.67	6.78	6.82	6.89	6.91
$P_o = 0.1$	6.62	6.68	6.78	6.82	6.91	6.94
$P_o = 0.2$	6.62	6.68	6.78	6.82	6.90	6.94
$P_o = 0.3$	6.63	6.68	6.79	6.83	6.91	6.95
$P_o = 0.4$	6.63	6.68	6.79	6.83	6.91	6.95
$P_o = 1$	6.63	6.68	6.79	6.83	6.91	6.94

All values are in dB.

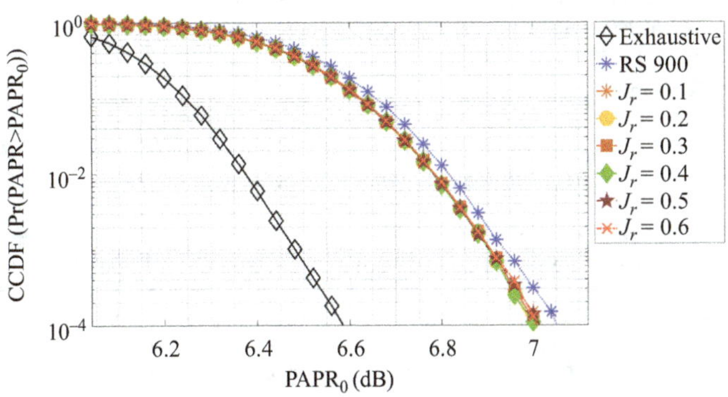

Figure 6.4 PAPR reduction performance comparison of GOBBO–PTS algorithm with different jumping rate values (total view)

and it is difficult to find the best performance. The detailed view in Figure 6.5 shows that for jumping rates 0.4 and 0.2, we obtain a performance slightly better than the others. Table 6.2 reports the comparative results for different jumping rates and different CCDF values. One may observe that the better obtained values are for jumping rates 0.2 and 0.4. However, results are all very close and the only differences are for small CCDF value of 5×10^{-3} and 5×10^{-4}. So, jumping rate is quite robust for values less or equal to 0.6.

Next, we evaluate the combination of different opposition probability P_o and jumping rate j_r values. We select the two best values found for opposition probability of 0.05 and 0.1. For jumping rate, we select three values 0.2, 0.3, and 0.4. The corresponding CCDF curves are plotted in Figures 6.6 and 6.7. We may notice from the detailed view that for $P_o = 0.05, j_r = 0.3$ and $P_o = 0.10, j_r = 0.4$ results seem better. The combination $P_o = 0.05, j_r = 0.3$ performs better for low CCDF values, while for higher values the other combination is better. This is also apparent from Table 6.3 where the results for different CCDF values are listed.

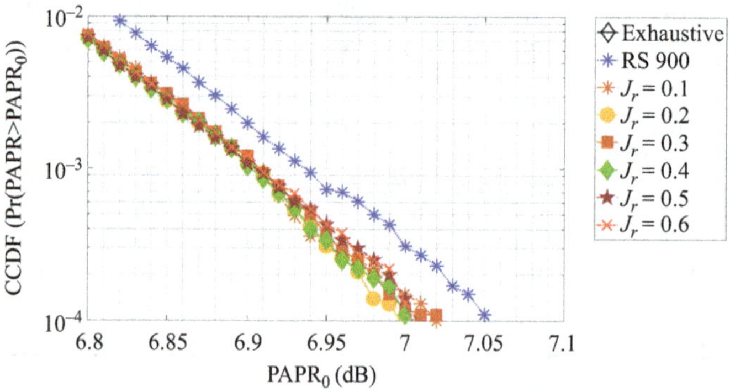

Figure 6.5 PAPR reduction performance comparison of GOBBO–PTS algorithm with different jumping rate values (detailed view)

Table 6.2 PAPR reduction performance comparison of GOBBO–PTS algorithm with different jumping rate values for different CCDF values

Method	**1.00e−01**	**5.00e−02**	**1.00e−02**	**5.00e−03**	**1.00e−03**	**5.00e−04**
$J_r = 0.1$	6.63	6.68	6.79	6.83	6.91	6.93
$J_r = 0.2$	6.63	6.68	6.79	6.83	6.91	6.94
$J_r = 0.3$	6.63	6.68	6.79	6.83	6.91	6.95
$J_r = 0.4$	6.63	6.68	6.79	6.82	6.91	6.94
$J_r = 0.5$	6.63	6.68	6.79	6.82	6.91	6.95
$J_r = 0.6$	6.63	6.68	6.79	6.83	6.91	6.95

All values are in dB.

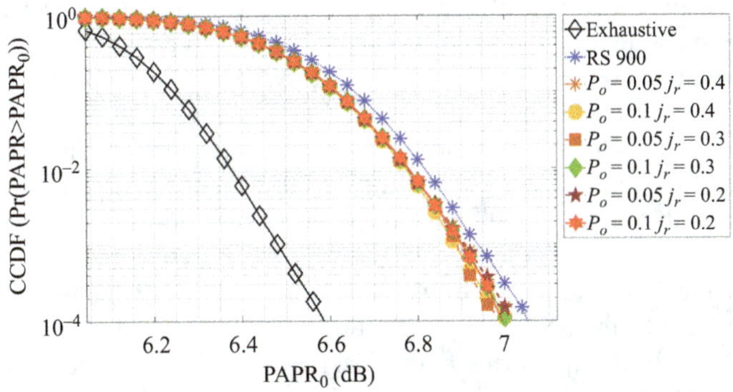

Figure 6.6 PAPR reduction performance comparison of GOBBO–PTS algorithm with different opposition probability and jumping rate values (total view)

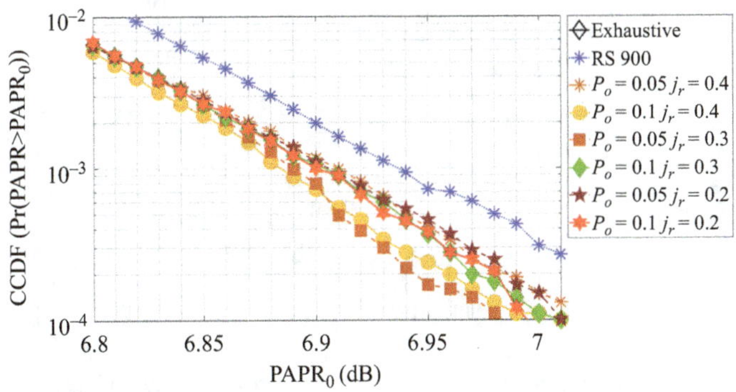

Figure 6.7 PAPR reduction performance comparison of GOBBO–PTS algorithm with different opposition probability and jumping rate values (detailed view)

Table 6.3 PAPR reduction performance comparison of GOBBO–PTS algorithm with different opposition probability and jumping rate values for different CCDF values

Method	**1.00e−01**	**5.00e−02**	**1.00e−02**	**5.00e−03**	**1.00e−03**	**5.00e−04**
$P_o = 0.05,\ j_r = 0.4$	6.62	6.67	6.78	6.82	6.91	6.94
$P_o = 0.10,\ j_r = 0.4$	6.62	6.67	6.78	6.81	6.89	6.92
$P_o = 0.05,\ j_r = 0.3$	6.62	6.67	6.78	6.82	6.89	6.91
$P_o = 0.10,\ j_r = 0.3$	6.62	6.68	6.78	6.82	6.91	6.94
$P_o = 0.05,\ j_r = 0.2$	6.62	6.67	6.78	6.82	6.91	6.95
$P_o = 0.10,\ j_r = 0.3$	6.62	6.68	6.78	6.82	6.91	6.94

All values are in dB.

Additionally, we evaluate the GOBBO–PTS, as well the BBO–PTS performance using different values of population size and iterations. In cases, the total number of function evolutions remains constant and equal $NP \times G$. The population size is set to 30 when we change the iteration number. The same way, the iteration number is set to 30 when we vary the population size. Thus, there are two different numbers of objective–function evaluations 600 and 1,200. Moreover, we compare with the RS method with 600 and 1,200 random phase factors. Figures 6.8 and 6.9 show the comparative results. It is obvious that the more the objective–function evaluations, the smaller the PAPR value. Also, we notice that both GOBBO and BBO clearly perform better than the RS methods in both cases of 600 and 1, 200 evaluations. It appears that GOBBO outperforms the original BBO. The GOBBO performance seems to be dependent only on the number of objective–function evaluations. GOBBO with 40 iterations and population size 30 performs quite similarly as GOBBO with 30

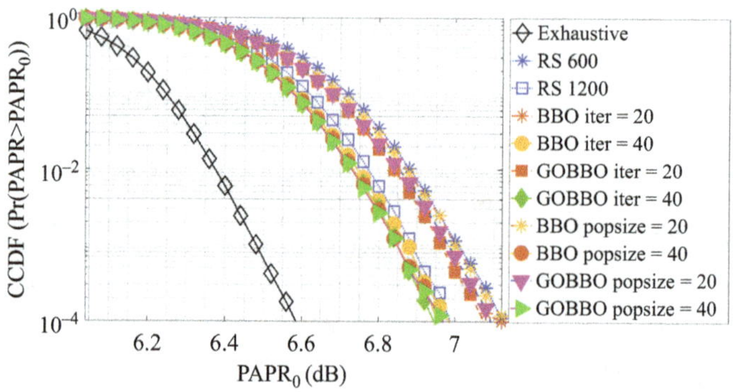

Figure 6.8 PAPR reduction performance comparison of GOBBO–PTS algorithm with different iterations and population size values (total view)

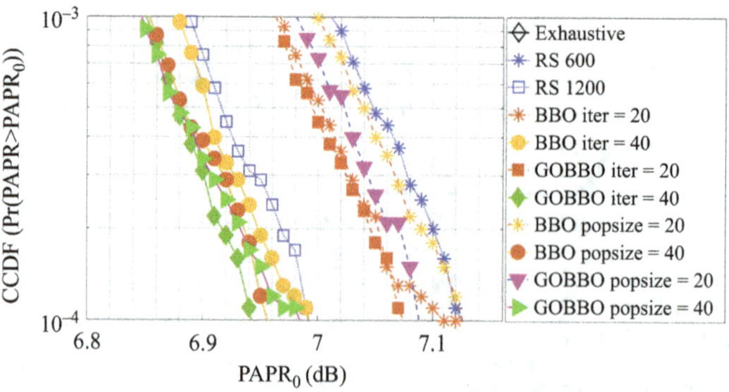

Figure 6.9 PAPR reduction performance comparison of GOBBO–PTS algorithm with different iterations and population size values (detailed view)

Table 6.4 PAPR reduction performance comparison of GOBBO–PTS algorithm with different iterations and population size values for different CCDF values

Method	1.00e−01	5.00e−02	1.00e−02	5.00e−03	1.00e−03	5.00e−04
BBO iter = 20	6.68	6.74	6.84	6.89	6.97	7.01
BBO iter = 40	6.59	6.64	6.75	6.79	6.88	6.91
GOBBO iter = 20	6.68	6.74	6.84	6.88	6.97	7.00
GOBBO iter = 40	6.58	6.63	6.74	6.78	6.86	6.88
BBO popsize = 20	6.70	6.76	6.88	6.92	7.00	7.05
BBO popsize = 40	6.58	6.64	6.74	6.78	6.86	6.89
GOBBO popsize = 20	6.68	6.74	6.86	6.90	6.99	7.03
GOBBO popsize = 40	6.58	6.63	6.74	6.77	6.85	6.88

All values are in dB.

iterations and population size 40. This is also evident from Table 6.4 that holds the comparative results for different CCDF values.

Overall, we will use the values of $P_o = 0.10, j_r = 0.4$ for GOBBO control parameters next. Since the number of objective–function evaluations is the only metric that affects performance, we will use a constant value of 900 for next simulations.

6.1.4 Comparison with other methods

In this section, we compare the GOBBO–PTS and the BBO–PTS with different methods. Thus, we also apply the ant colony optimization (ACO) [36], the artificial bee colony (ABC) [37], the Gbest-guided ABC (GABC) [38], the grey wolf optimizer (GWO) [39], the shuffled frog leaping algorithm (SFLA) [40], and the binary genetic algorithm (BGA) [41]. We set the population size, *NP*, to 30 and the maximum number of iterations, *G*, to 30 for all EAs. Thus, the computational complexity of all EAs is $NP \times G = 900$. The computational complexity of the exhaustive search for this case is $W^{M-1} = 32{,}768$. We also apply two specific PTS heuristics, the iterative flipping algorithm for PTS (IPTS) [4] with search complexity $(M-1)W = 30$, and the gradient descent (GD) method [10] with search complexity $C^r_{M-1} W^r I = C^2_{15} 2^2 3 = 1{,}260$.

The BBO–PTS control parameters in all simulations are given later. For all BBO algorithms, the habitat modification probability, P_{mod}, is set to one, and the maximum mutation rate, $m_{\max}$, is set equal to 0.005. The maximum immigration rate I, and the maximum emigration rate E are both set to one. The elitism parameter p is set to two. For GOBBO, the jumping rate is set to 0.4 and the opposition probability is set to 0.1. In the ABC algorithm, the *limit* parameter is set to 5. Moreover, in the GABC the nonnegative constant C is set to 1.5. For ACO, the initial pheromone value τ_0 is set to 10^{-6}, the pheromone update constant Q is set to 20, the exploration constant q_0 is set to 1, the global pheromone decay rate ρ_g is 0.9, the local pheromone decay rate ρ_l is 0.5, the pheromone sensitivity α is 1, and the visibility sensitivity is β

is 5. The population size of SFLA is also set to 30. The number of groups in SFLA is set to 5; therefore, each group consists of six frogs. For the BGA, the crossover probability is set to 0.9, and the mutation probability to 0.05.

Figures 6.10 and 6.11 depict the comparison between the CCDF by GOBBO–PTS and other PTS reduction techniques. We notice that GOBBO–PTS and BBO–PTS emerge as the best algorithms. For high CCDF values, both perform identically; however, for low CCDF values GOBBO–PTS clearly performs better as it is apparent from Figure 6.11. Table 6.5 lists the comparative results for different CCDF values. For Pr ($PAPR > PAPR_0$) $= 10^{-3}$, the PAPR of the original OFDM-transmitted signal is 11.16 dB. If we use exhaustive search in all combinations, the PAPR value reduces to 6.49 dB. GOBBO–PTS obtained the next best result of 6.89 dB, while very close

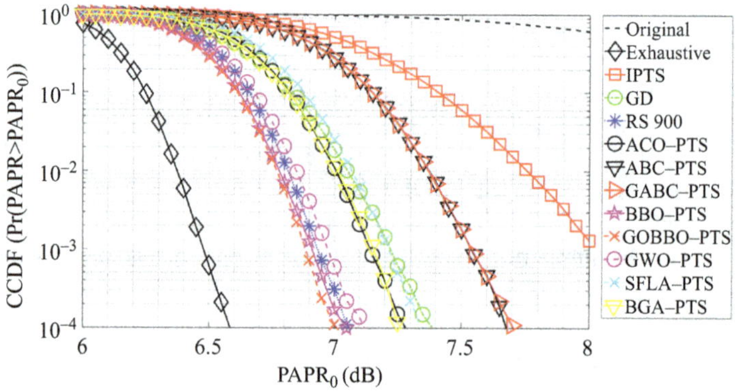

Figure 6.10 PAPR reduction performance comparison of the GOBBO–PTS algorithms with other PTS schemes for NP = 30, G = 30 (total view)

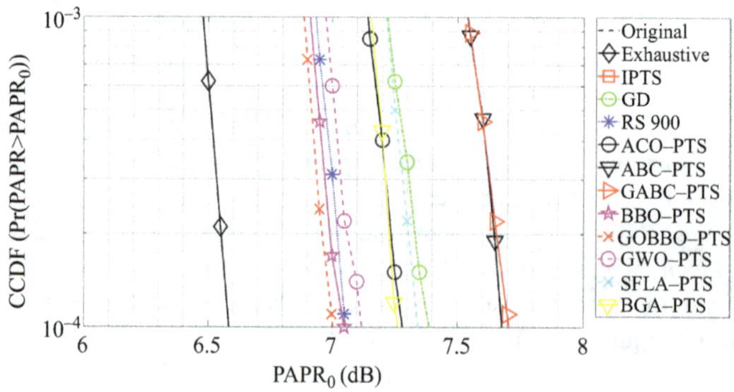

Figure 6.11 PAPR reduction performance comparison of the GOBBO–PTS algorithms with other PTS schemes for NP = 30, G = 30 (detailed view)

Table 6.5 PAPR reduction performance comparison of the GOBBO–PTS algorithms with other PTS schemes for NP = 30, G = 30 for different CCDF values

Method	**1.00e−01**	**5.00e−02**	**1.00e−02**	**5.00e−03**	**1.00e−03**	**5.00e−04**
Original	9.28	9.64	10.34	10.61	11.16	11.43
Exhaustive	6.25	6.30	6.38	6.41	6.49	6.52
IPTS	7.42	7.54	7.77	7.86	8.04	8.11
GD	6.84	6.91	7.06	7.11	7.23	7.27
RS 900	6.66	6.72	6.82	6.86	6.94	6.99
ACO–PTS	6.82	6.89	7.01	7.05	7.15	7.19
ABC–PTS	7.16	7.23	7.38	7.43	7.55	7.60
GABC–PTS	7.15	7.23	7.38	7.43	7.55	7.60
BBO–PTS	6.62	6.68	6.79	6.83	6.92	6.95
GOBBO–PTS	6.62	6.67	6.78	6.81	6.89	6.92
GWO–PTS	6.68	6.74	6.85	6.89	6.98	7.02
SFLA–PTS	6.88	6.95	7.07	7.12	7.22	7.26
BGA–PTS	6.82	6.89	7.02	7.06	7.16	7.19

All values are in dB.

BBO–PTS reduces the PAPR value to 6.92 dB. Next, the RS obtained a result of 6.94 dB. All the other methods obtained results with higher PAPR values. GWO–PTS also performed well with values close to that of RS.

Overall, the BBO algorithms and especially GOBBO are well-suited algorithms for solving the PAPR reduction problem. Moreover, GOBBO outperformed the original BBO algorithm.

6.2 Antenna selection in MIMO systems

Over the last years, the need for higher data rates is constantly growing. This fact makes MIMO antenna systems vital at a greater degree. These MIMO systems are capable of achieving high data rates through spatial multiplexing gain and improved reliability via diversity gain [42–44].

However, when multiple antennas at the transmitter and the receiver side of any connection are utilized, this leads to an increase of hardware complexity and cost considering the analog radio frequency (RF) chains (i.e., amplifiers, analog-to-digital converters, and filters) per antenna. The authors in [45] have argued that one way to reduce costs for hardware will be to select a suitable subset of antennas from the set of the all antennas and then use only these selected antennas. In this way, the MIMO advantage is kept and the complexity is reduced.

It is even more important to select an antenna selection (AS) within the emerging massive MIMO and 5G technology [46–49]. Therefore, it is preferable to have a larger antenna number than the RF chains number in each link end and out of this antenna set to select the antenna subset to use. However, the reduction of the antenna number will lead evidently to performance deterioration. Thus, it is very important to properly select an algorithm that will obtain good results at the AS problem. This question translates to which antenna subset should be chosen from each link end in order to provide the maximum channel capacity. In other words, the principle of AS is to choose the best L_R out of K_R antennas located at the receiver side and the best L_T out of K_T antennas located at the transmitter side. Thus, the AS problem is a combinatorial optimization problem, where all the problem unknown variables are in binary space.

An obvious way to solve this problem would be to use an exhaustive search (Ex.Search) approach. This means to search over all possible combinations of antennas in each side. However, the drawback in this case would be the high computational complexity. Consequently, we can counterbalance this high computational complexity by finding suboptimal algorithms with performance as close as possible to the one provided by the ES.

Several approaches have appeared in the literature that address the AS problem [6,50–56]. Several papers exist in the literature aiming to fulfill the AS criterion that propose different suboptimal algorithms, for channel capacity maximization. A separable transmit–receive successive selection as well as a successive joint transmit–receive selection were proposed and compared in terms of spectral efficiency and outage probability. Later, a computationally efficient joint transmit–receive AS algorithm was presented in [6], having improved performance compared to the previous decoupling methods.

Apart from decoupling-based algorithms mentioned in the previous paragraph, EAs were also applied in the same problem. These approaches include genetic algorithms (GAs) [57–60], and binary particle swarm optimization (PSO) [61–63].

However, in most of these approaches the algorithms work in real space and map the real values to binary space. In [64], the authors apply algorithms that are better suited to work in binary spaces and combinatorial optimization problems like BBO [65]. They apply the original BBO algorithm as well nonlinear migration models from [66]. The results in [64] outperform other popular algorithms.

In this section, we will try different BBO variants as well as other emerging algorithms. More specifically, we will apply the chaotic BBO (CBBO) [67], the emerging algorithms monarch butterfly optimization (MBO) [68], moth search algorithm (MSA) [69], and the popular PSO and ACO.

6.2.1 MIMO system model

We assume a MIMO system that uses spatial multiplexing in a frequency nonselective fading wireless channel. This MIMO system has K_T antennas for transmission and K_R antennas for reception. Then, the complex $K_R \times 1$ signal vector $\mathbf{y}$ can be expressed with the following form:

$$\mathbf{y} = \sqrt{\frac{E_x}{N_T}}\mathbf{H}\mathbf{x} + \mathbf{z} \tag{6.9}$$

where $\mathbf{x} = [x_1, x_2, \ldots, x_{K_T}]^T$ is the complex $K_T \times 1$ transmitted symbol vector, E_x is the constant signal energy of each transmitted signal x_i, which is transmitted from the *i*th transmit antenna. Moreover, **H** is the $K_R \times K_T$ channel matrix, where the element h_{km} stands for the link gain from between the *k*th transmit antenna to the *m*th receive antenna. This link gain is assumed to be a zero-mean and unit variance (i.i.d. (independent and identically distributed)) complex Gaussian random variable. Additionally, $\mathbf{z} = [z_1, z_2, \ldots, z_{K_R}]^T$ represents the $K_R \times 1$ white complex Gaussian noise vector with zero-mean and covariance matrix $N_0 I_{K_R}$, where I_{K_R} is a $K_R \times K_R$ identity matrix.

We assume that (1) the channel state information is available at the receive side only and (2) the transmitted complex vector **x** is statistically independent, i.e., $E\{xx^H\} = I_{K_T}$, where $E\{\}$is the expected value operator. Therefore, the capacity of the deterministic MIMO channel can be represented by [43,45]

$$C = \log_2 \det\left(\mathbf{I}_{\mathbf{N_T}} + \frac{E_x}{K_T N_0}\mathbf{H^H H}\right) \tag{6.10}$$

where det () and $()^{\mathbf{H}}$ are the determinant and the Hermitian operator, respectively. Equation (6.10) is applicable when $K_R \geq K_T$ [44].

A typical topology of a MIMO system is depicted in Figure 6.12. If there are L_T and L_R RF chains at the transmit and receive sides, respectively, then it is $L_T \leq K_T$ and $L_R \leq K_R$. Therefore, the primary objective in the AS problem is to select the optimal antenna subset. In other words, it is to choose L_T out of K_T antennas at the transmitter side and choose L_R out of K_R antennas at the receiver side, in order to maximize the capacity expressed by [45]

$$\overline{C} = \log_2 \det\left(\mathbf{I}_{L_\mathbf{T}} + \frac{E_x}{L_T N_0}\overline{\mathbf{H}}^{\mathbf{H}}\overline{\mathbf{H}}\right) \tag{6.11}$$

where $\overline{\mathbf{H}}$ denotes an $L_R \times L_T$ subblock matrix of **H**. Taking under consideration that for a MIMO wireless system with $K_T \times K_R$ antennas there will be K_{select} different possibilities of selection of (L_R, L_T) antennas in operation given by

$$\mathrm{K}_{select} = C_{K_T, L_T} \times C_{K_R, L_R} = \binom{K_T}{L_T} \times \binom{K_R}{L_R} \tag{6.12}$$

The AS optimization problem is therefore defined

$$max\overline{C}(\mathbf{b}) = \log_2 \det\left(\mathbf{I}_{L_\mathbf{T}} + \frac{E_x}{L_T N_0}\overline{\mathbf{H}}^{\mathbf{H}}(\mathbf{b})\,\overline{\mathbf{H}(\mathbf{b})}\right) \tag{6.13}$$

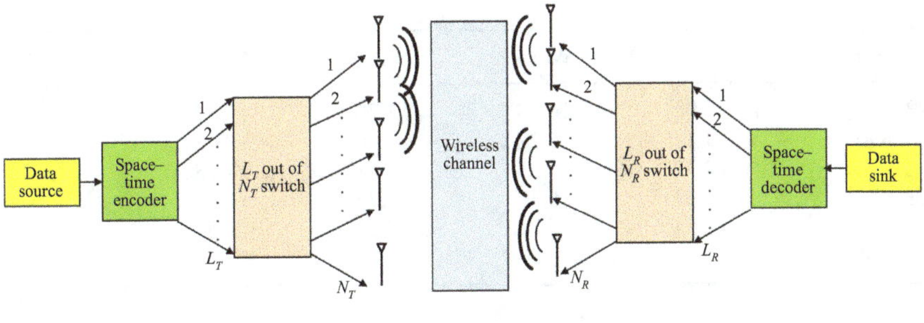

Figure 6.12 A typical MIMO system model that includes antenna selection

where $\mathbf{b} = [b_1, b_2, \ldots, b_{K_R}, b_{N_{R+1}}, \ldots, b_{K_R+K_T}]$ is a $K_T + K_R$ binary vector. If $b_i = 1$, then the antenna in position i is selected to transmit or receive depending on the position, otherwise the antenna does not operate. In the case of massive MIMO systems in 5G networks, the usage of the ES method is computationally intensive and therefore not practical.

6.2.2 CBBO algorithm selection

In this section, we evaluate the application of different CBBO variants to the AS problem. More precisely, we apply the original BBO, the CBBO0 (chaotic selection operator), CBBO1 (chaotic migration operator), CBBO2 (chaotic mutation operator), CBBO3 (selection/migration operators combined), and CBBO4 (selection/migration/mutation operators combined). In all BBO variants, we set the control parameters to $P_{\text{mod}} = 1$, $m_{\max} = 0.005$, and $I = 1, E = 1$. We assume a 16×16 MIMO system. We run Monte-Carlo simulations using 10,000 channel realizations in total. This means that in every simulation, a new random channel is generated. In order to evaluate the algorithms' performance, we calculate the mean/ergodic capacity obtained by each algorithm.

Table 6.6 shows the algorithms' comparative results in terms of ergodic capacity with $SNR = 20$ dB for $(L_T, L_R) = (2, 4)$. We notice that the best value of 19.12 bits/s/Hz is obtained by CBBO0, while CBBO2 and CBBO4 obtain the second best value of 19.02 bits/s/Hz. However, the best mean value is obtained by CBBO2. The comparison of the mean ergodic capacity value among the algorithms for different signal-to-noise ratio (SNR) values is presented in Table 6.7. We notice that CBBO2 clearly outperforms the other algorithms in four out of the five cases. This fact is also evident from the boxplot of the results in Figure 6.13. We notice that algorithms present similar width of interquartile range, thus a similar dispersion of values. However, it appears also that CBBO2 obtained the larger median value. The convergence rate graph for the $SNR = 20$ dB case is presented in Figure 6.14. All algorithms converge at similar speed, so no obvious advantage in terms of converge rate is found. However, CBBO2 converges at the higher ergodic capacity value.

Table 6.6 Comparative results of different CBBO algorithms of ergodic capacity with SNR = 20 dB values for $(L_T, L_R) = (2, 4)$

Algorithm	Best	Worst	Mean	Median	St. Dev.
BBO	18.96	**16.58**	17.70	**17.70**	0.44
CBBO0	**19.12**	16.22	17.66	17.68	0.46
CBBO1	18.77	16.23	17.58	17.58	0.46
CBBO2	19.02	16.39	**17.72**	17.69	**0.44**
CBBO3	18.98	16.11	17.52	17.53	0.45
CBBO4	19.02	16.04	17.53	17.55	0.44

Higher values are indicated in bold font.

Table 6.7 Comparative results of different CBBO algorithms of ergodic capacity values with different SNR values for $(L_T, L_R) = (2, 4)$

SNR (dB)	BBO	CBBO0	CBBO1	CBBO2	CBBO3	CBBO4
0	4.99	5.00	4.90	**5.02**	4.87	4.85
5	7.93	7.90	7.82	**7.95**	7.77	7.78
10	11.11	11.09	10.97	**11.12**	10.96	10.93
15	**14.39**	14.37	14.24	14.38	14.19	14.19
20	17.70	17.66	17.58	**17.72**	17.52	17.53

Higher values are indicated in bold font.

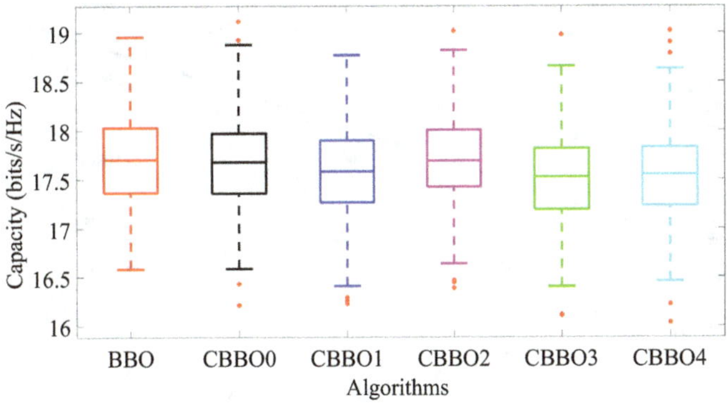

Figure 6.13 Ergodic capacity with SNR = 20 dB values for $(L_T, L_R) = (2, 4)$*. Boxplot of all algorithms results*

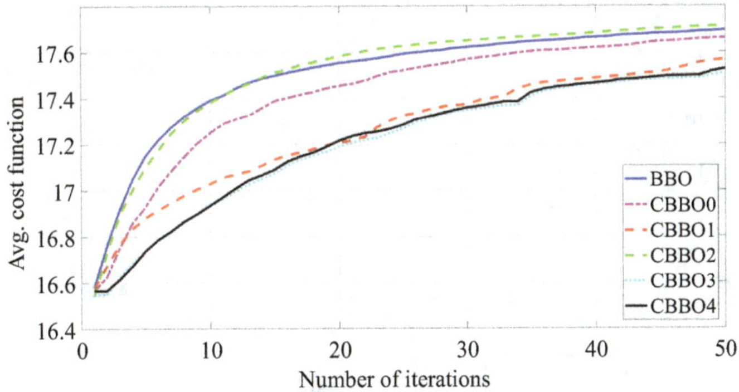

Figure 6.14 Ergodic capacity with SNR = 20 dB values for $(L_T, L_R) = (2, 4)$*. Convergence rate graph*

The ergodic capacity for different *SNR* values is depicted in Figure 6.15 as well the optimal ergodic capacity obtained by ES. Figure 6.16 shows a detailed view of the previous figure. It appears that CBBO2 derives the best result. Figures 6.17–6.21 show the CCDF of ergodic capacity with different *SNR* values for $(L_T, L_R) = (2, 4)$. For comparison purposes, the CCDF of the optimal result found by ES is also shown in figures. We notice that CBBO2 emerges as the best among the other algorithms. CBBO0 obtains the second best result.

Overall, based on the previous results, we select CBBO2 among all the CBBO variants to compare with other algorithms.

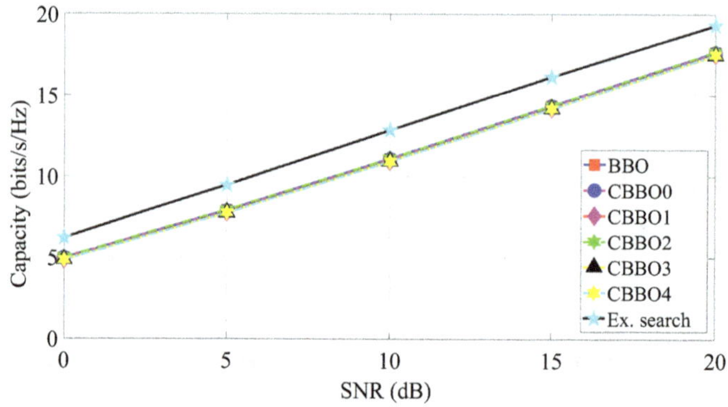

Figure 6.15 Ergodic capacity for different SNR values for $(L_T, L_R) = (2, 4)$ with CBBO algorithms (total view)

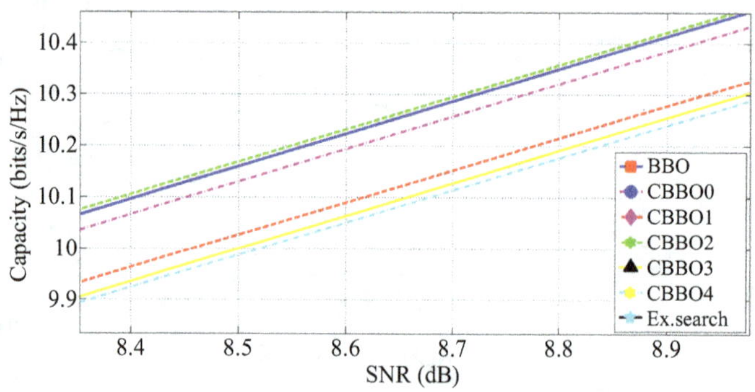

Figure 6.16 Ergodic capacity for different SNR values for $(L_T, L_R) = (2, 4)$ with CBBO algorithms (detailed view)

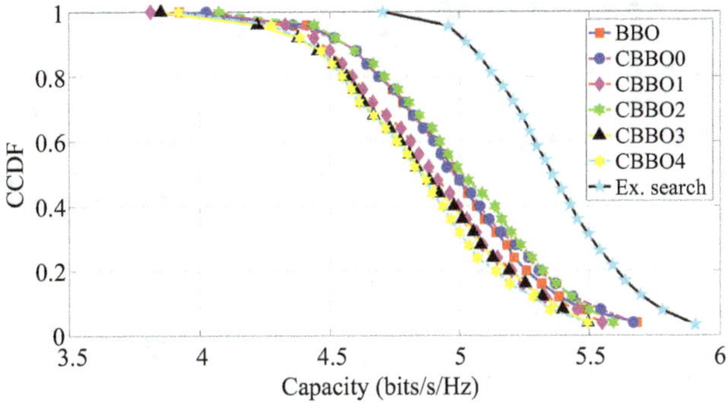

Figure 6.17 *CCDF of ergodic capacity with SNR = 0 dB values for $(L_T, L_R) = (2, 4)$*

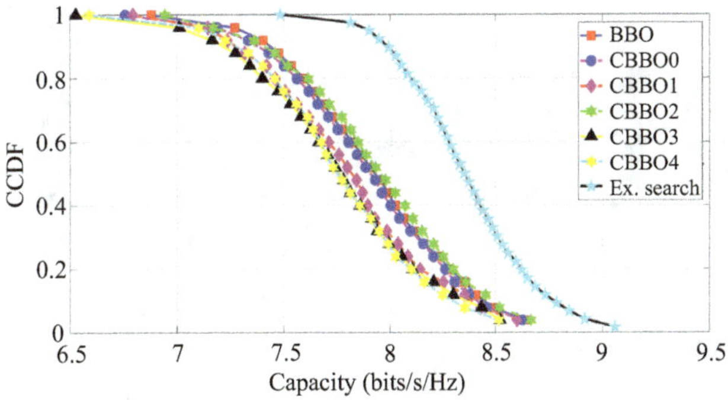

Figure 6.18 *CCDF of ergodic capacity with SNR = 5 dB values for $(L_T, L_R) = (2, 4)$*

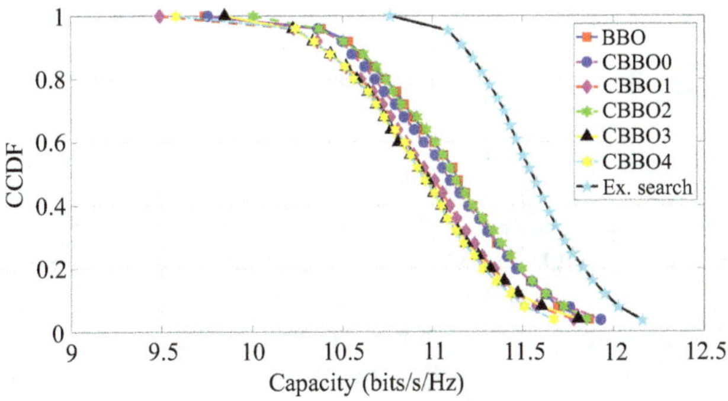

Figure 6.19 *CCDF of ergodic capacity with SNR = 10 dB values for $(L_T, L_R) = (2, 4)$*

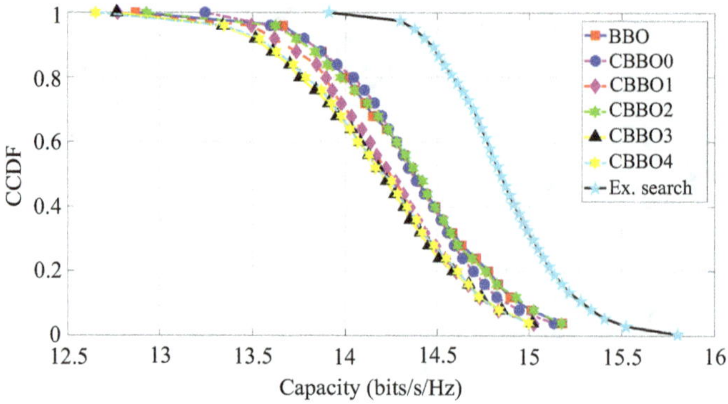

Figure 6.20 CCDF of ergodic capacity with SNR = 15 dB values for $(L_T, L_R) = (2, 4)$

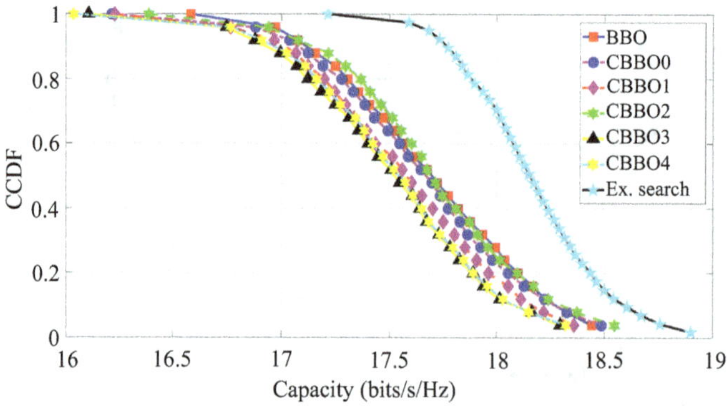

Figure 6.21 CCDF of ergodic capacity with SNR = 20 dB values for $(L_T, L_R) = (2, 4)$

6.2.3 Simulation results

This section presents the evaluation of all the algorithms. We apply the original BBO, CBBO2, MBO, PSO, ACO, and MSA to the channel capacity maximization problem. The objective–function we use is the one in (6.13). In BBO algorithms, we select the following values for the control parameters, $P_{\text{mod}} = 1$, $m_{\text{max}} = 0.005$, and $I = 1, E = 1$. For PSO, we set c_1 and c_2 to 2.05. Additionally, the inertia weight is linearly decreased starting from 0.9 to 0.4. For ACO, the initial pheromone value τ_0 is set to 1.0e−6, the pheromone update constant Q is set to 20, the exploration constant q_0 is set to 1, the global pheromone decay rate ρ_g is 0.9, the local pheromone decay rate ρ_l is 0.5, the pheromone sensitivity α is 1, and the visibility sensitivity β is 5. For MBO, the ratio $r_{pop} = 5/12$, and the migration period is set to 1.2, as it is suggested by the authors in [68]. In MSA, we set the control parameters as in [69] $\beta = 1.5, \phi = 0.618, S_{\max} = 1$. In all algorithms, we set the elitism parameter to two. Moreover, for all the algorithms, we set the population size $NP = 50$, and the maximum number of generations $G = 50$. Thus, a total number of 2,500 objective–function evaluations in each algorithm run.

We consider three AS scenarios, $(L_T, L_R) = \{(2,\ 4),\ (3,\ 5),\ (4,\ 6)\}$ for five different *SNR* values from 0 dB to 20 dB with a step of 5 dB. We calculate the mean/ergodic capacity and the outage capacity by running 10,000 simulations (different random channels) for each case.

Table 6.8 reports the comparative results. We notice that MSA performs better than the other algorithms in 10 out of the 15 cases. MBO outperforms the other algorithms in 3 out of the 15 cases. CBBO2 obtains the best result in two cases.

In all figures that follow, we plot the ES optimal values as well. Figures 6.22 and 6.23 show the ergodic capacity for $(L_T, L_R) = (2, 4)$. We can notice from the detailed view that MSA and MBO lines are almost identical, and MSA is slightly better. The next lines are the BBO and CBBO2, while PSO and ACO obtain the worst result. The ergodic capacity for $(L_T, L_R) = (3, 5)$ is depicted in Figures 6.24 and 6.25. We notice that for this case, the results from MSA, MBO, BBO, and CBBO2 are very

Table 6.8 Comparative results of different algorithms of ergodic capacity values with different SNR

SNR (dB)	CBBO2	BBO	MBO	MSA	PSO	ACO
$(L_T, L_R) = (2, 4)$						
0	5.02	4.99	5.14	**5.19**	4.82	4.73
5	7.95	7.93	8.12	**8.14**	7.66	7.64
10	11.12	11.11	11.31	**11.34**	10.81	10.79
15	14.38	14.39	14.54	**14.62**	14.09	14.07
20	17.72	17.70	**17.92**	17.89	17.40	17.39
$(L_T, L_R) = (3, 5)$						
0	6.22	6.25	6.24	**6.30**	5.85	5.73
5	10.34	10.38	10.43	**10.46**	9.89	9.74
10	14.98	14.98	**15.10**	**15.10**	14.45	14.35
15	19.89	19.84	**19.99**	19.95	19.33	19.19
20	24.86	24.77	24.90	**24.96**	24.26	24.18
$(L_T, L_R) = (4, 6)$						
0	7.24	7.26	7.20	**7.26**	6.72	6.57
5	12.44	**12.50**	12.36	12.44	11.74	11.51
10	**18.51**	18.50	18.46	18.46	17.74	17.50
15	**24.99**	24.87	24.89	24.96	24.22	23.92
20	31.52	31.48	31.45	**31.55**	30.75	30.50

Bold font shows the larger values.

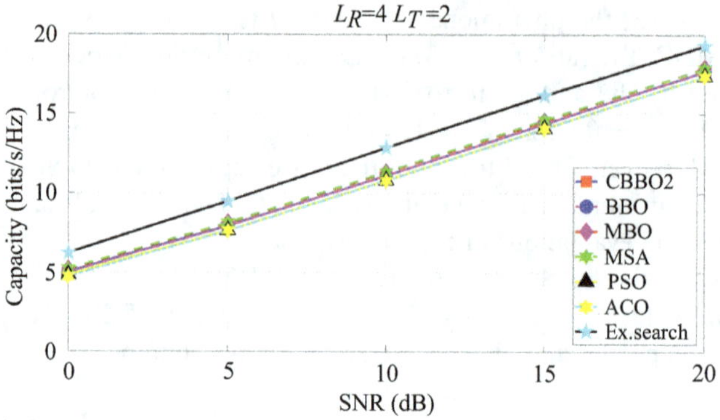

Figure 6.22 Ergodic capacity for different SNR values for $(L_T, L_R) = (2, 4)$ with different algorithms (total view)

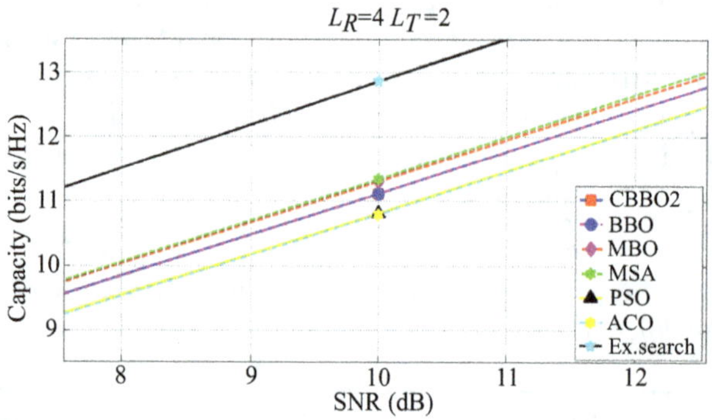

Figure 6.23 Ergodic capacity for different SNR values for $(L_T, L_R) = (2, 4)$ with different algorithms (detailed view)

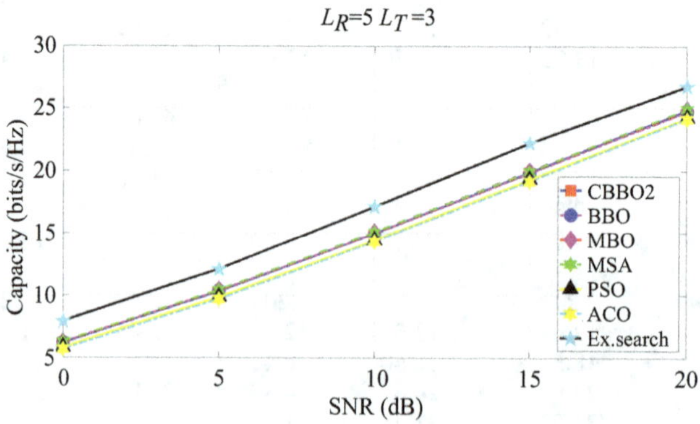

Figure 6.24 Ergodic capacity for different SNR values for $(L_T, L_R) = (3, 5)$ with different algorithms (total view)

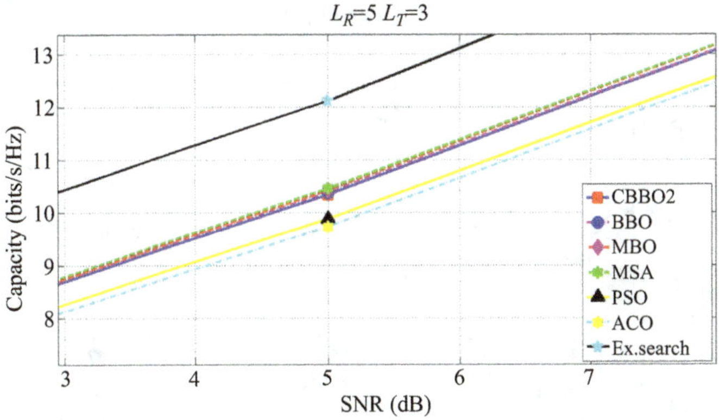

Figure 6.25 Ergodic capacity for different SNR values for $(L_T, L_R) = (3, 5)$ *with different algorithms (detailed view)*

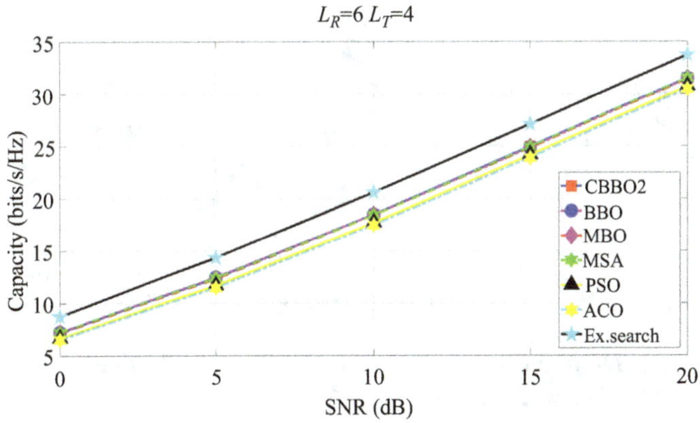

Figure 6.26 Ergodic capacity for different SNR values for $(L_T, L_R) = (4, 6)$ *with different algorithms (total view)*

close. Figures 6.26 and 6.27 present the ergodic capacity for the $(L_T, L_R) = (4, 6)$ case. Similarly with the previous case, the lines obtained from MSA, MBO, BBO, and CBBO2 are very close.

Moreover, we obtain the CCDF plots of the ergodic capacity for all cases. Figure 6.28 illustrates the CCDF for the $(L_T, L_R) = (2, 4)$ case. We can notice that MSA and MBO perform clearly better than the other algorithms. MBO is better for low CCDF values. BBO and CBBO2 are the next algorithms that perform well. CBBO2 seems to perform slightly better than the original BBO. Both PSO and ACO perform worse than the other algorithms. The CCDF plots of the ergodic capacity for $(L_T, L_R) = (3, 5)$ is shown in Figure 6.29. For this case, it appears that MSA is

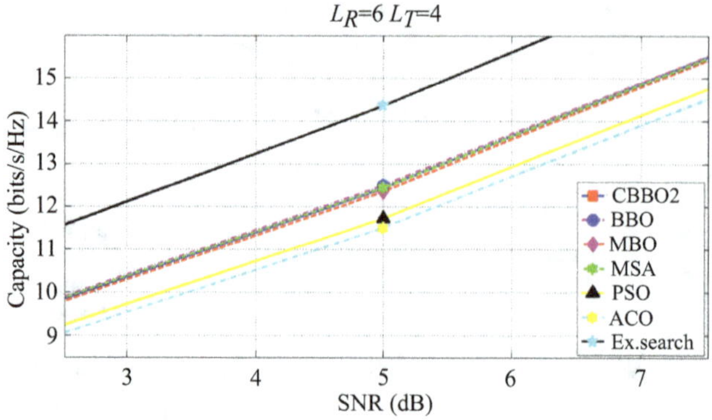

Figure 6.27 *Ergodic capacity for different SNR values for* $(L_T, L_R) = (4, 6)$ *with different algorithms (detailed view)*

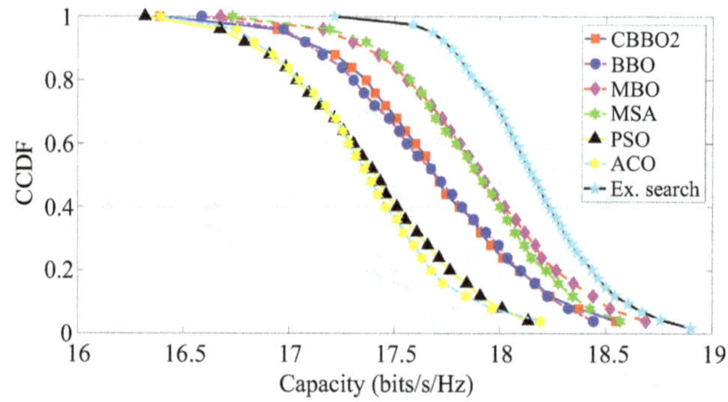

Figure 6.28 *CCDF of ergodic capacity with SNR = 20 dB values for* $(L_T, L_R) = (2, 4)$

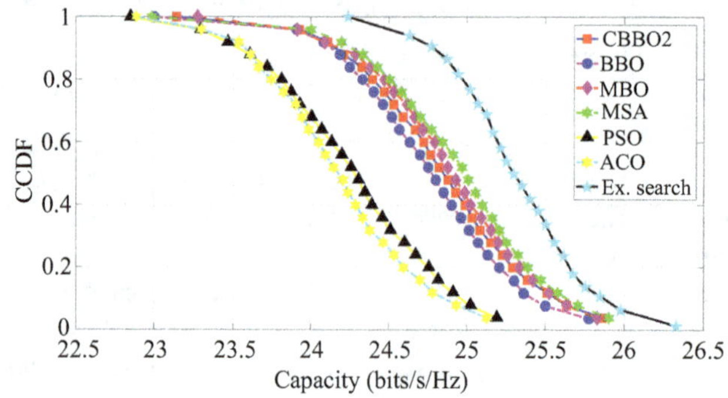

Figure 6.29 *CCDF of ergodic capacity with SNR = 20 dB values for* $(L_T, L_R) = (3, 5)$

the algorithm with the best performance. MBO and CBBO2 are the second and third algorithms, respectively. Additionally, Figure 6.30 depicts the CCDF plots of the ergodic capacity for $(L_T, L_R) = (4, 6)$. It is evident that MSA outperforms the other algorithms. MBO, CBBO2, and the original BBO perform almost identically. PSO clearly outperforms ACO in this case.

Next, we obtain the 1% outage capacity versus *SNR* plots for all three cases. Figures 6.31 and 6.32 show the 1% outage capacity for the $(L_T, L_R) = (2, 4)$ case. One may observe from the detailed view that MSA clearly outperforms the other algorithms. MBO is second, while CBBO2 and BBO perform almost identically. The 1% outage capacity for the $(L_T, L_R) = (3, 5)$ is plotted in Figures 6.33 and 6.34.

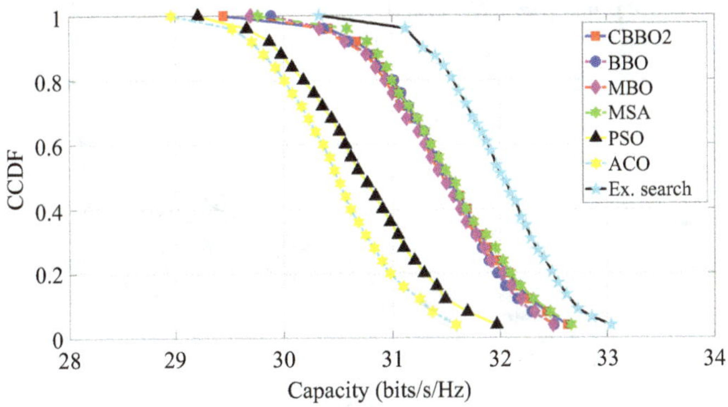

Figure 6.30 CCDF of ergodic capacity with SNR = 20 dB values for $(L_T, L_R) = (4, 6)$

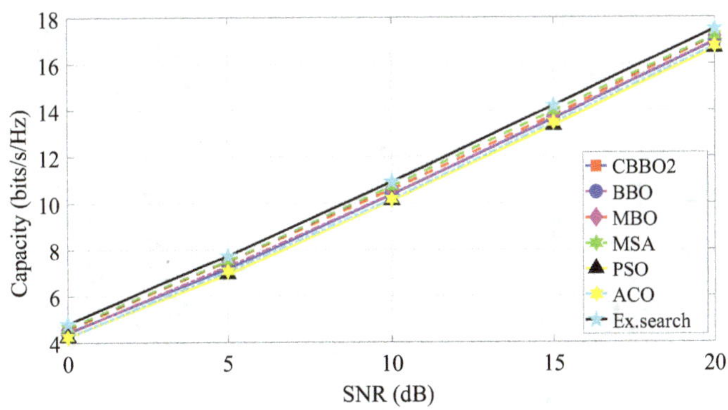

Figure 6.31 1% Outage capacity versus SNR for $(L_T, L_R) = (2, 4)$ with different algorithms (total view)

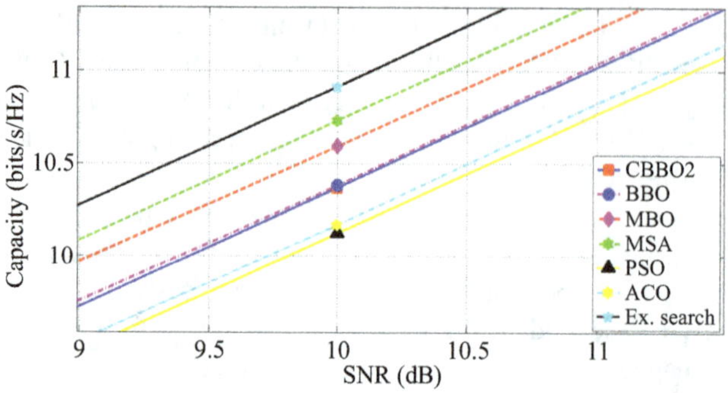

Figure 6.32 *1% Outage capacity versus SNR for* $(L_T, L_R) = (2, 4)$ *with different algorithms (detailed view)*

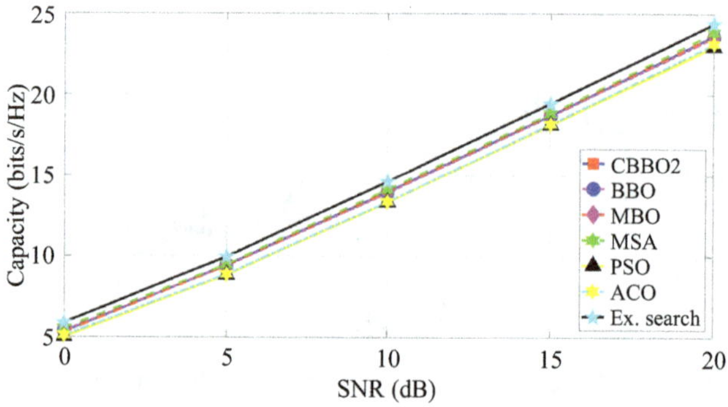

Figure 6.33 *1% Outage capacity versus SNR for* $(L_T, L_R) = (3, 5)$ *with different algorithms (total view)*

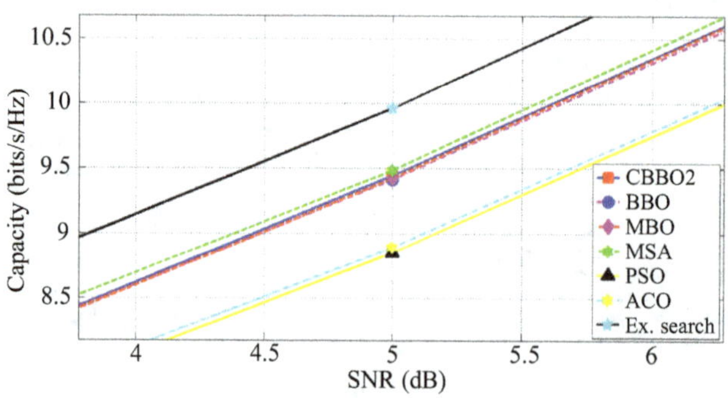

Figure 6.34 *1% Outage capacity versus SNR for* $(L_T, L_R) = (3, 5)$ *with different algorithms (detailed view)*

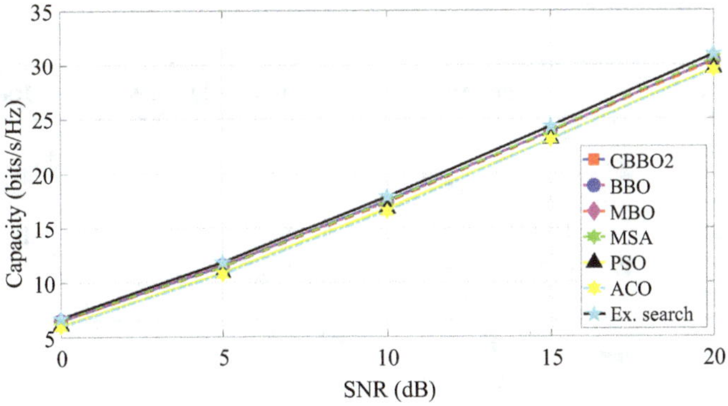

Figure 6.35 1% Outage capacity versus SNR for $(L_T, L_R) = (4, 6)$ with different algorithms (total view)

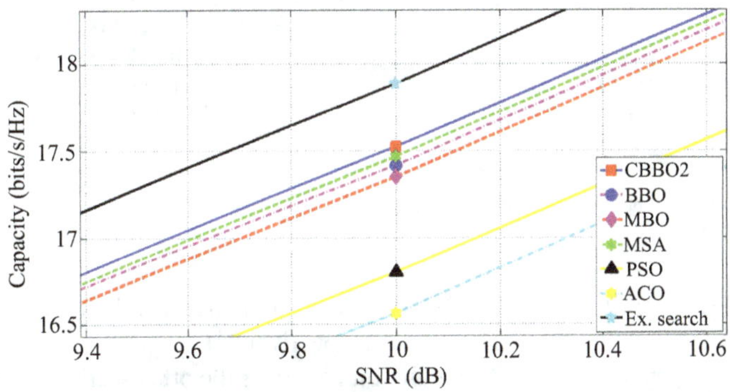

Figure 6.36 1% Outage capacity versus SNR for $(L_T, L_R) = (4, 6)$ with different algorithms (detailed view)

We notice that MSA obtains the best performance very close to MBO, CBBO2, and BBO. Finally, in Figures 6.35 and 6.36 we plot the 1% outage capacity for the $(L_T, L_R) = (4, 6)$ case. For this case, CBBO2 performs better than the other algorithms, and MSA is the second best in performance.

6.2.3.1 Nonparametric statistical tests

Moreover, in order to compare the algorithms' performance, we have conducted nonparametric statistical tests. These are the Friedman test and the Wilcoxon signed-rank test, which have been used as a metric for the performance evaluation of EAs [70–72]. We perform these tests using the data from the results reported in Table 6.8. Table 6.9 reports the ranking results according to Friedman test. One may observe

Table 6.9 Average rankings achieved by Friedman test

Method	**Average rank**	**Normalized values**	**Rank**
CBBO2	2.87	1.87	3
BBO	3.13	2.04	4
MBO	2.47	1.61	2
MSA	**1.53**	**1.00**	**1**
PSO	5.00	3.26	5
ACO	6.00	3.91	6

Table 6.10 Wilcoxon signed-rank test between MSA and the other algorithms

MSA versus	***p*-Values**
CBBO2	**0.012085**
BBO	**8.54E−04**
MBO	**1.15E−02**
PSO	**6.10E−05**
ACO	**6.10E−05**

The bold font indicates values below significant level.

that the MSA algorithm ranked first and clearly outperforms the other algorithms. MSA and CBBO2 rank second and third, respectively.

In order to examine if MSA actually outperforms the other methods, we perform the Wilcoxon signed-rank test, the significance of which is set to 0.05. Table 6.10 holds all the *p*-values found for the case of MSA versus the other algorithms. We observe that for all algorithms, the *p*-values are below the significance level (0.05). Thus, we can accept that a significant difference exists between MSA and the other algorithms. MSA proves to be significantly better than all the other methods.

6.3 Cognitive radio engine design

Cognitive radio (CR) systems have been proposed as an effective solution to the problem of the shortage of available spectrum [73]. According to measurements, the real problem is based on the fact that the current licensed spectrum is not fully utilized. Mitola suggested that CR systems have the ability to monitor and detect the conditions of their operating environment, and reconfigure their transmission parameters in order to provide the best service that satisfies the user's demands [74,75].

The CR system gets the information about its operating environment by the environmental parameters that also determine the effectiveness and the accuracy of the decisions made by the system. Moreover, one may notice that in order to determine the possible configurations of the system, we must define the performance objectives that will be satisfied by it. Although the primary objective of dynamic spectrum access is to improve spectrum utilization efficiency, there are additional objectives that should be satisfied like the bit-error-rate (BER) minimization, the throughput maximization, the power consumption minimization, and the interference minimization.

It has been proven several times in the literature that EAs can be applied to the problem of reconfiguring the transmission parameters according to certain performance objectives. The optimization of CR systems has been performed in the literature using GAs [76–80], PSO [81,82], or ACO [83]. Additionally, multi-objective EAs have also been used to the CR problem [84,85]. In [86], the authors apply BBO, while in [87] a real-coded BBO algorithm is utilized. The authors in [88] present a survey of all these techniques.

6.3.1 Problem formulation

The basic concept of a cognitive engine is real-time tuning of transmission parameters according to the environment information. Thus, at first the environmental and the transmission parameters need to be correctly set. The environmental parameters represent the inputs of the cognitive engine, so that the engine can decide about the transmission parameters (Figure 6.37). Thus, we can express the optimization problem as the search for the best possible transmission parameters that are subject to the known environmental parameters. In this section, we use the same environmental parameters as in [77,87]. These are the SNR, BER, and background noise power (N). The transmission parameters are the decision variables of the CR engine that need to be found. Table 6.11 holds the transmission parameters [86]. One may observe in Table 6.11 that the modulation type used in our system is binary PSK (BPSK), for

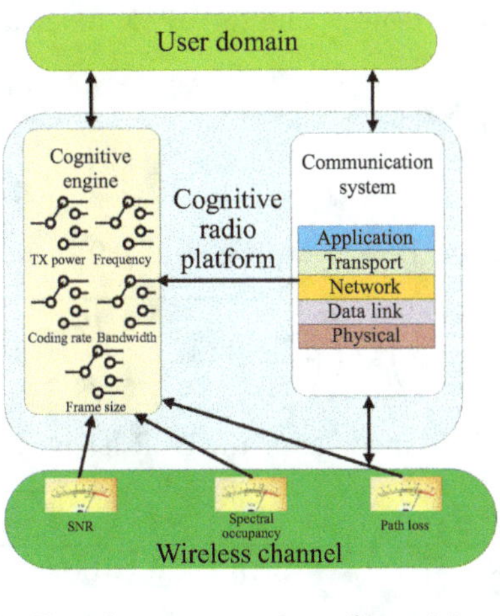

Figure 6.37 Visual representation of cognitive radio engine

Table 6.11 Transmission parameters

Parameter name	Description	Value	
		Min	**Max**
Transmission power	Amount of transmitted power	0.158 mW	2.51 mW
Modulation index	Number of symbols for the current modulation	2	256
Bandwidth	Bandwidth of the transmitted signal in Hz	2 MHz	32 MHz
TDD	Percentage of transmit time	25%	100%
Symbol rate	Number of symbols per second	125 ksps	1 Msps

modulation index = 2, quadrature PSK, for *modulation index* = 4, and QAM, for *modulation index* > 4.

We use the same five objective–functions as in [86,87]. We calculate the total fitness from the sum of the five objective–functions using suitable weight values. Each one of the single objective fitness functions for a CR system with L-independent subcarriers is defined next.

We define the power consumption minimization by

$$F_{\text{min-}power}(\bar{P}) = \frac{\sum_{i=1}^{L} P_i}{L \times P_{\max}} \tag{6.14}$$

where P_i is the transmitted power on the ith subcarrier, $P_{\max}$ is the maximum available transmit power, and $\bar{P} = (P_1, P_2, \ldots, P_L)$ is the vector of the transmitted power. The objective–function of BER minimization is expressed as

$$F_{\text{min-}ber}(\bar{P}, \bar{M}) = \frac{\log_{10}(0.5)}{\log_{10}\left(\frac{\sum_{i=1}^{L} Pbe_i(M_i, P_i)}{L}\right)} \tag{6.15}$$

where M_i is the modulation index on the ith subcarrier, $\bar{M} = (M_1, M_2, \ldots, M_L)$ is the vector of the modulation indices, and for BPSK modulation the BER is defined as [89]

$$Pbe_i(P_i) = Q\left(\sqrt{\frac{P_i}{N_0}}\right) \tag{6.16}$$

Also for M-ary PSK, the BER is expressed as [89]

$$Pbe_i(P_i, M_i) = \left(\frac{2}{\log_2 M_i}\right) Q\left(\sqrt{2\log_2(M_i) \times \frac{P_i}{N_0} \times \sin\frac{\pi}{M_i}}\right) \tag{6.17}$$

Additionally, for M-ary QAM modulation the BER is given by [89]

$$Pbe_i(P_i, M_i) = \left(\frac{4}{\log_2 M_i}\right)\left(1 - \frac{1}{\sqrt{M_i}}\right) Q\left(\sqrt{\frac{3\log_2(M_i)\, P_i}{(M_i - 1)\, N_0}}\right) \tag{6.18}$$

We assume that $N_0 = -104$ dBm as set in previous chapters in order to compare results. The objective–function of throughput maximization is defined as [77]

$$F_{\text{max-}throughput}\left(\bar{M}\right) = 1 - \frac{\log_2\left(\frac{\sum_{i=1}^{L} M_i}{L}\right)}{\log_2\left(M_{\max}\right)} \tag{6.19}$$

where $M_{\max}$ is the maximum modulation index. The objective–function of interference minimization is given by [86]

$$F_{\text{min-}interference}\left(\bar{P}, \bar{B}, \bar{TDD}\right) = \frac{\sum_{i=1}^{L}\left[\left(P_i \times B_i \times TDD_i\right) - \left(P_{\min} \times B_{\min}\right)\right]}{L\left(P_{\max} \times B_{\max} \times 100\right)} \tag{6.20}$$

where B_i and TDD_i are the bandwidth and the time-division duplexing on the ith subcarrier, respectively, $\bar{B} = (B_1, B_2, \ldots, B_L)$, $\overline{TDD} = (TDD_1, TDD_2, \ldots, TDD_L)$ are the corresponding vectors of bandwidth and time-division duplexing, respectively, and $B_{\min}$ and $B_{\max}$ are the minimum and maximum bandwidths available. The objective–function of spectral efficiency maximization is expressed as [86]

$$F_{\text{max-}specteff}\left(\bar{M}, \bar{B}, \bar{R}_s\right) = 1 - \frac{\sum_{i=1}^{L} \frac{M_i R_{s,i} B_{\min}}{M_{\max} R_{s,\max} B_i}}{L} \tag{6.21}$$

where $R_{s,i}$ is the symbol rate on the ith subcarrier, $R_s = (R_{s,1}, R_{s,2}, \ldots, R_{s,L})$ is the corresponding vector of the symbol rate, and $R_{s,\max}$ is the maximum symbol rate.

If we apply a weighted sum approach in order to fulfill all five objectives, then the total objective function is defined by [77]

$$F_{total}(\bar{x}) = \sum_{i=1}^{k} w_i F_i(\bar{x}) \tag{6.22}$$

where k is the number of objective–functions, and $w_1, w_2, \ldots, w_k$, $0 \le w_i \le 1$ for $i = 1, 2, \ldots k$ are suitable weight values subject to

$$w_1 + w_2 + \cdots + w_k = 1 \tag{6.23}$$

Therefore, we can express the five objective fitness functions for a CR system with N-independent subcarriers [77] using

$$\begin{aligned} F_{multicarrier}\left(\bar{P}, \bar{M}, \bar{B}, \overline{TDD}, \bar{R}_s\right) = {} & w_1 F_{\text{min-}power}(\bar{P}) + w_2 F_{\text{min-}ber}(\bar{P}, \bar{M}) \\ & + w_3 F_{\text{max-}throughput}\left(\bar{M}\right) \\ & + w_4 F_{\text{min-}interference}\left(\bar{P}, \bar{B}, \overline{TDD}\right) \\ & + w_5 F_{\text{max-}specteff}\left(\bar{M}, \bar{B}, \bar{R}_s\right) \end{aligned} \tag{6.24}$$

The weights w_1, w_2, w_3, w_4, w_5 define the search direction of the EA and must conform to the limitations of (6.23). Table 6.12 lists five different weight vectors that represent

Table 6.12 Weighting values

Scenario	**Weight vector** [t] $[w_1, w_2, w_3, w_4, w_5]$
Minimize power consumption	$[0.65, 0.05, 0.15, 0.10, 0.05]$
Minimize bit-error-rate	$[0.05, 0.70, 0.05, 0.05, 0.15]$
Maximize throughput	$[0.05, 0.10, 0.70, 0.10, 0.05]$
Minimize interference	$[0.05, 0.05, 0.15, 0.70, 0.05]$
Maximize spectral efficiency	$[0.05, 0.15, 0.10, 0.05, 0.70]$
Balanced mode (equal weights)	$[0.2, 0.2, 0.2, 0.2, 0.2]$

different scenarios of the CR engine configuration. Each weight vector emphasizes on different performance objectives. This fact leads the EAs to a solution set that is related to a specific scenario. We have chosen the same weights as in [87], where these weight values were found on the basis of a trial and error procedure.

6.3.2 Numerical results

We will compare results for multiple subcarrier systems for four cases. The carrier numbers are 128, 256, 512, 1, 024, respectively. We apply the popular algorithms PSO, differential evolution (DE), evolutionary strategy (ES) [90], and the emerging nature-inspired algorithms the elephant herding optimization (EHO) [91], the MBO [68], and the MSA [69]. The problem dimension in each is $5 \times N_{car}$, where N_{car} is the number of subcarriers.

We set the following control parameters. For PSO, we set c_1 and c_2 to 2.05. Additionally, the inertia weight is linearly decreased starting from 0.9 to 0.4. For DE, we use a mutation control parameter $F = 0.5$, and a crossover constant is $CR = 0.5$. For the ES, we set as in [65] $\lambda = 10$ offspring each generation, and standard deviation $\sigma = 1$ for changing solutions. For the EHO, we used five clans and the parameters $\alpha = 0.5$ and $\beta = 0.1$. For MBO, the ratio $r_{pop} = 5/12$, and the migration period is set to 1.2, as it is suggested by the authors in [68]. In MSA, we set the control parameters as in [69] $\beta = 1.5, \phi = 0.618, S_{\max} = 1$. The population size is set to 100, and the maximum number of iterations is set to 200 for all algorithms. The elitism size is set to two for all algorithms.

Table 6.13 holds the algorithms' comparative results for different weight vectors for 128 carriers. We observe that DE and MSA perform better than the other algorithms for this case. The boxplot for the balanced mode case is plotted in Figure 6.38. We notice that MSA obtained the results with the smaller dispersion of values, while DE is second. The convergence plot for this case is depicted in Figure 6.39. MSA converges faster than the other algorithms.

The comparative results for different weight vectors for 256 carriers are reported in Table 6.14. MSA performs better in three cases, while MBO in two. DE performs better in one case. Figure 6.40 shows the boxplot for the balanced mode case.

Table 6.13 Comparative results of the total fitness values for different weight vectors for 128-carrier system

Scenario	EHO	PSO	DE	MBO	ES	MSA
Min power mode	0.1857	0.3765	0.1110	0.1030	0.3072	**0.0924**
Min BER mode	0.1991	0.2166	0.2063	0.2169	0.2172	**0.1885**
Max throughput mode	0.2350	0.2319	**0.1227**	0.1066	0.1720	0.1708
Min interference mode	0.1826	0.3572	0.0923	0.3859	0.2921	**0.0913**
Max spectral efficiency mode	0.7336	0.6812	**0.5128**	0.6995	0.5781	0.6428
Balanced mode	0.3803	0.4158	**0.2621**	0.3346	0.3660	0.2653

The smaller values are in bold font.

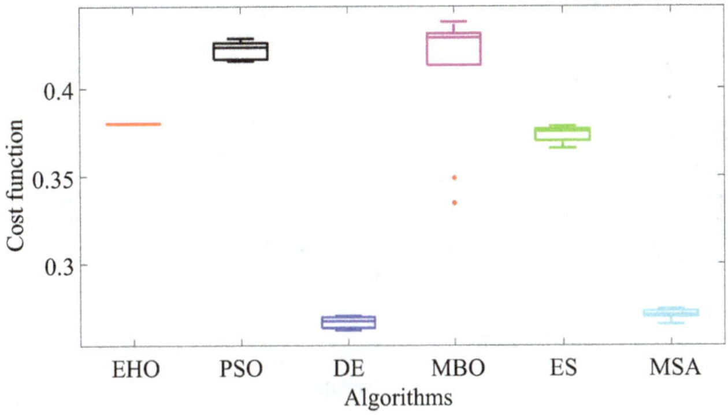

Figure 6.38 Balanced mode (equal weights) 128 subcarriers case. Boxplot of all algorithms results

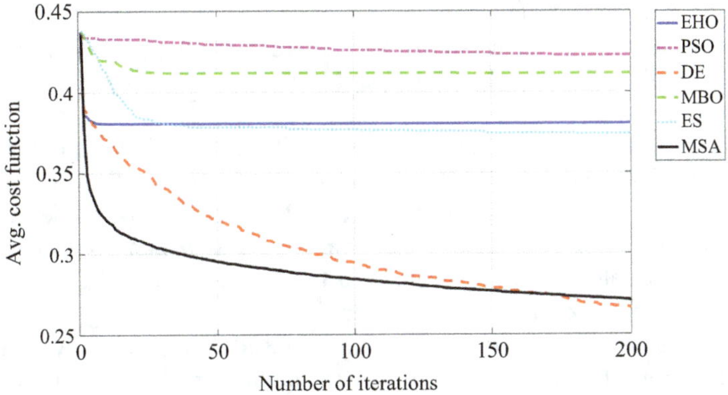

Figure 6.39 Balanced mode (equal weights) 128 subcarriers case. Convergence rate graph

Table 6.14 Comparative results of the total fitness values for different weight vectors for 256-carrier system

Scenario	EHO	PSO	DE	MBO	ES	MSA
Min power mode	0.1857	0.4018	0.1779	**0.0940**	0.3526	0.1007
Min BER mode	0.1992	0.2211	0.2296	0.2208	0.2274	**0.1948**
Max throughput mode	0.2445	0.2451	0.1373	**0.1235**	0.1883	0.2016
Min interference mode	0.1826	0.3813	0.1522	0.4055	0.3280	**0.1025**
Max spectral efficiency mode	0.7383	0.7080	**0.5683**	0.7087	0.6015	0.6727
Balanced mode	0.3803	0.4313	0.3061	0.3021	0.3798	**0.2802**

The smaller values are in bold font.

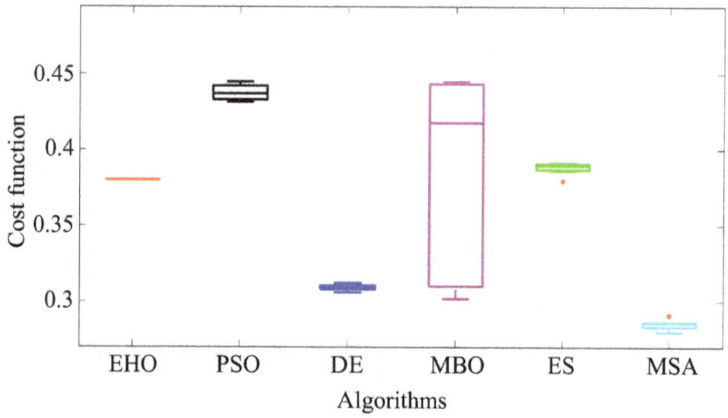

Figure 6.40 Balanced mode (equal weights) 256 subcarriers case. Boxplot of all algorithms results

We observe that most of the algorithms present a small dispersion of values, while MBO obtained the larger dispersion of values. Similar to previous case, the convergence rate graph in Figure 6.41 shows that MSA is the faster algorithm in terms of convergence.

Table 6.15 lists the comparative results for different weight vectors for 512 carriers. The results show that no algorithm clearly outperforms all, MSA and MBO are better in two cases. In boxplot of Figure 6.42 of the balanced mode case, MSA obtained the best result in terms of smaller values. Figure 6.43 plots the corresponding convergence rate graph, MSA obtains a fast convergence and DE continues to converge up to the maximum number of iterations.

Finally, Table 6.16 reports the comparative results for different weight vectors for 1,024 carriers. This is the problem with the higher dimension, in this case there are $5 \times 1{,}024 = 5{,}120$ decision variables. MSA is clearly the best algorithm in three out of the six cases. DE performs well in two cases. The boxplot for the balanced mode case is depicted in Figure 6.44. We notice that all algorithms obtained results

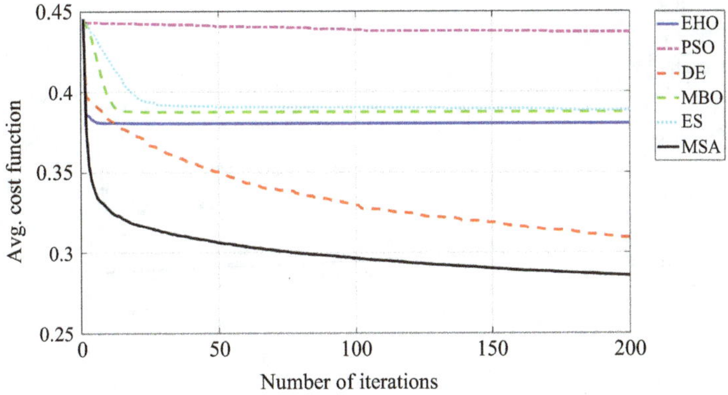

Figure 6.41 Balanced mode (equal weights) 256 subcarriers case. Convergence rate graph

Table 6.15 Comparative results of the total fitness values for different weight vectors for 512-carrier system

Scenario	**EHO**	**PSO**	**DE**	**MBO**	**ES**	**MSA**
Min power mode	0.1857	0.4210	0.2462	0.1187	0.3816	**0.1099**
Min BER mode	**0.1993**	0.2241	0.2388	0.2224	0.2396	0.2034
Max throughput mode	0.2505	0.2518	0.1456	**0.1176**	0.1939	0.2218
Min interference mode	0.1826	0.3971	0.2259	0.4231	0.3609	**0.1149**
Max spectral efficiency mode	0.7437	0.7197	**0.6030**	0.7164	0.6224	0.6970
Balanced mode	0.3804	0.4367	0.3361	**0.3007**	0.4010	0.2945

The smaller values are in bold font.

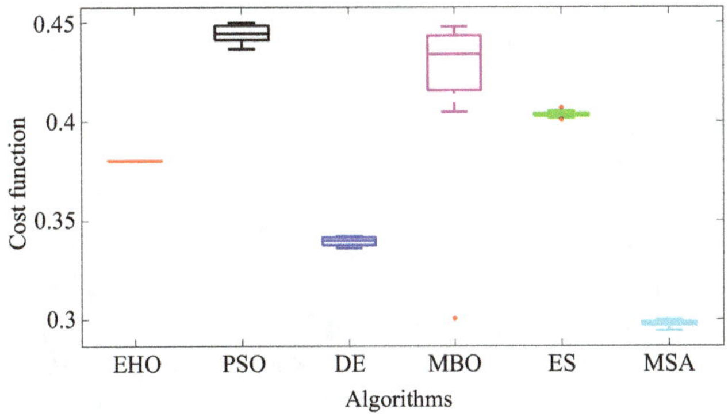

Figure 6.42 Balanced mode (equal weights) 512 subcarriers case. Boxplot of all algorithms results

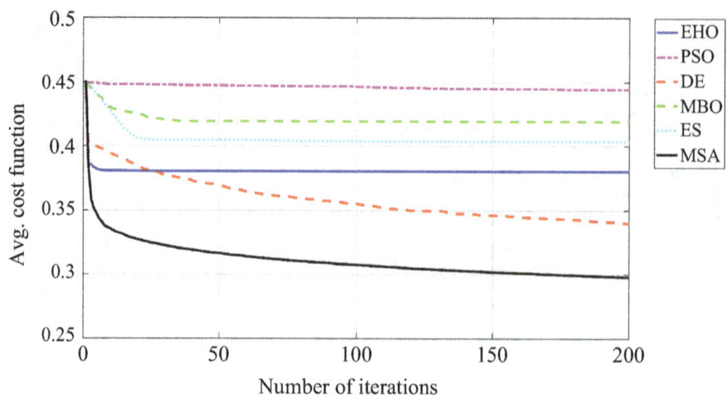

Figure 6.43 Balanced mode (equal weights) 512 subcarriers case. Convergence rate graph

Table 6.16 Comparative results of the total fitness values for different weight vectors for 1,024-carrier system

Scenario	EHO	PSO	DE	MBO	ES	MSA
Min power mode	0.1857	0.4328	0.3012	0.3240	0.3955	**0.1198**
Min BER mode	**0.1993**	0.2291	0.2515	0.2284	0.2487	0.2197
Max throughput mode	0.2624	0.2556	**0.1534**	0.1686	0.2034	0.2393
Min interference mode	0.1826	0.4121	0.2802	0.3879	0.3694	**0.1258**
Max spectral efficiency mode	0.7465	0.7227	**0.6307**	0.7150	0.6435	0.7162
Balanced mode	0.3804	0.4412	0.3583	0.4289	0.4049	**0.3027**

The smaller values are in bold font.

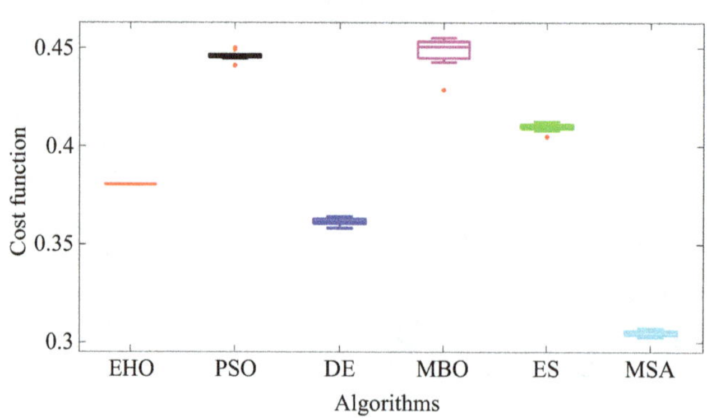

Figure 6.44 Balanced mode (equal weights) 1,024 subcarriers case. Boxplot of all algorithms results

with small dispersion of values. The convergence rate graph in Figure 6.45 reveals that MSA presents the faster convergence speed.

The previous results have obtained a number of solutions for each case. In order to select the best compromised solution, a fuzzy decision maker (described in Chapter 2) was applied. Figures 6.46–6.49 illustrate the decision variables in all subcarriers for each case of the best compromised solution. Table 6.17 shows the objective–function

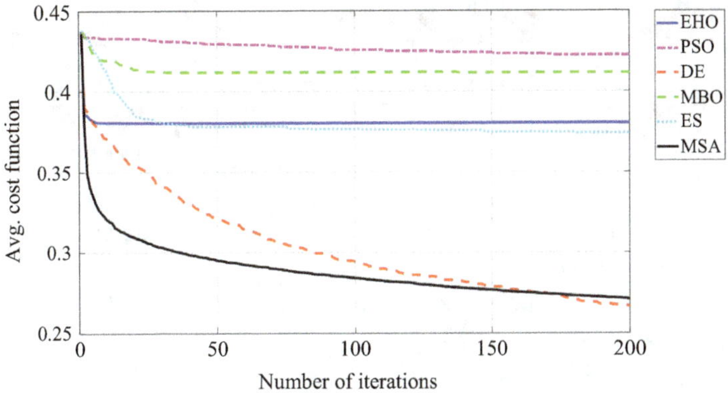

Figure 6.45 Balanced mode (equal weights) 128 subcarriers case. Convergence rate graph

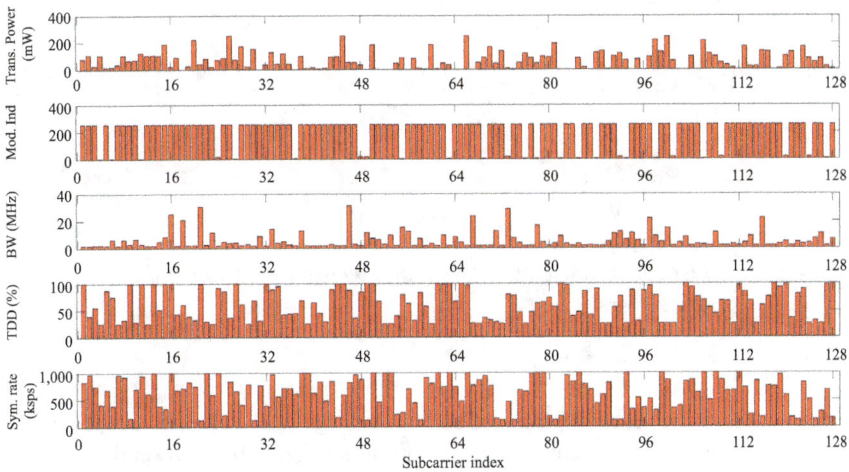

Figure 6.46 Best compromised solution found by fuzzy decision maker for 128 subcarriers

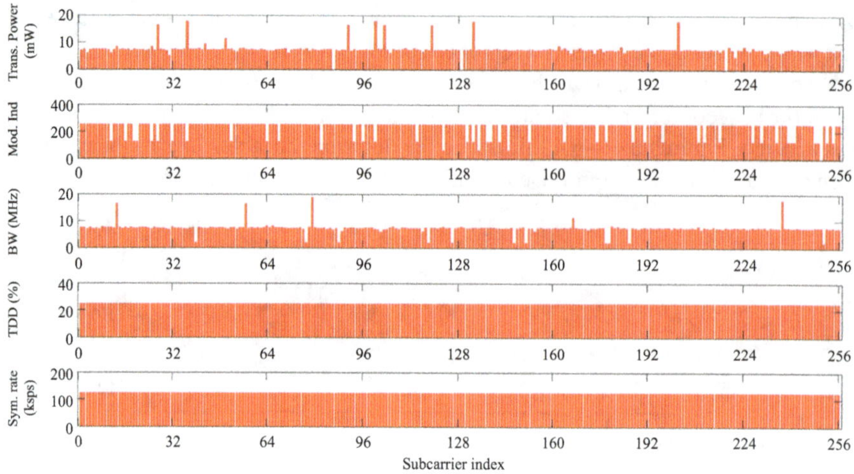

Figure 6.47 Best compromised solution found by fuzzy decision maker for 256 subcarriers

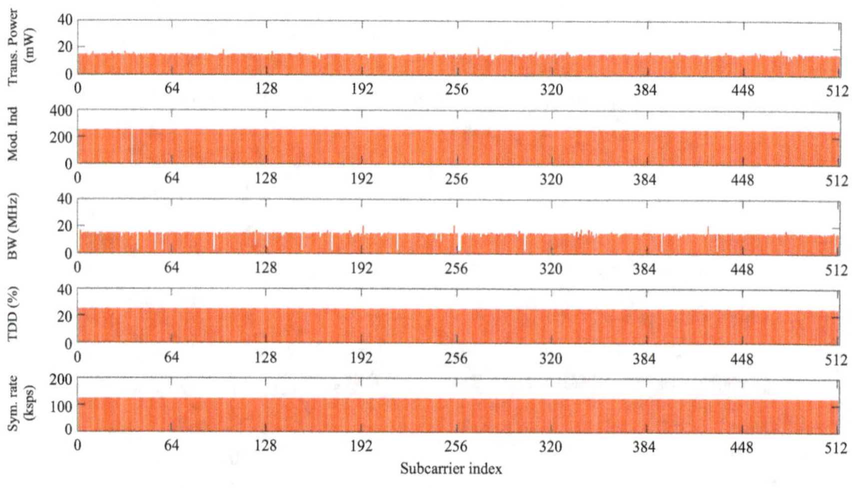

Figure 6.48 Best compromised solution found by fuzzy decision maker for 512 subcarriers

values in case of the best compromised solutions as well as the algorithm that obtained the result. We can observe that the three solutions are found by MBO and one by DE.

In order to evaluate the algorithms' performance, nonparametric statistical tests were conducted on the basis of the results of all cases. A first indication of the algorithm that outperforms the other is shown in Table 6.18. We notice that out of the 24

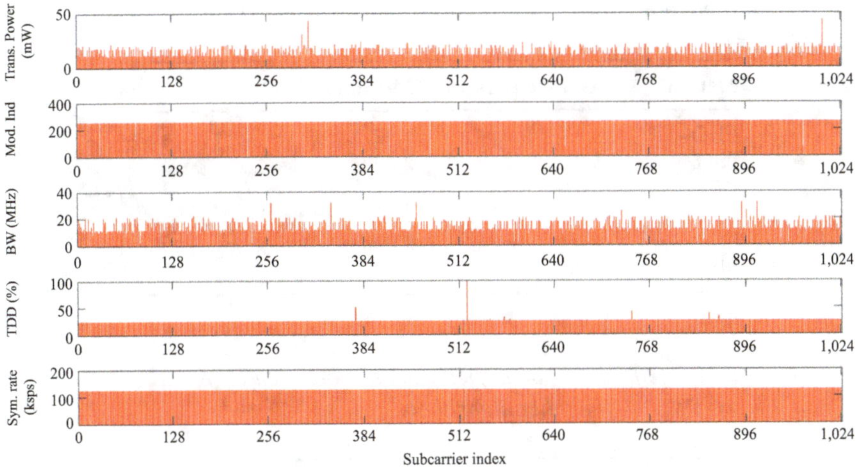

Figure 6.49 Best compromised solution found by fuzzy decision maker for 1,024 subcarriers

Table 6.17 Best compromised solutions for the cases of 128, 256, 512, and 1,024 subcarriers

Carriers	Min power	Min BER	Max Thr.	Min int.	Max s.e.	Algorithm
128	2.804E−01	1.933E−01	9.605E−01	1.25E−01	3.28E−01	DE
256	3.078E−02	1.223E−01	9.823E−01	1.69E−01	3.38E−02	MBO
512	5.981E−02	6.902E−02	9.997E−01	3.94E−01	1.98E−02	MBO
1,024	5.570E−02	8.112E−02	9.985E−01	3.66E−01	2.06E−02	MBO

Table 6.18 Algorithm ranking first and second for all cases

Algorithm	Ranking first	Ranking second
EHO	2	5
PSO	0	0
DE	6	6
MBO	4	5
ES	0	4
MSA	**12**	4

The smaller values are in bold font.

cases MSA was the first in 12, and second in 4. This result is better quantified in Table 6.19 that lists the average rankings by Friedman test. We observe that MSA ranks first and DE second. Moreover, in order to test if the differences in the algorithms' performance are significant, we perform the Wilcoxon signed-rank test between MSA

Table 6.19 Average rankings achieved by Friedman test

Method	Average rank	Normalized values	Rank
EHO	3.83	1.92	4
PSO	5.21	2.60	5
DE	2.63	1.31	2
MBO	3.33	1.67	3
ES	4.00	2.00	6
MSA	**2.00**	**1.00**	**1**

Table 6.20 Wilcoxon signed-rank test between MSA and the other algorithms

MSA versus	*p*-Values
EHO	**5.200E−05**
PSO	**1.600E−05**
DE	2.509E−01
MBO	**2.260E−02**
ES	**7.050E−04**

The bold font indicates values below significant level.

and the other algorithms. The significance level is set to 0.05. Table 6.20 lists the results of the Wilcoxon signed-rank test. We notice that MSA proves to be significantly better than EHO, PSO, MBO, and ES. However, MSA is not significantly better than DE.

6.4 Spectrum allocation in cognitive radio networks

The need for a dynamic way of spectrum assignment steams from the increasing demand of free spectrum channels and the usual underutilization of the licensed spectrum bands. CR networks (CRNs) [73] are the most prominent solution for this type of problem. The main advantage of the CRNs is the fact that they have the ability to adapt to their environment through sensing mechanisms so that they can fully use the available spectrum channels [73,92–96]. The basic idea is to allow secondary users (SUs) to take advantage of channels assigned to primary users (PUs) without causing any interference. There are a huge number of approaches in the literature that address the problem of spectrum sharing in CRs. Figure 6.50 depicts an example deployment of a CRN with distributed secondary and PUs. In relation to the network architecture, the solutions are classified into centralized and distributed. Moreover, in relation to the spectrum allocation behavior, they are classified into cooperative or noncooperative and in relation to spectrum access technique into overlay or underlay

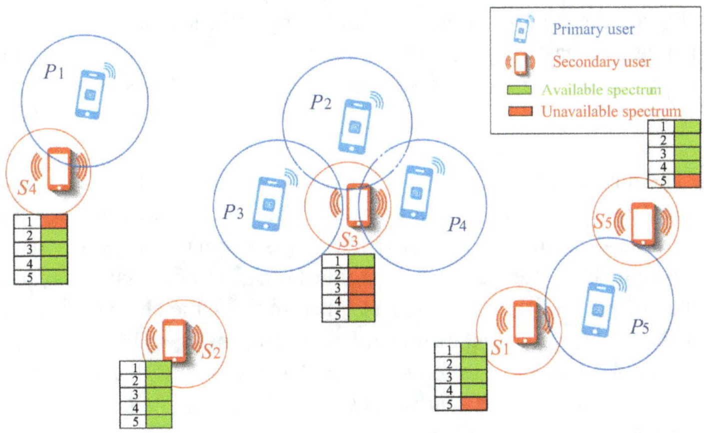

Figure 6.50 Spectrum allocation in cognitive radio networks example deployment

[97]. These solutions for dynamic spectrum access among others include game theory [98], pricing and auction mechanisms [99,100], local bargaining [101], and graph coloring [102].

In this section, we study a centralized approach as in [103], where a central node calculates an allocation assignment based on global knowledge [102,104]. The authors in [102] show that the spectrum allocation optimization problem is an *NP*-hard one.

EAs are once more a proper optimization method for solving the spectrum allocation problem [104–108]. In [104], a quantum GA and a PSO are deployed. In [106], the authors use a binary harmony search algorithm, while the authors in [107–109] apply a chemical reaction optimisation (CRO) algorithm. In [103], the authors apply successfully a CBBO algorithm.

6.4.1 Problem formulation

In this section, we briefly exhibit the general spectrum allocation model given in [102]. Moreover, in [103], the authors apply successfully a CBBO algorithm to the same problem.

As it stated earlier, here we assume a centralized approach. Thus, there is a main server that collects the information about all the channel assignments in the network. First, the main server gathers all the required data about power, spectrum, and interference from both primary and SUs. Afterward, the central server makes decisions on channel assignments based on an assignment algorithm. Then, the main server broadcasts the channel assignments on a free preset channel. This centralized approach has several limitations in a real network. The first limitation is that a free channel for broadcasting the channel assignments to the SUs is needed. Probably, this channel will be in a licensed band and thus will be of fixed and limited bandwidth. Another implication of this fact is that it limits the available bandwidth for network control messages, as the number of the network users increases.

Additionally, another serious limitation is the fact that the server processing complexity will increase at least polynomially as the number of connected users increases. Therefore, as network size increases, this in consequence leads to a computational bottleneck.

We assume a CRN that consists of N SUs that are competing for K spectrum channels. The number of PUs that have constant coverage radius ($d_p = const$) is also K. Moreover, the SUs have a coverage radius that ranges between a minimum and maximum value $d_{\min}$ and $d_{\max}$. This radius adjusts dynamically in order to avoid interference with PUs. We have the rule that an SU's transmission cannot overlap with the transmission of the PU that is assigned to the same channel. Thus, each SU n, where $1 \leq n \leq N$, can adjust its transmission range $d_S(n,m)$ by tuning its transmit power on channel m to avoid interference with PUs. We assume that a secondary user n can use the same channel m as a nearby PU k; only if: $d_S(n,m) \leq Dist(n,k) - d_P(k,m)$ where $Dist(n,k)$ is the distance between n and k.

In general, interference range d_S is limited by the predefined minimum and maximum transmit power, i.e., $[d_{\min}, d_{\max}]$. The SUs are placed in different locations and the technology employed in different bands and user requirements. Thus, different users will grasp different available spectrum. Each SU identifies its position with reference to the PUs' position. There are three possible scenarios and these are depicted in Figure 6.51.

- The SU is in the protection range of the PU. Thus, the SU cannot be assigned to the channel that the PU uses (Figure 6.51a).
- The SU is located outside the protection range of the PU, and it is $d_S < d_{\min}$; again the SU cannot be assigned to the channel that the PU uses (Figure 6.51b).
- The SU is located outside the protection range of the PU and $d_S > d_{\min}$; then this SU may be assigned to this channel, with a transmission range of $d_S(N,K)$ (Figure 6.51c).

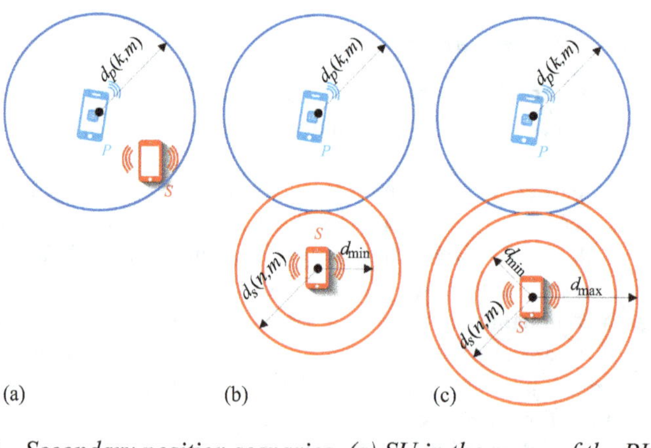

Figure 6.51 Secondary position scenarios. (a) SU in the range of the PU, (b) SU outside the range of the PU $d_s < d_{min}$ (c) SU outside the range of the PU $d_s > d_{min}$

In relation to these rules, each SU adapts its set of distances d_S with the set of available channels for it. This adjustment is subject to the constraint that two SUs using the same channel cannot overlap.

We may define the following matrices that should be computed on the basis of the locations of PUs and SUs. These are [102–104] as follows:

- *Channel availability matrix*: $L = \left\{l_{n,k} \mid l_{n,k} \in \{0,1\}\right\}_{N\times K}$ is an $N \times K$ binary matrix that represents which channel is available for every SU: $l_{n,k} = 1$ if and only if channel k is available at user n, otherwise $l_{n,k} = 0$.
- *Channel reward matrix*: $S = \left\{s_{n,k}\right\}_{N\times K}$ is an $N \times K$ matrix that denotes the channel reward (maximum bandwidth or throughput) that can be obtained by SU n, using channel m.
- *Interference constraint matrix*:
  $I = \left\{i_{n,p,k} \mid i_{n,p,k} \in \{0,1\}\right\}_{N\times N\times K}$ is an $N \times N \times K$ binary matrix, representing the interference constraints among SUs. When both users n and p use channel k, the interference between them can be expressed as

$$i_{n,p,k} = \begin{cases} 1 & \text{if users } n \text{ and } k \text{ interfere with each other} \\ 0 & \text{otherwise} \end{cases} \tag{6.25}$$

- *Conflict free channel assignment matrix*: $A = \{\alpha_{n,k} \mid \alpha_{n,k} \in \{0,1\}, \alpha_{n,k} \leq l_{n,k}\}_{N\times K}$ is an $N \times K$ binary matrix that represents the following assignment:

$$a_{n,k} = \begin{cases} 1 & \text{if channel } k \text{ is allocated to SU } n \\ 0 & \text{otherwise} \end{cases} \tag{6.26}$$

Moreover, the conflict free assignment should meet all the interference constraints given by C:

$$\alpha_{n,k} + \alpha_{p,k} \leq 1, \text{ if } i_{p,n,k} = 1, \forall n, p \leq N, k \leq M \tag{6.27}$$

Additionally, the user reward vector that each SU gets for a given channel assignment is expressed as

$$R = \left\{ r_n = \sum_{k=1}^{K} a_{n,k} s_{n,k} \right\}_{N\times 1} \tag{6.28}$$

The main objective in the spectrum allocation problem is to maximize the network utilization $U(R)$. This can be expressed mathematically with

$$A^* = \underset{A\in\Lambda_{L,I}}{\operatorname{argmax}}\, U(R) \tag{6.29}$$

where A^* denotes the optimal free channel assignment matrix, and $\Lambda(L,I)_{N,K}$ denotes the set of conflict free channel assignments for given L and I. We can express the three main objective–functions as [102–104]

- *Maximum–sum reward (MSR)*:

$$U_{sum}(R) = \sum_{n=1}^{N} r_n \tag{6.30}$$

which maximizes the total spectrum utilization in the system regardless of fairness.

- *Max–min reward (MMR)*:

$$U_{min}(R) = \min_{0 \le n \le N} r_n \tag{6.31}$$

which maximizes the spectrum utilization at the bottleneck user or the user with the least allotted spectrum.

- *Max–proportional fair (MPF)*:

$$U_{fair}(R) = \left(\prod_{n=1}^{N} r_n \right)^{1/N} \tag{6.32}$$

which addresses fairness for single hop flows.

The encoding of the matrix L is an issue that needs to be decided for solving the problem using EAs. Figure 6.52 illustrates a possible way to represent the structure of a vector (member of the population), for $N = 5$, and $K = 5$. Similarly, to [104], we can decrease the numbers of bits for each vector by selecting only the one bits. We observe that encoding all the elements requires 25 bits, while the encoding of only the elements with ones of matrix L needs 9 bits. Thus, in order to derive the objective–function value for each possible solution vector, we can map the vector to the channel assignment matrix, as it is shown in Figure 6.52.

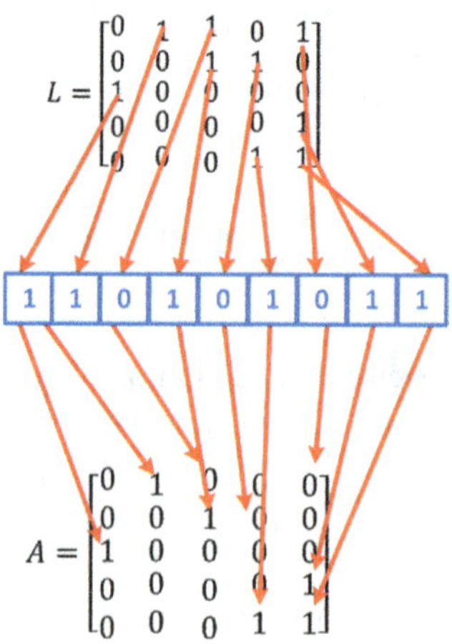

Figure 6.52 Structure of a sample member of the population

6.4.2 Simulation results

In the spectrum allocation case, we will apply the following algorithms; a simple GA, PSO, ACO, and the emerging algorithms MBO [68] and the GCMBO (MBO algorithm enhanced with greedy strategy and self-adaptive crossover operator) [110]. We generate random topologies with dimensions 20×20. We assume that the coverage radius of PUs is fixed at $d_p = 2$. Moreover, we make the assumption that the coverage radius of SUs is between the values $d_{\min} = 1$ and $d_{\max} = 4$. We generate 50 random topologies with dimensions 20×20 and we obtain the mean values.

The selection of an algorithm control parameters plays an important role and is usually problem dependent. Generally, we select control parameters for all algorithms that commonly perform well regardless of the characteristics of the problem to be solved. For the GA, we use single-point crossover with crossover probability 0.9, and the mutation probability is set to 0.1. For PSO, we set c_1 and c_2 to 2.05. Additionally, the inertia weight is linearly decreased starting from 0.9 to 0.4. For ACO, the initial pheromone value τ_0 is set to 1.0e−6, the pheromone update constant Q is set to 0, the exploration constant q_0 is set to 0.2, the global pheromone decay rate ρ_g is 0.2, the local pheromone decay rate ρ_l is 0.05, the pheromone sensitivity α is 1, and the visibility sensitivity β is 2. For MBO and GCMBO, the ratio $r_{pop} = 5/12$, and the migration period is set to 1.2, as it is suggested by the authors in [68]. We set the population size and the maximum number of iterations to the same values for all algorithms. These are 20 and 300 (as in [103]), respectively. We consider five different cases for different combinations of SUs and PUs. These are for $(N, K) = \{(4, 4), (4, 12), (12, 8), (8, 12), (16, 18)\}$.

Table 6.21 lists the average objective–function results for all the algorithms. We have a total of $5 \times 3 = 15$ problem cases. We notice that GCMBO clearly outperforms

Table 6.21 Average reward values obtained by all algorithms

Case (*N*, *K*)	Function	PSO	GA	ACO	GCMBO	MBO
(4, 4)	MSR	144.34	**152.80**	143.24	151.45	150.23
(4, 4)	MMR	16.65	24.31	23.87	**24.92**	24.33
(4, 4)	MPF	23.97	**34.91**	32.95	33.73	33.34
(4, 12)	MSR	425.70	430.03	425.66	**436.58**	393.23
(4, 12)	MMR	46.82	76.44	75.68	**87.29**	73.96
(4, 12)	MPF	80.34	103.53	97.48	**105.60**	96.26
(12, 8)	MSR	477.46	508.51	497.37	**526.78**	446.54
(12, 8)	MMR	3.97	11.74	15.66	**15.72**	12.71
(12, 8)	MPF	5.82	35.07	29.67	**34.43**	26.50
(8, 12)	MSR	606.24	**640.33**	588.13	626.97	540.76
(8, 12)	MMR	15.67	**43.97**	42.40	42.29	34.11
(8, 12)	MPF	37.47	**70.91**	59.70	67.89	57.29
(16, 18)	MSR	1186.35	**1243.31**	1115.58	1198.24	995.60
(16, 18)	MMR	4.34	24.03	21.52	**28.13**	18.16
(16, 18)	MPF	20.57	**66.94**	48.63	62.63	45.62

The larger values are in bold.

the other algorithms in 8 out of the 15 cases, while the GA performs better in 7 cases. For topologies with a small number of users GCMBO performs better, while for topologies with a larger user number the GA is better and GCMBO is second best. MBO also performs well, slightly worse than GCMBO. PSO and ACO obtain the worst performance. We can observe better these facts in Table 6.22 where the rankings of the algorithms coming first and second are listed. Clearly GCMBO and GA are the algorithms that perform better than the others. ACO outperforms PSO and obtains the second best in two cases.

Moreover, we compare the algorithms in terms of nonparametric statistical tests. The rankings according to Friedman test are reported in Table 6.23. We notice that GCMBO obtained the best rank of 1.60; however, GA is very close with rank 1.67. PSO failed to find good results in this case, and its performance is poor. The Wilcoxon signed-rank test between GCMBO and the other algorithms is presented in Table 6.24, where the p-values below 0.05 (the significant level) are shown in boldface. We observe that GCMBO for the spectrum allocation optimization problem is significantly better than all algorithms except GA.

Next, we select the case of $N = 18, K = 16$ and present the corresponding boxplots and convergence rate graphs for this case. Figures 6.53–6.55 depict the boxplots for this case. We observe that for the MSR case, both ACO and MBO obtained results with the smaller interquartile range. GA obtained better results than MSA. However, the results are different for the MMR case. We may observe that GCMBO

Table 6.22 Algorithm ranking first and second for all cases

Algorithm	Ranking first	Ranking second
PSO	0	0
GA	7	5
ACO	0	2
GCMBO	8	7
MBO	0	1

Table 6.23 Average rankings achieved by Friedman test

Method	Average rank	Normalized values	Rank
PSO	4.47	2.79	5
GA	1.67	1.04	2
ACO	3.33	2.08	3
GCMBO	**1.60**	**1.00**	**1**
MBO	3.93	2.46	4

The smaller values are in bold font.

Table 6.24 Wilcoxon signed-rank test between GCMBO and the other algorithms

GCMBO versus	***p*-Values**
PSO	**6.100E−05**
GA	9.517E−01
ACO	**3.660E−04**
MBO	**6.100E−05**

The bold font indicates values below significant level.

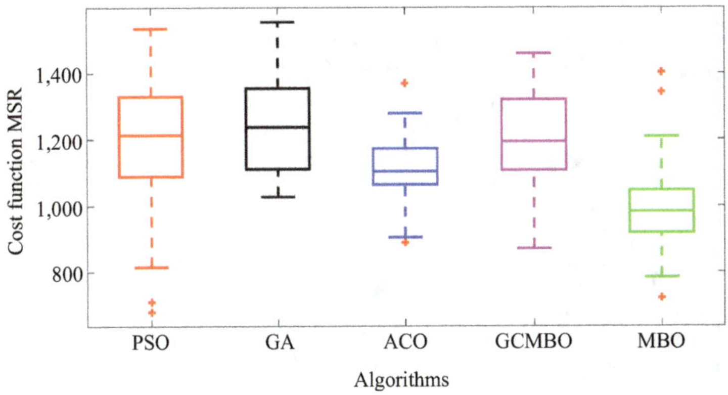

Figure 6.53 Maximum–sum reward. Boxplot of obtained data for $N = 16, K = 18$ users

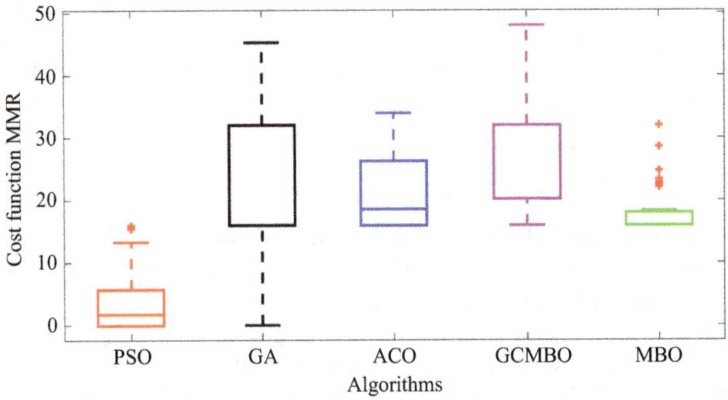

Figure 6.54 Max–min reward. Boxplot of obtained data for $N = 16, K = 18$ users

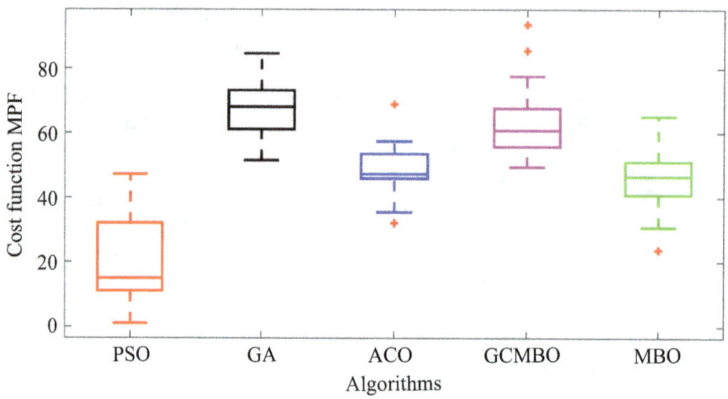

Figure 6.55 Max–proportional fair. Boxplot of obtained data for $N = 16, K = 18$ users

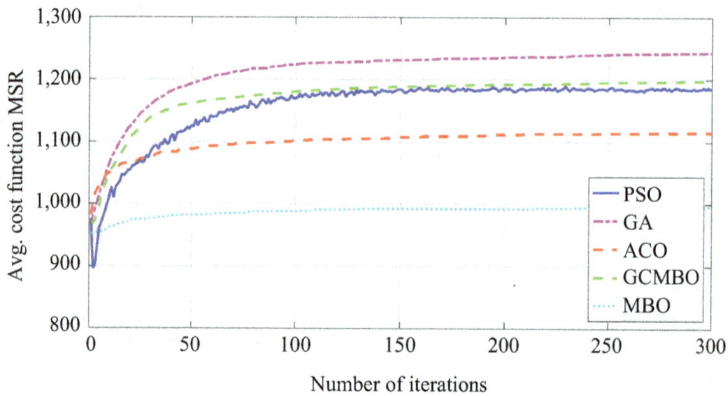

Figure 6.56 Maximum–sum reward. Convergence rate plot for $N = 16, K = 18$ users

obtained the best results. However, for this case MBO and PSO found results with smaller dispersion of values. For the MPF case, almost all algorithms obtain similar interquartile range. GCMBO obtained the best result, while GA is second best.

Figures 6.56–6.58 show the convergence rate graphs. In all cases, the algorithms converge at a similar speed. The only difference is with the PSO algorithm in MMR and MPF functions, where it fails to converge fast and the average convergence lines seem like noise.

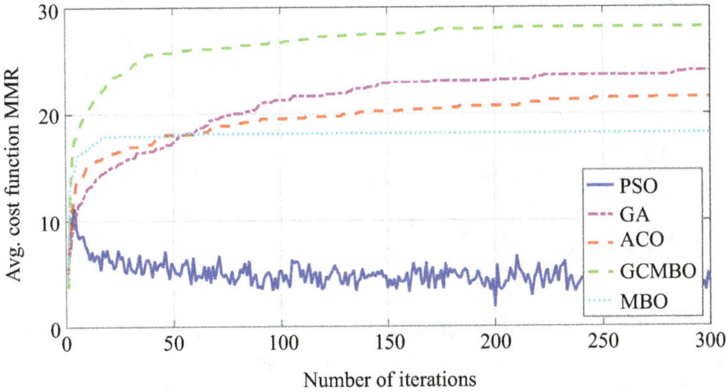

Figure 6.57 *Max–min reward. Convergence rate plot for $N = 16, K = 18$ users*

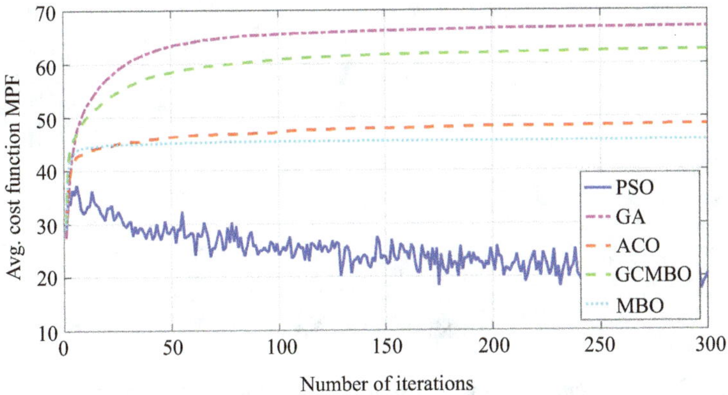

Figure 6.58 *Max–proportional fair. Convergence rate plot for $N = 16, K = 18$ users*

6.4.3 *Asymptotic behavior*

In this section, we study the asymptotic behavior of the three objective–functions. Thus, we run the GCMBO algorithm using 50 independent runs of each objective–function, for different PU, SU, and $d_{\max}$ numbers. In the first case, we set the number of SUs fixed to $N = 8$, and we increase the number of PUs from 10 to 50. Figures 6.59–6.61 depict the asymptotic graphs for each case. We may observe that in all three cases, the objective–function value increases. This can be expected since as the number of PUs increase, there are more available channels for the SUs. Thus, it becomes more probable that an SU will be assigned a free channel.

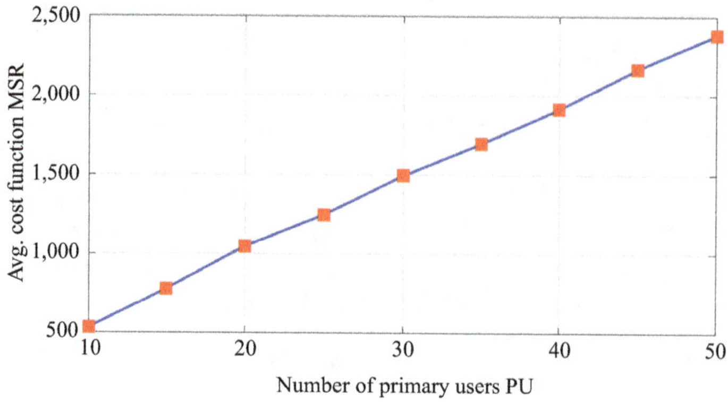

Figure 6.59 Maximum–sum reward. Asymptotic behavior for $N = 8$ secondary users

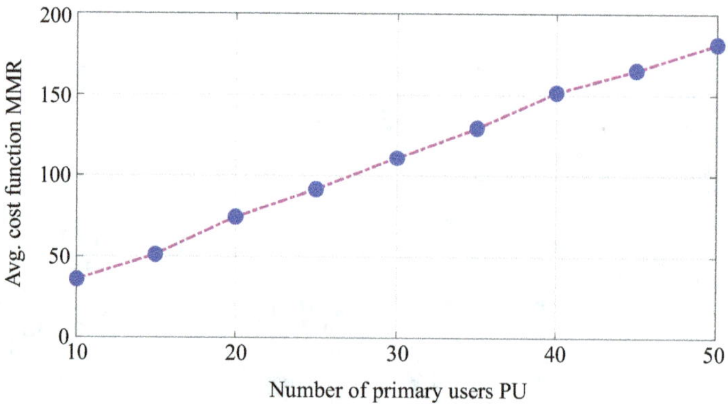

Figure 6.60 Max–min reward. Asymptotic behavior for $N = 8$ secondary users

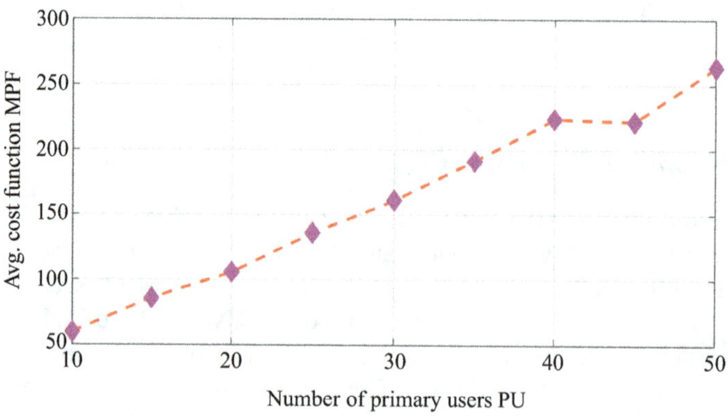

Figure 6.61 Max–proportional fair. Asymptotic behavior for $N = 8$ secondary users

Next, we study the effect of increasing the SUs' number. We set the number of PUs fixed to $K = 10$ and the number of SU is increased from 10 to 20. This is illustrated in Figures 6.62–6.64. We may observe that the MMR function value is reducing and approaches zero as the number of secondary increases. The same asymptotic behavior can also be noticed for the MPF function. This can be explained in the following way. The more the number of SUs increase, the fewer the available channels. Additionally, we notice that the MSR function value increases exponentially as the number of SUs increase.

Finally, we keep the numbers of PUs and SUs fixed to $K = 10, N = 8$ and we increase maximum coverage range d_{max} from 2 m to 8 m. Figures 6.65–6.67 plot the asymptotic graphs for this case. We notice that all three objective–function values

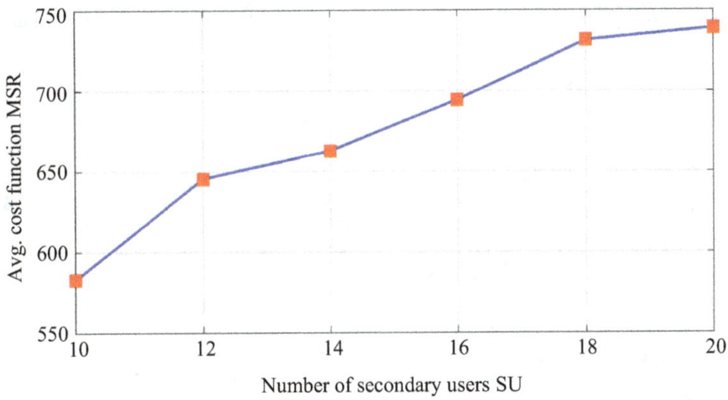

Figure 6.62 Maximum–sum reward. Asymptotic behavior for K = 10 primary users

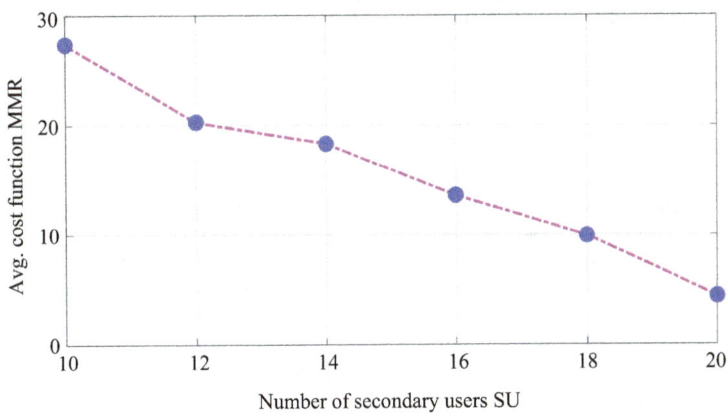

Figure 6.63 Max–min reward. Asymptotic behavior for K = 10 primary users

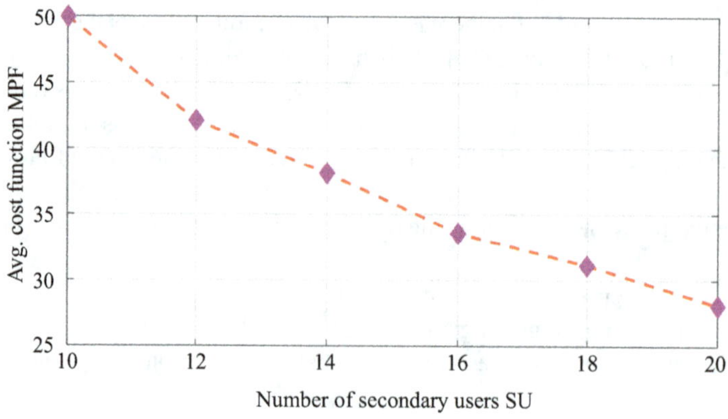

Figure 6.64 Max–proportional fair. Asymptotic behavior for K = 10 primary users

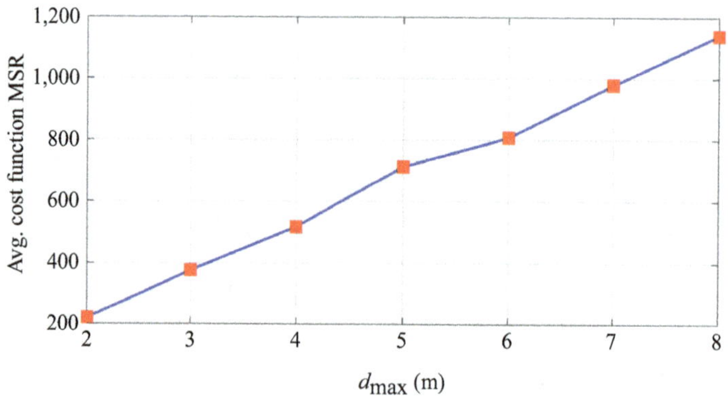

Figure 6.65 Maximum–sum reward. Asymptotic behavior for K = 10 primary and N = 8 secondary users for different d_{max} values

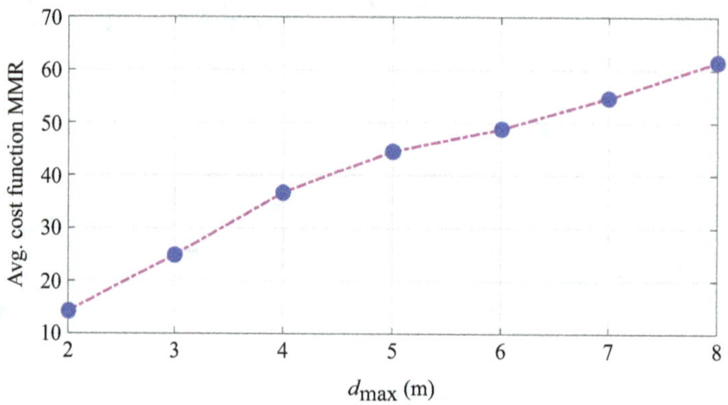

Figure 6.66 Max–min reward. Asymptotic behavior for K = 10 primary and N = 8 secondary users for different d_{max} values

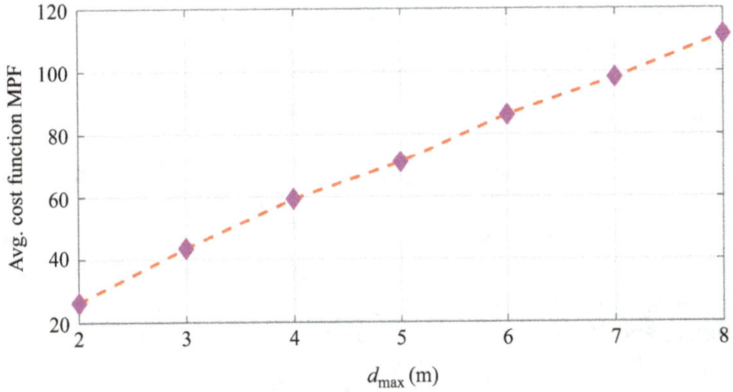

Figure 6.67 Max–proportional fair. Asymptotic behavior for K = 10 primary and N = 8 secondary users for different d_{max} values

increase almost linearly as d_{max} increases. The larger the coverage range the more possible for an SU to be located outside the protection range of a PU but at range less than the d_{max}, so that channel assignments can be made.

6.5 Optimization of wireless sensor networks

A set of wireless sensors (few tens or even hundreds) that work together to monitor an area gathers data from specific applications form a wireless sensor network (WSN). There a huge number of applications that WSNs are utilized like indoor/outdoor environmental monitoring, power monitoring, military target tracking, seismic sensing, healthcare and human activity monitoring [111–113]. If the detection scheme is decentralized, then each sensor node transmits to the central node (called *fusion center*) a summary of its own observations after some preliminary local processing, and then the central node chooses one of a few hypotheses to make the final decision. Therefore, the fusion center does not have direct access to the raw observation data. Thus, the data rate requirements of the decentralized scheme are much lower than in a centralized scheme (where all nodes send directly all the information to the fusion center), because only a few bits of the observation data are transmitted [114]. This problem of distributed detection and fusion under constraints has attracted a lot of attention in the literature [115–117].

In this section, we assume that there is a WSN in which the fusion center is decentralized, and it works through a binary hypothesis testing problem, and the observations of local nodes are correlated. Thus, in this case we may define as the optimization goal the minimization of the total power, while the detection error probability at the fusion center lies below a certain threshold, by the optimal allocation of the total power resources. The use of EAs to this problem has already been addressed in the literature using PSO [118], a hybrid BBO–DE algorithm [119], and a new hybrid TLBO–Jaya algorithm [120].

6.5.1 System model

We assume that a WSN (using a decentralized detection scheme) has L spatially distributed sensor nodes and a fusion center as it is shown in Figure 6.68. Then, we may denote the observed signal vector as $\mathbf{x} = [x_1, x_2, \ldots, x_L]^T$. Moreover, we assume that two possible local observation states exist. The signal may be present or absent. Thus, each node tests the binary hypothesis problem under the hypotheses H_0 (no signal) and H_1 (signal present). The prior probabilities of the hypotheses are $P(H_0) = \pi_0$ and $P(H_1) = \pi_1$.

Therefore, we consider the problem of detecting a constant signal in additive white Gaussian noise, under the assumption that the local observation z_k is obtained at the kth node, then this can be formulated as [118,119]

$$\begin{aligned} H_0 &: z_k = v_k, \quad k = 1, 2, \ldots, L \\ H_1 &: z_k = x_k + v_k, \quad k = 1, 2, \ldots, L \end{aligned} \tag{6.33}$$

We assume that the additive local noise, v_k, is zero-mean Gaussian with variance σ_v^2. The signal x_k is a known positive constant signal, so that $x_k = m, \ k = 1, 2, \ldots, L,$

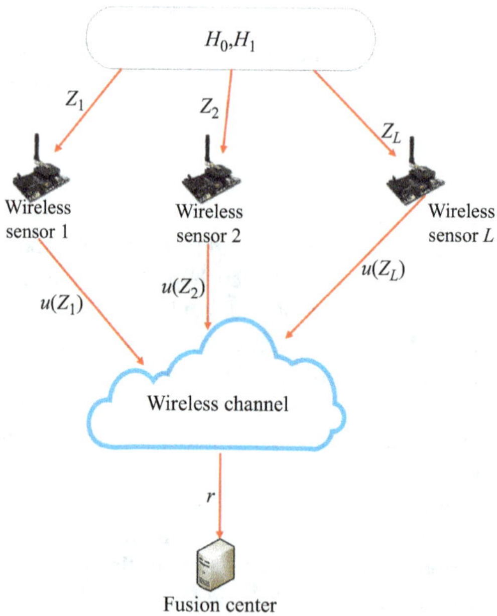

Figure 6.68 Decentralized binary hypothesis detection problem

for all sensors. Additionally, we may define the local observation of SNR, $\gamma = m^2/\sigma_v^2$. This observation in a vector form is expressed

$$\mathbf{z} = \mathbf{x} + \mathbf{v} \tag{6.34}$$

where $\mathbf{v} = [v_1, v_2, \ldots, v_L]^T$ is a zero-mean Gaussian vector with covariance matrix $\boldsymbol{\Sigma}_\mathbf{v}$. Each sensor node makes a local decision $u_k(z_k)$. Moreover, we consider that each sensor retransmits an amplified version of its observation to the fusion center [119]

$$u_k(z_k) = g_k z_k, \; k = 1, 2, \ldots, L \tag{6.35}$$

where g_k denotes the gain at the kth node that it is subject to attenuation and fading. Therefore, the received signal r_k that comes from the kth node is arrived at the fusion center under the two hypotheses [118,119]

$$\begin{aligned} H_0 : r_k &= n_k; \; k = 1, 2, \ldots, L \\ H_1 : r_k &= h_k g_k x_k + n_k; \; k = 1, 2, \ldots, L \end{aligned} \tag{6.36}$$

where h_k denotes the channel fading coefficient, g_k denotes the gain, $n_k = h_k g_k v_k + w_k$ is the effective noise at the fusion center with zero-mean, and w_k is the receiver noise that is considered to be a sequence of i.i.d. components of zero-mean with variance σ_w^2. Then, the covariance matrix of the effective noise vector $\mathbf{n}$ can be expressed as

$$\boldsymbol{\Sigma}_\mathbf{n} = h_k g_k \boldsymbol{\Sigma}_\mathbf{v} g_k h_k + \boldsymbol{\Sigma}_\mathbf{w} \tag{6.37}$$

where $\boldsymbol{\Sigma}_\mathbf{w} = \sigma_\mathbf{w}^2 \mathbf{I}$ and $\mathbf{I}$ are the receiver noise covariance matrix, and the $L \times L$ identity matrix, respectively. In reality, the sensor observations are correlated. We assume that adjacent nodes are equally spaced in straight line at a distance, d, then the correlation of noise samples from nodes i and j is $\rho^{d|i-j|}$, with $|\rho| \leq 1$. Furthermore, $\boldsymbol{\Sigma}_\mathbf{v}$ can be written in a form of a symmetric Hermitian Toeplitz matrix [118,119]

$$\Sigma_v = \sigma_v^2 \begin{bmatrix} 1 & \rho^d & \cdots & \rho^{d(L-2)} & \rho^{d(L-1)} \\ \rho^d & 1 & \cdots & \rho^{d(L-3)} & \rho^{d(L-2)} \\ \vdots & \vdots & \cdots & \vdots & \vdots \\ \rho^{d(L-1)} & \rho^{d(L-2)} & \cdots & \rho^d & 1 \end{bmatrix} \tag{6.38}$$

It is written in vector notation as

$$\mathbf{r} = \mathbf{A}\mathbf{x} + \mathbf{n} \tag{6.39}$$

where $\mathbf{r} = [r_1, r_2, \ldots, r_L]^T$ denotes the received information vector, $\mathbf{n} = [n_1, n_2, \ldots, n_L]^T$ denotes the noise vector, and $\mathbf{A} = diag\,(h_1 g_1, h_2 g_2, \ldots, h_L g_L)$.

The observations $\mathbf{r}$ at the fusion center are distributed as follows [118,119]:

$$\begin{aligned} H_0 &: \mathbf{r} \sim N\,(0, \boldsymbol{\Sigma}_\mathbf{n}) \\ H_1 &: \mathbf{r} \sim N\,(\mathbf{Am}, \boldsymbol{\Sigma}_\mathbf{n}) \end{aligned} \tag{6.40}$$

The noise in the fusion center covariance matrix is [118,119]

$$\boldsymbol{\Sigma}_\mathbf{n} = \mathbf{A}\boldsymbol{\Sigma}_\mathbf{v} A + \sigma_\mathbf{w}^2 \mathbf{I} \tag{6.41}$$

where $\mathbf{m}$ is the L-length vector with all components equal to m. If we consider the threshold, $\tau = \pi_0/\pi_1$ (we assume minimum probability of error Bayesian fusion), then the optimal Bayesian decision rule is formulated mathematically as [118]

$$\delta(\mathbf{r}) = \begin{cases} 1 \text{ if } T(\mathbf{r}) \geq \ln \tau \\ 0 \text{ if } T(\mathbf{r}) < \ln \tau \end{cases} \tag{6.42}$$

A threshold rule on the log-likelihood ratio (LLR) of the observation vector is the optimal procedure for deciding between the two hypotheses. Therefore, we may write the LLR for the detection problem as [119]

$$T(\mathbf{r}) = m\mathbf{e}^{\mathrm{T}} f A \Sigma_{\mathbf{n}}^{-1} \mathbf{r} - \frac{1}{2} m^2 \mathbf{e}^{\mathrm{T}} A \boldsymbol{\Sigma}_{\mathbf{n}}^{-1} \mathbf{A}\mathbf{e} \tag{6.43}$$

The distribution of the LLR for the detection problem under the two hypotheses is given by [118,119]

$$H_0 : T(\mathbf{r}) \sim N\left(-\frac{1}{2} m^2 \mathbf{e}^{\mathrm{T}} A \boldsymbol{\Sigma}_{\mathbf{n}}^{-1} \mathbf{A}\mathbf{e}, m^2 \mathbf{e}^{\mathrm{T}} A \boldsymbol{\Sigma}_{\mathbf{n}}^{-1} \mathbf{A}\mathbf{e}\right) \tag{6.44}$$

$$H_1 : T(\mathbf{r}) \sim N\left(\frac{1}{2} m^2 \mathbf{e}^{\mathrm{T}} A \boldsymbol{\Sigma}_{\mathbf{n}}^{-1} \mathbf{A}\mathbf{e}, m^2 \mathbf{e}^{\mathrm{T}} A \boldsymbol{\Sigma}_{\mathbf{n}}^{-1} \mathbf{A}\mathbf{e}\right) \tag{6.45}$$

According to the literature [118–120], we can further make the assumption that the two hypotheses are equally probable than the threshold is, $\tau = 1$. The probability of fusion error (i.e., the fusion center chooses H_1 when H_0 is true, or H_0 when H_1 is true) is

$$P_e = Q\left(\frac{1}{2}\sqrt{m^2 \mathbf{e}^{\mathrm{T}} A \boldsymbol{\Sigma}_{\mathbf{n}}^{-1} \mathbf{A}\mathbf{e}}\right) \tag{6.46}$$

where $Q(.)$ denotes the Gaussian Q-function. Thus, the optimal power allocation problem is to obtain a set of L optimal sensor gains, $\mathbf{g} = (g_1, g_2, \ldots, g_L)$, that minimizes the total power, and to keep the probability of fusion error under a specified threshold, ε, [119], i.e.,

$$\begin{aligned} \min \ f(\mathbf{g}) &= \sum\nolimits_{\ell=1}^{L} g_\ell^2 \\ \zeta(\mathbf{g}) &= Q\left(\frac{1}{2}\sqrt{m^2 \mathbf{e}^{\mathrm{T}} A \boldsymbol{\Sigma}_{\mathbf{n}}^{-1} \mathbf{A}\mathbf{e}}\right) - \varepsilon \leq 0 \\ \psi_l(\mathbf{g}) &= -g_\ell \leq 0, \ l = 1, 2, \ldots, L \end{aligned} \tag{6.47}$$

We can combine the previous objective and constraint functions in one objective–function using a penalty method. As in [120], we use a dynamically modified penalty function. This method is reported in [121] that it was found to be a more effective approach than using static penalty functions. Thus, we can express the objective–function as

$$F(\mathbf{g}) = f(\mathbf{g}) + iter\left[\sum_{i=0}^{L} \theta(q_i(\mathbf{g}))\, q_i(\mathbf{g})^{\lambda(q_i(\mathbf{g}))}\right]$$

$$q_i(\mathbf{g}) = \begin{cases} \max\{0, \zeta(\mathbf{g})\}, & \text{if } i = 0 \\ \max\{0, \psi_i(\mathbf{g})\}, & \text{otherwise} \end{cases} \tag{6.48}$$

$$\lambda(x) = \begin{cases} 1 \text{ if } x < 1 \\ 2 \text{ otherwise} \end{cases}$$

$$\theta(x) = \begin{cases} 10 \quad \text{if } x \le 0.1 \\ 100 \text{ if } \; 0.1 < x \le 1 \\ 300 \text{ otherwise} \end{cases}$$

where *iter* denotes the algorithm's current iteration number, $\theta(q_i(\mathbf{g}))$ is a multistage assignment function and $\lambda(q_i(\mathbf{g}))$ represents the power of the penalty function.

6.5.2 Numerical results

We apply several EAs to different WSN configurations, these are the PSO, and the emerging nature-inspired algorithms; the GWO [39], the salp swarm algorithm (SSA) [122], the gravitational search algorithm (GSA) [123], and the hybrid PSOGSA [124].

For PSO, we set c_1 and c_2 to 2.05. Additionally, the inertia weight is linearly decreased starting from 0.9 to 0.4. For GSA and PSOGSA, we set $G_0 = 100$, and $\alpha = 20$. Moreover, for PSOGSA we use, according to [124], $c_1' = 0.5$, $c_2' = 1.5$, w is selected randomly within [0, 1]. GWO and SSA do not require the setting of any control parameter. For all algorithm, we set the population size to 100 and the maximum number of iterations is set to 600; thus, a total number of 60,000 objective–function evaluation for each algorithm run. All algorithms run for 50 independent trials for each case. The population is initialized randomly at the beginning of each algorithm, according to the lower and upper bounds we set. The stopping criterion is the number of objective–function evaluations. The initial position (solution) in each dimension of the population is within the range [0,10].

We use the objective–function (cost function) defined in (6.48). Additionally, we make the assumption that the channel fading coefficients h_i follow a Rayleigh distribution with unit mean, and without loss of generality, they are ranked in descending order $h_1 \ge h_2 \ge \cdots \ge h_L$.

First, we perform simulations with two different sensor node numbers, $L = \{15, 25\}$, different fusion error threshold $\varepsilon = \{0.001, 0.05, 0.01, 0.1\}$ values for each node number, and different correlation factor $\rho = \{0, 0.01, 0.1, 0.5\}$ values. The *SNR* is set fixed to 10 dB. Therefore, we obtain eight different optimization cases for each ρ value.

Table 6.25 Algorithms' comparative results for L = 15 sensors, γ = 10 dB

ρ	ε	GSA	GWO	PSO	SSA	PSOGSA
0	0.1	4.367E+00	3.085E+00	3.081E+00	3.085E+00	**3.078E+00**
	0.05	8.156E+00	6.719E+00	6.695E+00	6.700E+00	**6.676E+00**
	0.01	1.984E+01	1.873E+01	1.856E+01	1.859E+01	**1.851E+01**
	0.001	4.497E+01	4.408E+01	4.376E+01	4.372E+01	**4.365E+01**
0.01	0.1	4.531E+00	3.099E+00	3.102E+00	3.100E+00	**3.092E+00**
	0.05	8.139E+00	6.762E+00	6.734E+00	6.728E+00	**6.715E+00**
	0.01	2.000E+01	1.886E+01	1.869E+01	1.869E+01	**1.864E+01**
	0.001	4.547E+01	4.441E+01	4.417E+01	4.426E+01	**4.410E+01**
0.1	0.1	4.330E+00	3.240E+00	3.249E+00	3.243E+00	**3.235E+00**
	0.05	8.360E+00	7.141E+00	7.099E+00	7.089E+00	**7.076E+00**
	0.01	2.109E+01	1.996E+01	1.994E+01	1.992E+01	**1.988E+01**
	0.001	4.974E+01	4.897E+01	4.865E+01	4.858E+01	**4.855E+01**
0.5	0.1	4.965E+00	3.953E+00	3.960E+00	3.958E+00	**3.952E+00**
	0.05	9.992E+00	9.093E+00	9.103E+00	9.074E+00	**9.056E+00**
	0.01	3.133E+01	3.066E+01	3.061E+01	3.056E+01	**3.052E+01**
	0.001	1.257E+02	1.250E+02	1.244E+02	1.243E+02	**1.242E+02**

The smaller values are in bold.

Table 6.25 holds the algorithms' comparative results for $L = 15$ sensors. The first line of Table 6.25 lists the results for $\rho = 0$, i.e., the uncorrelated case. We observe that PSOGSA clearly outperforms all algorithms in all cases. GWO, PSO, and SSA obtain perform in general well very close to the result from PSOGSA. The same conclusion is conducted from Table 6.26 that reports the algorithms' comparative results for $L = 25$ sensors. We notice again that PSOGSA emerges as the best algorithm. The problem now requires finding the values of 25 unknowns, and the differences between the algorithms are more clear.

In order to study the dependence of the total network power with increasing *SNR* value, we obtain Figures 6.69–6.72 for $L = 15$ sensors and fusion error threshold $\varepsilon = 0.1$. We notice that the total network power (objective–function value) is decreasing exponentially as the *SNR* increases. This can be expected as when the local *SNR* is high, it is not required to turn on all nodes but rather a relatively smaller number of nodes to achieve the same performance. Therefore, the total system power is decreasing. All the algorithms except GSA seem to converge to the same value as the *SNR* increases. This final function value seems to converge to a value close to 2, and that is independent of the correlation ρ. The GSA final value is a bit higher near 3.

Next, we evaluate the algorithms' performance when the number of nodes is very large. We set $L = 300$ sensors, and $\rho = 0.1$ for two different cases of fusion error threshold. In the case, we set the population size to 300 for all algorithms. Figure 6.73 shows the boxplot for the $\varepsilon = 0.1$ case. We notice that in this case, GWO performs better than the other algorithms, while PSOGSA is second in performance. If we

Table 6.26 Algorithms' comparative results for L = 25 sensors, $\gamma = 10\,dB$

ρ	ε	GSA	GWO	PSO	SSA	PSOGSA
0	0.1	7.843E+00	9.392E−01	9.712E−01	9.412E−01	**9.343E−01**
	0.05	8.034E+00	1.959E+00	1.988E+00	1.957E+00	**1.949E+00**
	0.01	1.273E+01	6.279E+00	6.318E+00	6.237E+00	**6.212E+00**
	0.001	2.527E+01	1.761E+01	1.748E+01	1.739E+01	**1.731E+01**
0.01	0.1	6.586E+00	9.417E−01	9.772E−01	9.409E−01	**9.372E−01**
	0.05	7.444E+00	1.974E+00	1.990E+00	1.971E+00	**1.964E+00**
	0.01	1.189E+01	6.337E+00	6.355E+00	6.316E+00	**6.282E+00**
	0.001	2.516E+01	1.782E+01	1.772E+01	1.763E+01	**1.756E+01**
0.1	0.1	7.751E+00	9.746E−01	9.905E−01	9.719E−01	**9.654E−01**
	0.05	8.169E+00	2.109E+00	2.140E+00	2.104E+00	**2.095E+00**
	0.01	1.342E+01	7.112E+00	7.080E+00	7.009E+00	**6.958E+00**
	0.001	2.715E+01	2.049E+01	2.036E+01	2.019E+01	**2.009E+01**
0.5	0.1	8.205E+00	1.156E+00	1.184E+00	1.146E+00	**1.139E+00**
	0.05	8.428E+00	3.099E+00	3.120E+00	3.082E+00	**3.067E+00**
	0.01	1.752E+01	1.303E+01	1.305E+01	1.291E+01	**1.284E+01**
	0.001	5.539E+01	5.144E+01	5.117E+01	5.111E+01	**5.068E+01**

The smaller values are in bold.

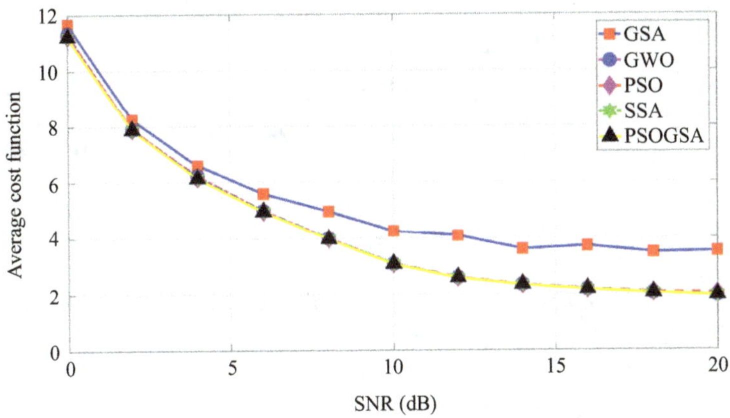

Figure 6.69 SNR values versus average cost function for L = 15 sensors, $\varepsilon = 0.1$, $\rho = 0$

set the fusion error threshold to $\varepsilon = 0.05$, the results are very similar. Figure 6.74 depicts the boxplot for this case. The problem becomes more difficult with smaller fusion error threshold. One may observe that GWO reaches a lower value with smaller dispersion of values. The performance of PSOGSA comes next.

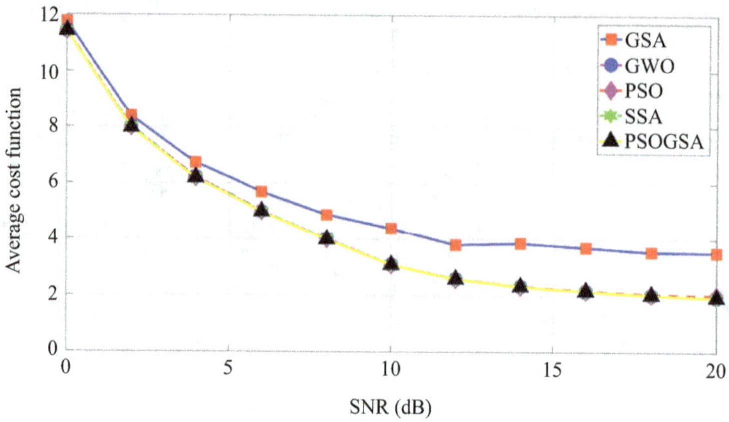

Figure 6.70 SNR values versus average cost function for L = 15 sensors, ε = 0.1, ρ = 0.01

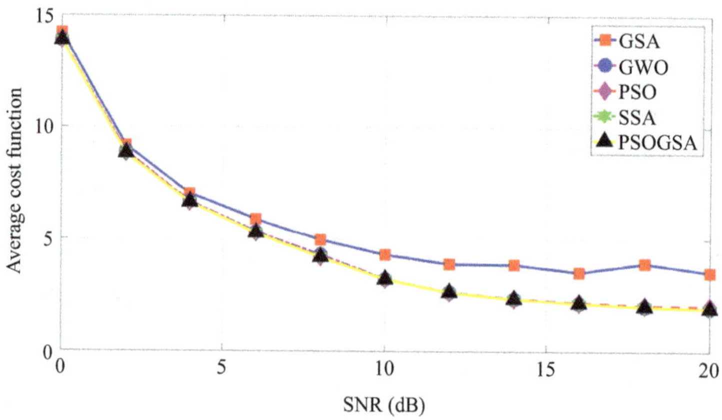

Figure 6.71 SNR values versus average cost function for L = 15 sensors, ε = 0.1, ρ = 0.1

The corresponding convergence rate graphs for this case are illustrated in Figures 6.75 and 6.76. From both figures we see a similar converge speed. GWO is the faster and requires less than 100 iterations to reach its final value, while PSOGSA requires more than 200 iterations. GSA is as fast as GWO, however the final value is larger than that of the other algorithms. SSA in both cases requires about 300 iterations to converge, and PSO converges slower than the other algorithms requiring the maximum number of iterations.

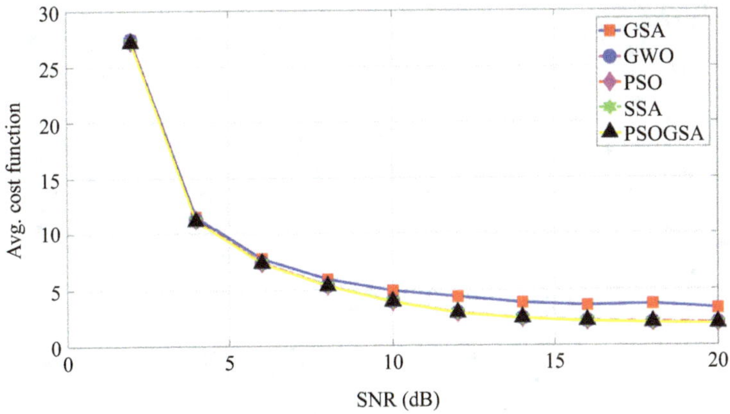

Figure 6.72 SNR values versus average cost function for L = 15 sensors, ε = 0.1, ρ = 0.5

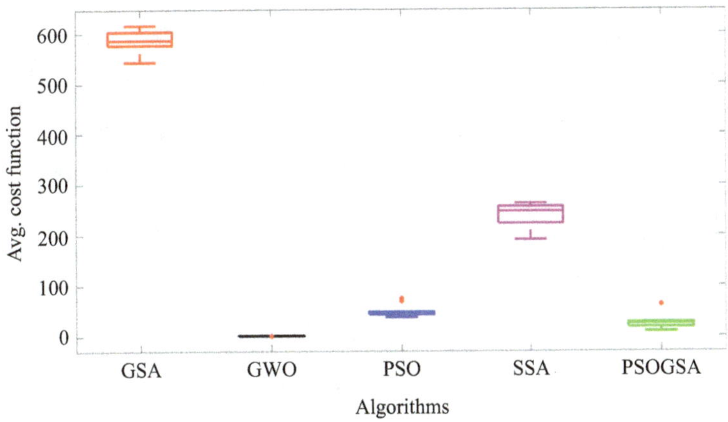

Figure 6.73 Boxplot of all algorithms results for L = 300 sensors, ε = 0.1, ρ = 0.1

Finally, we study the effect of asymptotically increasing the sensor node number for $\rho = 0.1$ (correlated observations) and different values of fusion error threshold. Figure 6.77 shows this asymptotic behavior of the total network power as the node number increases. The results are all obtained with GWO that performed well with large node number previously. We notice that as the node number becomes very large, the total network power reaches a specific value that remains more or less unchanged. This value is different for different values of fusion error threshold. For larger fusion

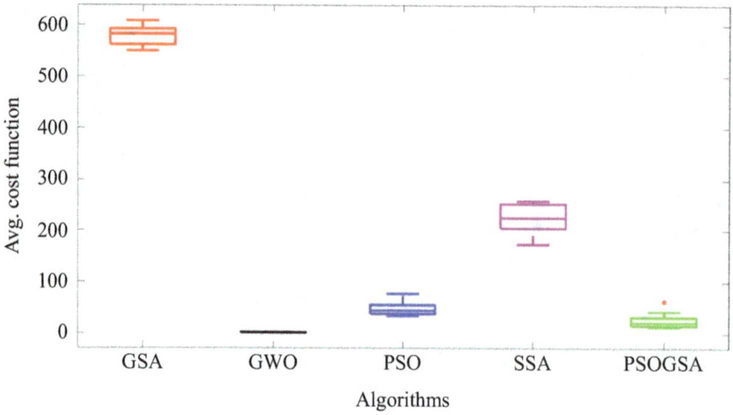

Figure 6.74 Boxplot of all algorithms results for L = 300 sensors, ε = 0.05, ρ = 0.1, g = 10 dB

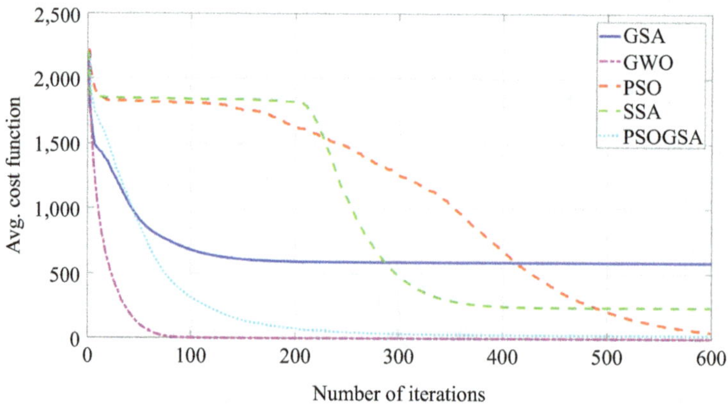

Figure 6.75 Convergence rate graph for L = 300 sensors, ε = 0.1, ρ = 0.1, g = 10 dB

error threshold, the total network power is larger, while for smaller values the total network power values are very close. This can be explained by the fact that in order to obtain a given fusion error threshold, only a small number of active sensors are required. Thus, as the number of sensors increase the number of active sensors is near a certain value, then the total network power converges to a specific value.

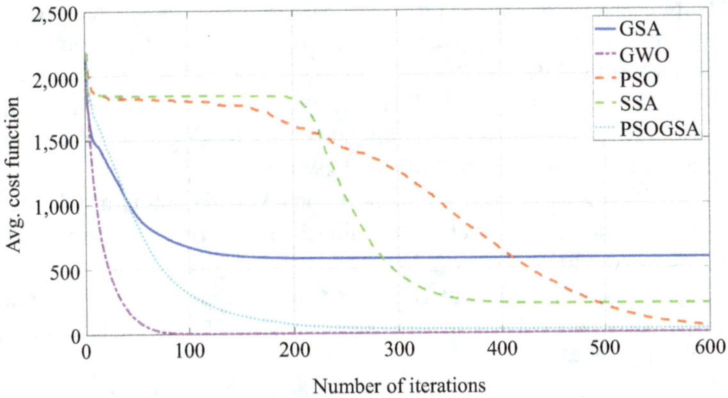

Figure 6.76 Convergence rate graph for $L = 300$ sensors, $\varepsilon = 0.05$, $\rho = 0.1$, $g = 10$ dB

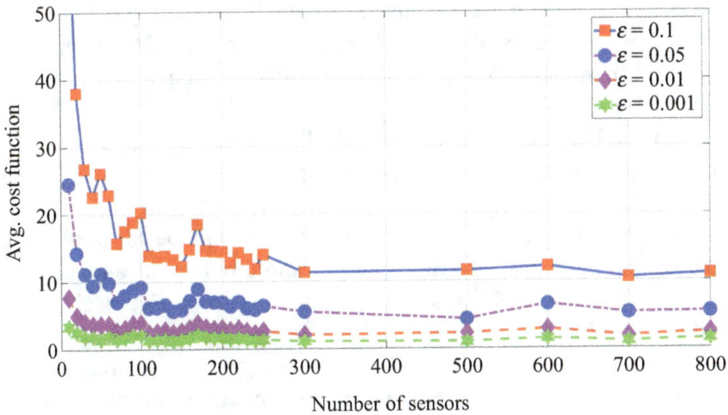

Figure 6.77 Sensor numbers versus average cost function for $\rho = 0.1$, and different ε values

References

[1] Hara S and Prasad R. Overview of multicarrier CDMA. IEEE Communications Magazine. 1997;35(12):126–133.

[2] van Nee R and Prasad R. OFDM for Wireless Multimedia Communications. Artech House Publishers; 2000.

[3] Helaly T, Dansereau R, and El-Tanany M. BER performance of OFDM signals in presence of nonlinear distortion due to SSPA. Wireless Personal Communications. 2012;64(4):749–760.

[4] Cimini Jr LJ and Sollenberger NR. Peak-to-average power ratio reduction of an OFDM signal using partial transmit sequences. IEEE Communications Letters. 2000;4(3):86–88.

[5] Alavi A, Tellambura C, and Fair I. PAPR reduction of OFDM signals using partial transmit sequence: An optimal approach using sphere decoding. IEEE Communications Letters. 2005;9(11):982–984.

[6] Chen JC. Partial transmit sequences for PAPR reduction of OFDM signals with stochastic optimization techniques. IEEE Transactions on Consumer Electronics. 2010;56(3):1229–1234.

[7] Cho YJ, No JS, and Shin DJ. A new low-complexity PTS scheme based on successive local search using sequences. IEEE Communications Letters. 2012;16(9):1470–1473.

[8] Duanmu C and Chen H. Reduction of the PAPR in OFDM systems by intelligently applying both PTS and SLM algorithms. Wireless Personal Communications. 2014;74(2):849–863.

[9] Gao J, Wang J, and Wang B. Improved particle swarm optimization for PAPR reduction of OFDM systems. In: 2010 International Conference on Networking, Sensing and Control, ICNSC 2010; 2010. p. 621–624.

[10] Han SH and Lee JH. PAPR reduction of OFDM signals using a reduced complexity PTS technique. IEEE Signal Processing Letters. 2004;11(11):887–890.

[11] Hasan MM. VLM precoded SLM technique for PAPR reduction in OFDM systems. Wireless Personal Communications. 2013;73(3): 791–801.

[12] Hung HL. Using evolutionary computation technique for trade-off between performance peak-to average power ration reduction and computational complexity in OFDM systems. Computers and Electrical Engineering. 2011;37(1):57–70.

[13] Hung HL and Huang YF. Peak to average power ratio reduction of multicarrier transmission systems using electromagnetism-like method. International Journal of Innovative Computing, Information and Control. 2011; 7(5 A):2037–2050.

[14] Hung HL, Huang YF, Yeh CM, *et al.* Performance of particle swarm optimization techniques on paper reduction for OFDM systems. In: Conference Proceedings – IEEE International Conference on Systems, Man and Cybernetics; 2008. p. 2390–2395.

[15] Jiang T, Xiang W, Richardson PC, *et al.* PAPR reduction of OFDM signals using partial transmit sequences with low computational complexity. IEEE Transactions on Broadcasting. 2007;53(3):719–724.

[16] Kwon OJ and Ha YH. Multi-carrier PAP reduction method using sub-optimal PTS with threshold. IEEE Transactions on Broadcasting. 2003;49(2):232–236.

[17] Lim DW, Heo SJ, No JS, *et al.* A new PTS OFDM scheme with low complexity for PAPR reduction. IEEE Transactions on Broadcasting. 2006;52(1): 77–82.

[18] Taspinar N, Kalinli A, and Yildirim M. Partial transmit sequences for PAPR reduction using parallel tabu search algorithm in OFDM systems. IEEE Communications Letters. 2011;15(9):974–976.

[19] Tellambura C. Improved phase factor computation for the PAR reduction of an OFDM signal using PTS. IEEE Communications Letters. 2001;5(4):135–137.

[20] Varahram P and Ali BM. A low complexity partial transmit sequence for peak to average power ratio reduction in OFDM systems. Radioengineering. 2011;20(3):677–682.

[21] Varahram P, Mohammady S, and Ali BM. A robust peak-to-average power ratio reduction scheme by inserting dummy signals with enhanced partial transmit sequence in OFDM systems. Wireless Personal Communications. 2013;72(2):1125–1137.

[22] Wang Y, Chen W, and Tellambura C. A PAPR reduction method based on artificial bee colony algorithm for OFDM signals. IEEE Transactions on Wireless Communications. 2010;9(10):2994–2999.

[23] Wang L and Tellambura C. Cross-entropy-based sign-selection algorithms for peak-to-average power ratio reduction of OFDM systems. IEEE Transactions on Signal Processing. 2008;56(10 I):4990–4994.

[24] Weng CE, Chang CW, Chen CH, *et al.* Novel low-complexity partial transmit sequences scheme for PAPR reduction in OFDM systems using adaptive differential evolution algorithm. Wireless Personal Communications. 2013;71(1):679–694.

[25] Wen JH, Lee SH, Huang YF, *et al.* A suboptimal PTS algorithm based on particle swarm optimization for PAPR reduction in OFDM systems. EURASIP Journal on Wireless Communications and Networking. 2008; 2008(14).

[26] Yang L, Chen RS, Soo KK, *et al.* An efficient sphere decoding approach for PTS assisted PAPR reduction of OFDM signals. AEU – International Journal of Electronics and Communications. 2007;61(10):684–688.

[27] You YH, Jeon WG, Paik JH, *et al.* Low-complexity PAR reduction schemes using SLM and PTS approaches for OFDM-CDMA signals. IEEE Transactions on Consumer Electronics. 2003;49(2):284–289.

[28] Yu X, Li S, Zhu Z, *et al.* An improved artificial bee colony-partial transmit sequence algorithm for PAPR reduction in OFDM systems. International Journal of Wireless and Mobile Computing. 2013;6(5):473–480.

[29] Zhang Y, Ni Q, and Chen HH. A new partial transmit sequence scheme using genetic algorithm for peak-to-average power ratio reduction in a multi-carrier code division multiple access wireless system. International Journal of Autonomous and Adaptive Communications Systems. 2009;2(1): 40–57.

[30] Zhang Y, Ni Q, Chen HH, *et al.* An intelligent genetic algorithm for PAPR reduction in a multi-carrier CDMA wireless system. In: IWCMC 2008 – International Wireless Communications and Mobile Computing Conference; 2008. p. 1052–1057.

[31] Kim SS, Kim MJ, and Gulliver TA. A new PTS for PAPR reduction by local search in GA. In: IEEE International Conference on Neural Networks – Conference Proceedings; 2006. p. 2370–2373.

[32] Shukla J, Joshi A, Bansal R, *et al.* PAPR reduction of OFDM systems using PTS with genetic algorithm at low computational complexity. In: International Conference on Recent Advances and Innovations in Engineering, ICRAIE 2014; 2014.

[33] Yao YC, Hung HL, and Wen JH. PAPR reduction in OFDM system using differential evolution-based partial transmit sequences scheme. In: Proceedings – International Conference on Machine Learning and Cybernetics. vol. 3; 2011. p. 1414–1419.

[34] Goudos SK. A novel generalized oppositional biogeography-based optimization algorithm: Application to peak to average power ratio reduction in OFDM systems. Open Mathematics. 2016;14(1):705–722.

[35] Rahnamayan RS, Tizhoosh HR, and Salama MMA. Opposition-based differential evolution. IEEE Transactions on Evolutionary Computation. 2008;12(1):64–79.

[36] Dorigo M, Maniezzo V, and Colorni A. Ant system: Optimization by a colony of cooperating agents. IEEE Transactions on Systems, Man, and Cybernetics, Part B: Cybernetics. 1996;26(1):29–41.

[37] Karaboga D and Basturk B. A powerful and efficient algorithm for numerical function optimization: Artificial bee colony (ABC) algorithm. Journal of Global Optimization. 2007;39(3):459–471.

[38] Zhu G and Kwong S. Gbest-guided artificial bee colony algorithm for numerical function optimization. Applied Mathematics and Computation. 2010;217(7):3166–3173.

[39] Mirjalili S, Mirjalili SM, and Lewis A. Grey wolf optimizer. Advances in Engineering Software. 2014;69:46–61.

[40] Eusuff M, Lansey K, and Pasha F. Shuffled frog-leaping algorithm: A memetic meta-heuristic for discrete optimization. Engineering Optimization. 2006;38(2):129–154.

[41] Wang Y, Chen W, and Tellambura C. Genetic algorithm based nearly optimal peak reduction tone set selection for adaptive amplitude clipping PAPR reduction. IEEE Transactions on Broadcasting. 2012;58(3): 462–471.

[42] Biglieri E, Calderbank R, Constantinides A, *et al.* MIMO Wireless Communications. New York: Cambridge University Press; 2007.

[43] Molisch AF and Win MZ. MIMO systems with antenna selection. IEEE Microwave Magazine. 2004;5(1):46–56.

[44] Telatar E. Capacity of multi-antenna Gaussian channels. European Transactions on Telecommunications. 1999;10(6):585–595.

[45] Molisch AF, Win MZ, Choi YS, *et al.* Capacity of MIMO systems with antenna selection. IEEE Transactions on Wireless Communications. 2005;4(4):1759–1771.

[46] Amadori PV and Masouros C. Interference-driven antenna selection for massive multiuser MIMO. IEEE Transactions on Vehicular Technology. 2016;65(8):5944–5958.

[47] Benmimoune M, Driouch E, Ajib W, *et al.* Feedback reduction and efficient antenna selection for massive MIMO system. In: 82nd IEEE Vehicular Technology Conference, VTC Fall 2015. Institute of Electrical and Electronics Engineers Inc.; 2015.

[48] Benmimoune M, Driouch E, Ajib W, *et al.* Novel transmit antenna selection strategy for massive MIMO downlink channel. Wireless Networks. 2017;23:2473–2484.

[49] Gao X, Edfors O, Tufvesson F, *et al.* Multi-switch for antenna selection in massive MIMO. In: 58th IEEE Global Communications Conference, GLOBECOM 2015. Institute of Electrical and Electronics Engineers Inc.; 2015.

[50] Chun N, Zhifeng T, Mehta NB, *et al.* Antenna selection for next generation IEEE 802.16 mobile stations. In: Proceedings of the IEEE International Conference on Communications; 2008.

[51] Gore DA and Paulraj AJ. MIMO antenna subset selection with space-time coding. IEEE Transactions on Signal Processing. 2002;50(10):2580–2588.

[52] Gorokhov A, Gore DA, and Paulraj AJ. Receive antenna selection for MIMO spatial multiplexing: Theory and algorithms. IEEE Transactions on Signal Processing. 2003;51(11):2796–2807.

[53] Sanayei S and Nosratinia A. Antenna selection in MIMO systems. IEEE Communications Magazine. 2004;42(10):68–73.

[54] Sanayei S and Nosratinia A. Capacity of MIMO channels with antenna selection. IEEE Transactions on Information Theory. 2007;53(11):4356–4362.

[55] Uchida D, Arai H, Inoue Y, *et al.* Antenna selection based on minimum eigenvalue in dual-polarized directional MIMO antenna. In: Proceedings of the 71st IEEE Vehicular Technology Conference (VTC 2010-Spring). IEEE; 2010. p. 16–19.

[56] Phan KT and Tellambura C. Receive antenna selection based on union-bound minimization using convex optimization. IEEE Signal Processing Letters. 2007;14(9):609–612.

[57] Karamalis PD, Skentos ND, and Kanatas AG. Selecting array configurations for MIMO systems: An evolutionary computation approach. IEEE Transactions on Wireless Communications. 2004;3(6):1994–1998.

[58] Lain JK. Joint transmit/receive antenna selection for MIMO systems: A real-valued genetic approach. IEEE Communications Letters. 2011;15(1):58–60.

[59] Lu HY and Fang WH. Joint transmit/receive antenna selection in MIMO systems based on the priority-based genetic algorithm. IEEE Antennas and Wireless Propagation Letters. 2007;6:588–591.

[60] Lu HY and Fang WH. Joint receive antenna selection and symbol detection for MIMO systems: A heterogeneous genetic approach. IEEE Communications Letters. 2009;13(2):97–99.

[61] Naeem M and Lee DC. Low-complexity joint transmit and receive antenna selection for MIMO systems. Engineering Applications of Artificial Intelligence. 2011;24(6):1046–1051.

[62] Naeem M and Lee DC. A joint antenna and user selection scheme for multiuser MIMO system. Applied Soft Computing Journal. 2014;23:366–374.

[63] Tohidi MS and Azmi P. Low-complexity throughput-based antenna selection method. Wireless Personal Communications. 2014;75(1):385–395.

[64] Fountoukidis KC, Kalialakis C, Psannis KE, *et al.* MIMO antenna selection using biogeography-based optimization with nonlinear migration models. International Journal of Communication Systems. 2018;31(17).

[65] Simon D. Biogeography-based optimization. IEEE Transactions on Evolutionary Computation. 2008;12(6):702–713.

[66] Guo W, Wang L, and Wu Q. An analysis of the migration rates for biogeography-based optimization. Information Sciences. 2014;254:111–140.

[67] Saremi S, Mirjalili S, and Lewis A. Biogeography-based optimisation with chaos. Neural Computing and Applications. 2014;25(5):1077–1097.

[68] Wang GG, Deb S, and Cui Z. Monarch butterfly optimization. Neural Computing and Applications. 2019;31:1995–2014.

[69] Wang GG. Moth search algorithm: A bio-inspired metaheuristic algorithm for global optimization problems. Memetic Computing. 2018;10(2):151–164.

[70] Derrac J, Garcia S, Molina D, *et al.* A practical tutorial on the use of nonparametric statistical tests as a methodology for comparing evolutionary and swarm intelligence algorithms. Swarm and Evolutionary Computation. 2011;1(1):3–18.

[71] García S, Fernández A, Luengo J, *et al.* Advanced nonparametric tests for multiple comparisons in the design of experiments in computational intelligence and data mining: Experimental analysis of power. Information Sciences. 2010;180(10):2044–2064.

[72] García S, Molina D, Lozano M, *et al.* A study on the use of non-parametric tests for analyzing the evolutionary algorithms' behaviour: A case study on the CEC'2005 Special Session on Real Parameter Optimization. Journal of Heuristics. 2009;15(6):617–644.

[73] Haykin S. Cognitive radio: Brain-empowered wireless communications. IEEE Journal on Selected Areas in Communications. 2005;23(2):201–220.

[74] Mitola Iii J and Maguire Jr GQ. Cognitive radio: Making software radios more personal. IEEE Personal Communications. 1999;6(4):13–18.

[75] Mitola J. Cognitive Radio: An Integrated Agent Architecture for Software Defined Radio. Ph.D Dissertation, Royal Institute of Technology, Kungliga Tekniska Hogskolan, Sweden; 2000.

[76] Hauris JF. Genetic algorithm optimization in a cognitive radio for autonomous vehicle communications. In: Proceedings of the 2007 IEEE International Symposium on Computational Intelligence in Robotics and Automation, CIRA 2007; 2007. p. 427–431.

[77] Newman TR. Multiple Objective Fitness Functions for Cognitive Radio Adaptation. Ph.D Dissertation, University of Kansas; 2008.

[78] Newman TR, Barker BA, Wyglinski AM, *et al.* Cognitive engine implementation for wireless multicarrier transceivers. Wireless Communications and Mobile Computing. 2007;7(9):1129–1142.

[79] Newman TR, Rajbanshi R, Wyglinski AM, *et al.* Population adaptation for genetic algorithm-based cognitive radios. Mobile Networks and Applications. 2008;13(5):442–451.

[80] Zhang Z and Xie X. Application research of evolution in cognitive radio based on GA. In: 2008 3rd IEEE Conference on Industrial Electronics and Applications, ICIEA 2008; 2008. p. 1575–1579.

[81] Tan X, Zhang H, and Hu J. A hybrid architecture of cognitive decision engine based on particle swarm optimization algorithms and case database. Annales des Telecommunications/Annals of Telecommunications. 2014;69(11–12):593–605.

[82] Yu Y, Tan X, Xie Y, *et al.* Cognitive radio decision engine based on binary chaotic particle swarm optimization. Journal of Information and Computational Science. 2013;10(12):3751–3761.

[83] Zhao N, Li S, and Wu Z. Cognitive radio engine design based on ant colony optimization. Wireless Personal Communications. 2012;65(1):15–24.

[84] Chen W, Li T, and Yang T. Intelligent control of cognitive radio parameter adaption: Using evolutionary multi-objective algorithm based on user preference. Ad Hoc Networks. 2015;26:3–16.

[85] Pradhan PM and Panda G. Pareto optimization of cognitive radio parameters using multiobjective evolutionary algorithms and fuzzy decision making. Swarm and Evolutionary Computation. 2012;7:7–20.

[86] Kaur K, Rattan M, and Patterh MS. Biogeography-based optimisation of cognitive radio system. International Journal of Electronics. 2014;101(1): 24–36.

[87] Paraskevopoulos A, Dallas PI, Siakavara K, *et al.* Cognitive radio engine design for IoT using real-coded biogeography-based optimization and fuzzy decision making. Wireless Personal Communications. 2017;97(2):1813–1833.

[88] Pradhan PM and Panda G. Comparative performance analysis of evolutionary algorithm based parameter optimization in cognitive radio engine: A survey. Ad Hoc Networks. 2014;17:129–146.

[89] Proakis JG. Digital Communications. 4th ed. New York, NY: McGraw-Hill; 1995.

[90] Mezura-Montes E and Coello CAC. A simple multimembered evolution strategy to solve constrained optimization problems. IEEE Transactions on Evolutionary Computation. 2005;9(1):1–17.

[91] Wang G, Deb S, and dos S Coelho L. Elephant herding optimization. In: 2015 3rd International Symposium on Computational and Business Intelligence (ISCBI); 2015. p. 1–5.

[92] Pandit S and Singh G. Throughput maximization with reduced data loss rate in cognitive radio network. Telecommunication Systems. 2013;57(2): 209–215.

[93] Sakran H and Shokair M. Hard and softened combination for cooperative spectrum sensing over imperfect channels in cognitive radio networks. Telecommunication Systems. 2013;52(1):61–71.

[94] Wang W, Wu K, Luo H, *et al.* Sensing error aware delay-optimal channel allocation scheme for cognitive radio networks. Telecommunication Systems. 2013;52(4):1895–1904.

[95] Goratti L, Baldini G, and Rabbachin A. An urn occupancy approach for cognitive radio networks in DTVB white spaces. Telecommunication Systems. 2014;56(2):229–244.

[96] Li X, Hu F, Zhang H, *et al.* Two-branch wavelet denoising for accurate spectrum sensing in cognitive radios. Telecommunication Systems. 2014;57(1):81–90.

[97] Akyildiz IF, Lee WY, Vuran MC, *et al.* NeXt generation/dynamic spectrum access/cognitive radio wireless networks: A survey. Computer Networks. 2006;50(13):2127–2159.

[98] Nie N and Comaniciu C. Adaptive channel allocation spectrum etiquette for cognitive radio networks. In: First IEEE International Symposium on New Frontiers in Dynamic Spectrum Access Networks, 2005. DySPAN 2005; 2005. p. 269–278.

[99] Huang J, Berry RA, and Honig ML. Auction-based spectrum sharing. Mobile Networks and Applications. 2006;11(3):405–418.

[100] Kloeck C, Jaekel H, and Jondral FK. Dynamic and local combined pricing, allocation and billing system with cognitive radios. In: First IEEE International Symposium on New Frontiers in Dynamic Spectrum Access Networks, 2005. DySPAN 2005; 2005. p. 73–81.

[101] Cao L and Zheng H. Distributed spectrum allocation via local bargaining. In: 2005 Second Annual IEEE Communications Society Conference on Sensor and Ad Hoc Communications and Networks, 2005. IEEE SECON 2005; 2005. p. 475–486.

[102] Peng C, Zheng H, and Zhao BY. Utilization and fairness in spectrum assignment for opportunistic spectrum access. Mobile Networks and Applications. 2006;11(4):555–576.

[103] Tegou TI, Tsiflikiotis A, Vergados DD, *et al.* Spectrum allocation in cognitive radio networks using chaotic biogeography-based optimisation. IET Networks. 2018;7(5).

[104] Zhao Z, Peng Z, Zheng S, *et al.* Cognitive radio spectrum allocation using evolutionary algorithms. IEEE Transactions on Wireless Communications. 2009;8(9):4421–4425.

[105] Ghasemi A, Jahromi AF, Masnadi-Shirazi MA, *et al.* Spectrum allocation based on artificial bee colony in cognitive radio networks. In: 6th International Symposium on Telecommunications (IST); 2012. p. 182–187.

[106] Hamza AS, Hamza HS, and El-Ghoneimy MM. In: Venkataraman H, Muntean GM, editors. Spectrum Allocation in Cognitive Radio Networks Using Evolutionary Algorithms. Dordrecht: Springer Netherlands; 2012.

[107] Lam AYS and Li VOK. Chemical reaction optimization for cognitive radio spectrum allocation. In: 2010 IEEE Global Telecommunications Conference GLOBECOM 2010; 2010. p. 1–5.

[108] Abbas N, Nasser Y, and Ahmad KE. Recent advances on artificial intelligence and learning techniques in cognitive radio networks. EURASIP Journal on Wireless Communications and Networking. 2015;2015(1).

[109] Lam AYS, Li VOK, and Yu JJQ. Power-controlled cognitive radio spectrum allocation with chemical reaction optimization. IEEE Transactions on Wireless Communications. 2013;12(7):3180–3190.

[110] Wang GG, Zhao X, and Deb S. A novel monarch butterfly optimization with greedy strategy and self-adaptive. In: Proceedings – 2015 2nd International Conference on Soft Computing and Machine Intelligence, ISCMI 2015; 2015. p. 45–50.

[111] Yick J, Mukherjee B, and Ghosal D. Wireless sensor network survey. Computer Networks. 2008;52(12):2292–2330.

[112] Nadeem A, Hussain MA, Owais O, *et al.* Application specific study, analysis and classification of body area wireless sensor network applications. Computer Networks. 2015;83:363–380.

[113] Hsia SC, Hsu SW, and Chang YJ. Remote monitoring and smart sensing for water meter system and leakage detection. IET Wireless Sensor Systems. 2012;2(4):402–408.

[114] Tsitsiklis JN. Decentralized detection. Advances in Statistical Signal Processing. 1993;2(2):297–344.

[115] Appadwedula S, Veeravalli VV, and Jones DL. Energy-efficient detection in sensor networks. IEEE Journal on Selected Areas in Communications. 2005;23(4):693–702.

[116] Jayaweera SK. Bayesian fusion performance and system optimization for distributed stochastic Gaussian signal detection under communication constraints. IEEE Transactions on Signal Processing. 2007;55(4): 1238–1250.

[117] Krasnopeev A, Xiao JJ, and Luo ZQ. Minimum energy decentralized estimation in a wireless sensor network with correlated sensor noises. EURASIP Journal on Wireless Communications and Networking. 2005; 2005(4):473–482.

[118] Wimalajeewa T and Jayaweera SK. Optimal power scheduling for correlated data fusion in wireless sensor networks via constrained PSO. IEEE Transactions on Wireless Communications. 2008;7(9):3608–3618.

[119] Boussaid I, Chatterjee A, Siarry P, *et al.* Hybridizing biogeography-based optimization with differential evolution for optimal power allocation in wireless sensor networks. IEEE Transactions on Vehicular Technology. 2011; 60(5):2347–2353.

[120] Tsiflikiotis A, Goudos SK, and Karagiannidis GK. Hybrid teaching-learning optimization of wireless sensor networks. Transactions on Emerging Telecommunications Technologies. 2017;28(11).

[121] Parsopoulos KE and Vrahatis MN. Particle swarm optimization method for constrained optimization problems. Intelligent Technologies–Theory and Application: New Trends in Intelligent Technologies. 2002;76: 214–220.

[122] Mirjalili S, Gandomi AH, Mirjalili SZ, *et al.* Salp swarm algorithm: A bio-inspired optimizer for engineering design problems. Advances in Engineering Software. 2017;114:163–191.

[123] Rashedi E, Nezamabadi-pour H, and Saryazdi S. GSA: A gravitational search algorithm. Information Sciences. 2009;179(13):2232–2248.

[124] Mirjalili S and Hashim SZM. A new hybrid PSOGSA algorithm for function optimization. In: 2010 International Conference on Computer and Information Application; 2010. p. 374–377.

Chapter 7
Design problems for 5G and beyond

This chapter presents design cases for 5G and beyond using EAs.

7.1 Multi-objective optimization in 5G massive MIMO wireless networks

The current fifth generation offers extremely wide spectrum and multi-gigabit-per-second data rates for mobile users. The massive multiple input–multiple output (MIMO) concept is one of critical technologies of 5G networks [1,2], where each base station (BS) is equipped with a large number of antennas and can provide service to multiple users, over the same time and frequency band. The network designer should take into account different requirements and objectives for the proper operation of 5G wireless networks with massive MIMO.

Thus, an efficient network optimization framework is required, which will have the ability to jointly consider all the conflicting 5G objectives. Such an optimization framework will obtain a set of possible solutions that belong to the Pareto front and have conflicting characteristics. This calls for a suitable decision maker, in order to obtain a best compromised solution. Toward this direction artificial intelligence (AI) techniques could be used, which, in general, play an important role toward future wireless networks and services [3]. Among others, evolutionary algorithms (EAs) belong to the core of AI paradigms.

The authors in [4] solve the multi-objective (MO) problem of 5G massive MIMO systems that considers the tradeoff between energy efficiency (EE) and spectral efficiency by transforming the MO problem into a single objective one and using weighted sum method (WSM). Additionally, an MO problem for 5G massive MIMO systems that consists of three conflicting objective functions, i.e., the average user rate, area rate, and EE, is reported in [5]. The authors in [5] solve the problem by using a scalarization (e.g., WSM, weighted product method, etc.) or a visualization method. Such an approach could be inefficient, especially as the number of objectives increases. Moreover, the output of these approaches and the achieved compromise among the multiple objectives is determined by the utilized weighting parameters, with their selection being a challenging problem. In [6] the authors apply two MOEAs, namely, the non-dominated sorting genetic algorithm-II (NSGA-II) [7] and the speed-constrained multi-objective particle swarm optimizer (SMPSO) [8].

In this section as in [6], we address the previous three-objective problem using a complete framework with MOEAs. In this chapter, we study the usage of more MOEAs to the same problem.

7.1.1 System model

A cellular network of M square cells of area $A = \alpha_{\max}^2$ (Figure 7.1), where the users are uniformly distributed in each cell at a minimum distance $\alpha_{\min}$ [5], is assumed. We define the vector $\bar{y} = (y_N, y_K, y_P)$, where y_N denotes the number of BS antennas per cell, y_K denotes the number of single antenna users per BS, and y_P is the transmission power per antenna in watt. The unknown variables y_N and y_P are bounded by

$$2 \leq y_N \leq Y_{\text{Nmax}} \tag{7.1}$$

and

$$0 \leq y_P \leq y_N Y_{\text{Pmax}} \tag{7.2}$$

respectively, where Y_{Nmax} denotes the maximum number of antennas in a BS, and Y_{Pmax} represents the maximum emitted power per BS antenna. Moreover, in regard to y_K, we may note that massive MIMO systems should satisfy the requirement $y_N \gg y_K$ [5], which is satisfied by setting the following constraint [5]:

$$1 \leq y_K \leq \frac{y_N}{2} \tag{7.3}$$

For reasons of simplification, we make the assumption of perfect channel state information for each BS and that zero-forcing (ZF) precoding is employed.

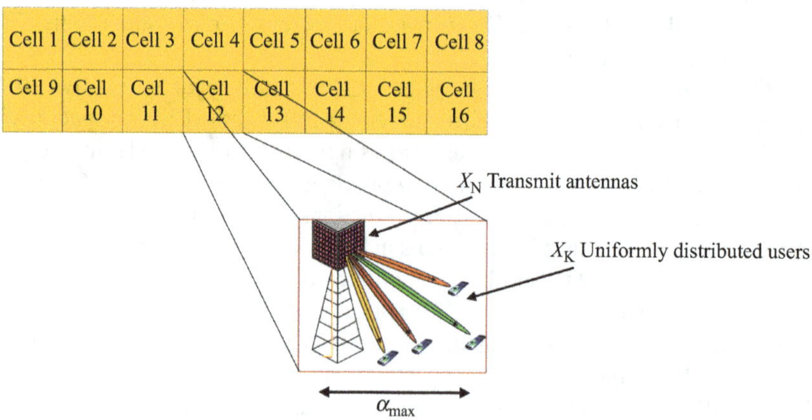

Figure 7.1 A sample 5G massive MIMO cellular network

Furthermore, we make the assumption that each user considers intercell interference as noise. Thus, the average user rate is expressed as [5]

$$f_1(\bar{y}) = W\left(1 - \frac{y_K}{W_c\tau_c}\right)\log_2\left(1 + \frac{\frac{y_P}{y_K}(y_N - y_K)}{\sigma^2\mu_1 + y_N\mu_2}\right) \tag{7.4}$$

where W is the transmission bandwidth, W_c is the coherence bandwidth, τ_c is the coherence time, and σ^2, μ_1, and μ_2 are the average noise power, inverse channel loss, and strength of intercell interference, respectively.

Additionally, we can express the total power consumption per cell as [5]

$$P_{\text{total}}(\bar{y}) = \frac{y_P}{\eta} + y_N C_N + y_K C_K + \frac{C_{\text{pre}}}{L_{\text{BS}}} + C_0 \tag{7.5}$$

where η is the efficiency of the power amplifiers at the BS, C_K denotes the hardware power per user, C_N represents the hardware power consumed per transmit antenna, C_0 denotes the static hardware power, and L_{BS} is the computational efficiency. Moreover, the floating-point operations per second (flops) required to compute ZF precoding are formulated as [5]

$$C_{\text{pre}} = 3y_K^2 y_N \frac{W}{W_c\tau_c} \tag{7.6}$$

Keeping in mind the above formulations, the average area rate and the EE are expressed as

$$f_2(\bar{y}) = \frac{y_K}{A} f_1(\bar{y}) \qquad \text{(bit/s/km}^2\text{)} \tag{7.7}$$

$$f_3(\bar{y}) = \frac{y_K f_1(\bar{y})}{P_{\text{total}(\bar{y})}} \qquad \text{(bit/J)} \tag{7.8}$$

respectively, where A is the area of each cell in km^2.

7.1.2 Multi-objective evolutionary algorithm-based solution

We can observe that the process of obtaining a solution for (7.9) is challenging. This is due to the multiple objectives that are also non-concave. This fact dramatically increases the complexity of the weighting method proposed in [5]. More accurately, because of the non-concavity property, if we want to find the Pareto front for each weighting vector, $\bar{w} = [w_1, w_2, w_3]$, a three-dimensional search is required. This can be time-consuming. Additionally, we may highlight that the complexity of the weighting method will become even higher, if we add more objectives. Therefore, a scalable and efficient framework is required.

MOEAs are suitable techniques to be applied to this particular problem.

7.1.3 Proposed optimization framework

This section focuses on the efficient and simultaneous maximization of the average user rate, the average area rate, and the EE. It is noted that these three objectives are mutually conflicting, which requires the use of MO optimization techniques [9]. The MO optimization problem that aims at simultaneously maximizing the aforementioned three metrics can be expressed as

$$\begin{aligned} &\max_{\bar{y}} \bar{f}(\bar{y}) = [f_1(\bar{y}), f_2(\bar{y}), f_3(\bar{y})] \\ &\textbf{s.t.} \quad (7.1), (7.2), (7.3) \end{aligned} \tag{7.9}$$

where $\bar{f}(\bar{y})$ is the vector of the objective functions. We denote Y and Z the search space and the objective space, respectively. Thus, using the mapping of $\bar{f} : Y \rightarrow Z$ each vector $\bar{y} \in Y$ corresponds to a vector $\bar{z} = \bar{f}(\bar{y}) \in Z$.

7.1.4 Numerical results

The simulation parameters used are reported in Table 7.1. We apply four different algorithms, namely, the NSGA-II [7], the non-dominated sorting genetic algorithm-III (NSGA-III) [10], the speed-constrained SMPSO [8], and the MOEA based on decomposition (MOEA/D) [11]. We run all algorithms for 20 independent trials. In all cases, we use 100 for population size and 250 for the maximum number of iterations.

We obtain the non-dominated solutions of all algorithm runs in order to form a Pareto front for each algorithm. Figures 7.2–7.4 (shown as examples) depict 2D cuts of the Pareto front. We notice that in this case it is difficult to reach a conclusion from these figures. We also obtain the 3D Pareto fronts. Figure 7.5 shows the Pareto fronts obtained by all algorithms in 3D objective space. We may observe that the Pareto front points obtained by NSGA-II seem to have a wider spread in 3D objective space. This is clearer from Figures 7.6–7.9 that depict the 3D Pareto front obtained by each algorithm. We notice that NSGA-II and SMPSO obtained a smoother distribution and a wide spread of non-dominated points, while NSGA-III and MOEAD obtained results with Pareto front in a smaller 3D space.

Moreover, Figures 7.10–7.13 illustrate the 3D decision space of each algorithm. We notice that NSGA-II the decision space points are in wider spread in the 3D decision space compared with those from the other algorithms.

Table 7.1 Simulation parameters [5]

Parameter	Value	Parameter	Value
C_N	0.5 W	C_K	0.3 W
C_0	10 W	L_{BS}	12.8 Gflops/W
η	0.31	α_{max}	250 m
α_{min}	35 m	W	10 MHz
σ^2	−100 dBm	μ_1	1.72×10^9
μ_2	0.54	W_c	200 kHz
τ_c	5 ms	M	16

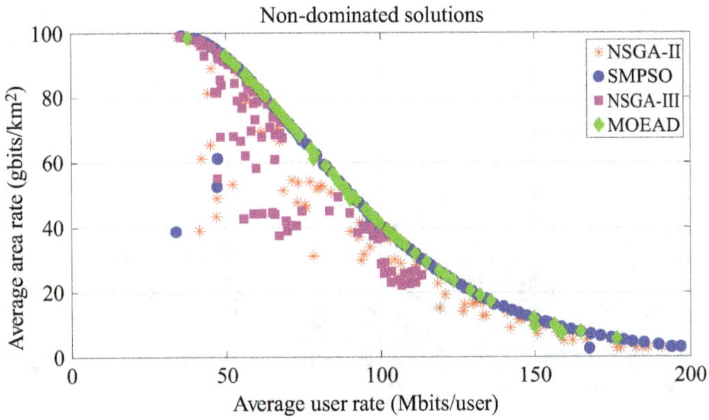

Figure 7.2 Comparative results 2D Pareto front of f_1 and f_2

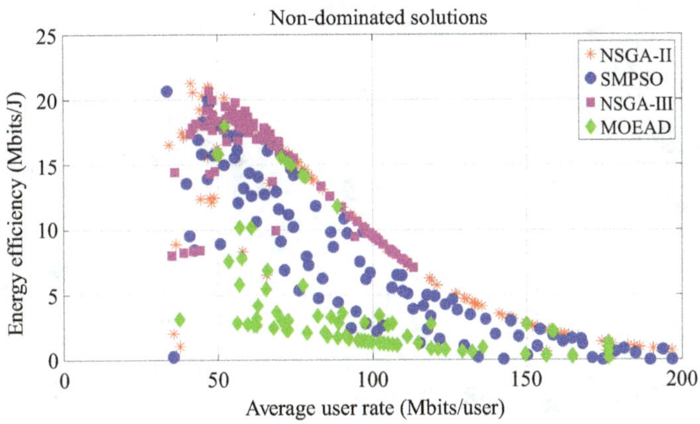

Figure 7.3 Comparative results 2D Pareto front of f_1 and f_3

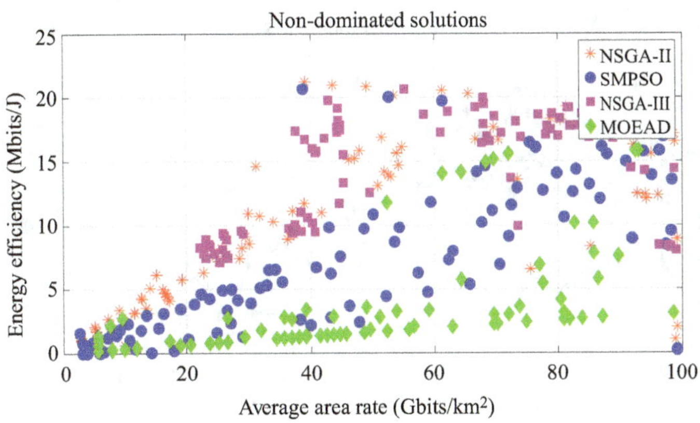

Figure 7.4 Comparative results 2D Pareto front of f_2 and f_3

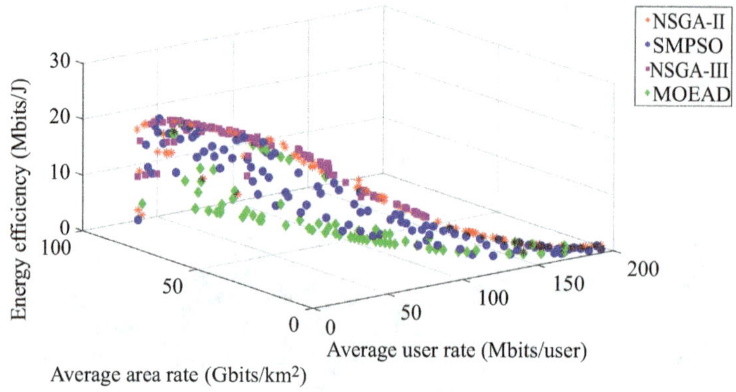

Figure 7.5 Comparative results 3D Pareto front by all algorithms

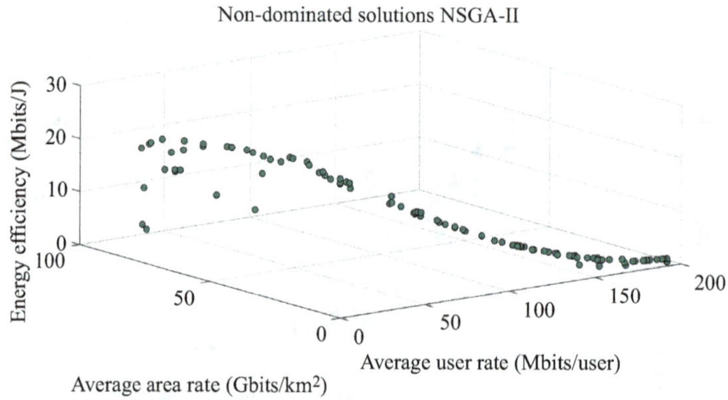

Figure 7.6 3D Pareto front obtained by NSGA-II

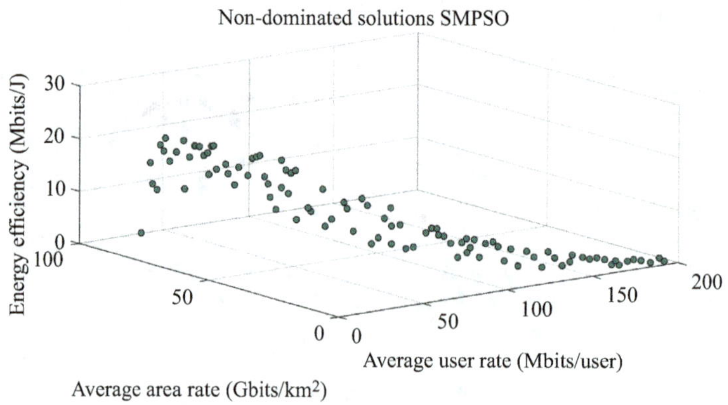

Figure 7.7 3D Pareto front obtained by SMPSO

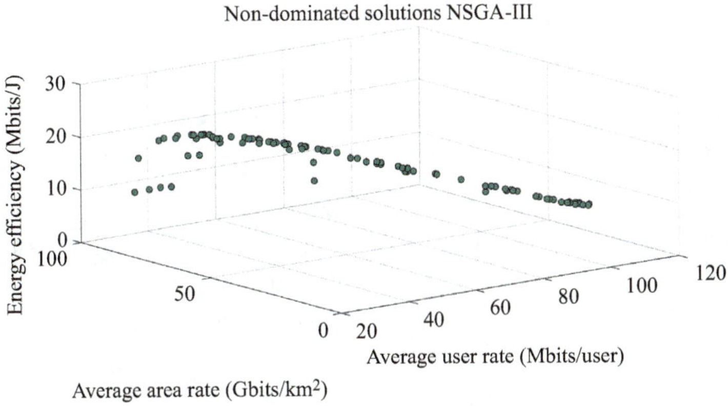

Figure 7.8 3D Pareto front obtained by NSGA-III

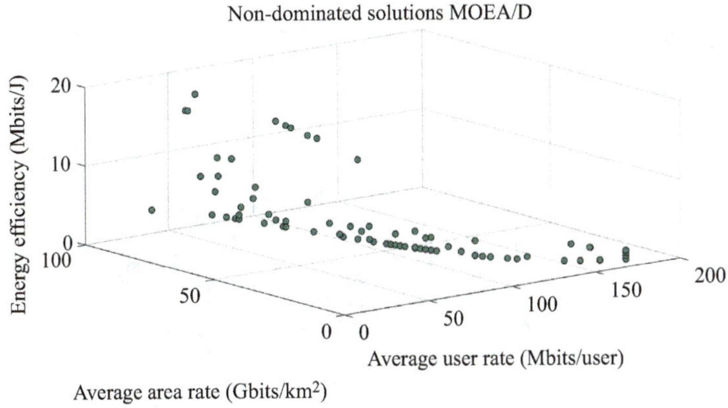

Figure 7.9 3D Pareto front obtained by MOEAD

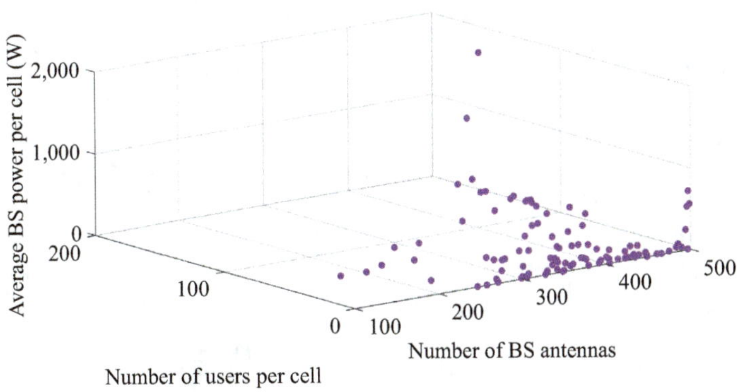

Figure 7.10 3D decision space of Pareto front obtained by NSGA-II

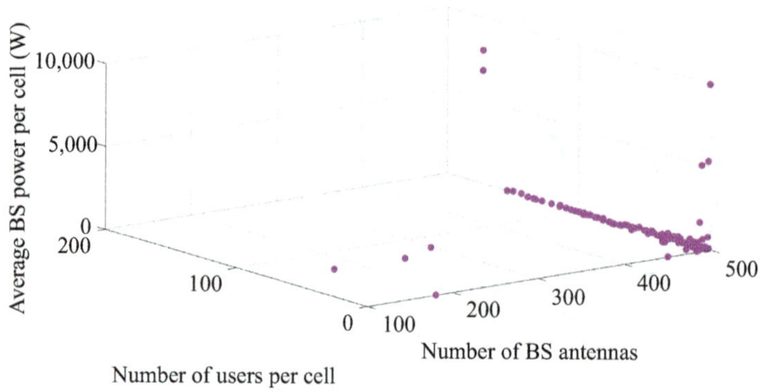

Figure 7.11 3D decision space of Pareto front obtained by SMPSO

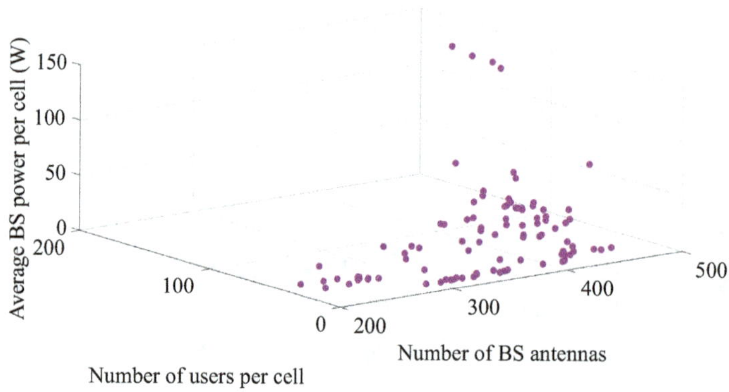

Figure 7.12 3D decision space of Pareto front obtained by NSGA-III

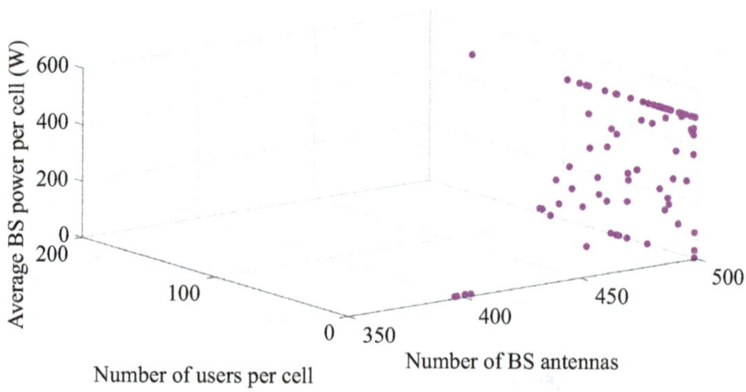

Figure 7.13 3D decision space of Pareto front obtained by MOEAD

In order to study how the algorithms evolve the objectives in each iteration, we plot convergence rate plots of each algorithm for two objectives: the average user rate and the average EE. Figures 7.14–7.17 plot these convergence rate plots. We notice from Figure 7.14 that both objectives evolve in conflict after the first 20 or more iterations. For SMPSO, it is more clear that the average EE objective increases its value after some iterations, while the average user rate reaches in less than 50 iterations to a higher value, that is deteriorating with iteration. For NSGA-III both objectives evolve within certain limits after the first 50 iterations. It is interesting to see how MOEAD deals with these objectives in Figure 7.17. We notice that MOEAD reaches a high value for both objectives at less than 20 iterations. These values remain unchanged for the rest of the iterations.

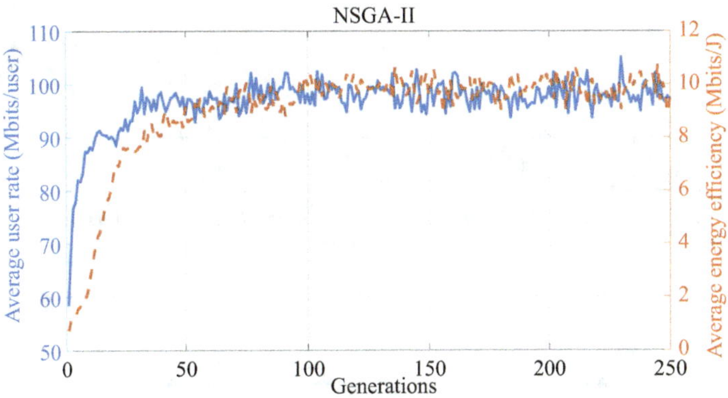

Figure 7.14 Average convergence rate by NSGA-II

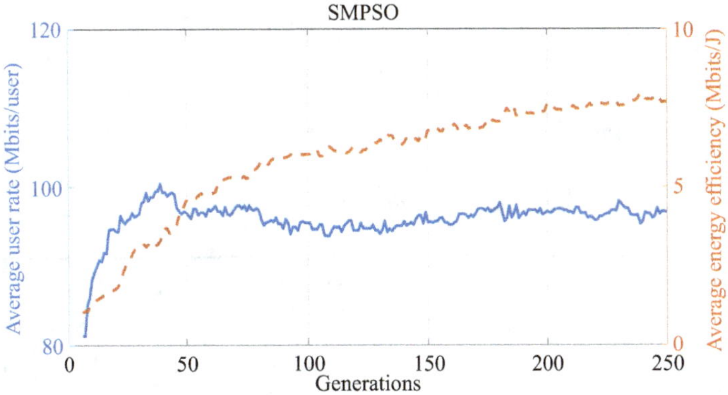

Figure 7.15 Average convergence rate by SMPSO

Additionally, we evaluate the four algorithms by performance indicators for MO algorithms. Table 7.2 reports the results. We observe that NSGA-II performs better that the NSGA-III and MOEAD in all three indicators. However, NSGA-II compared with SMPSO performs better in two indicators while SMPSO is better regarding the unary ε-indicator $QI_{\varepsilon 1}$. SMPSO also outperforms NSGA-III and MOEAD.

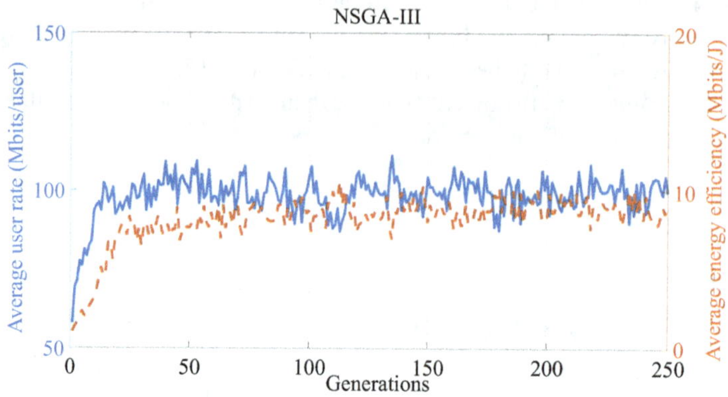

Figure 7.16 Average convergence rate by NSGA-III

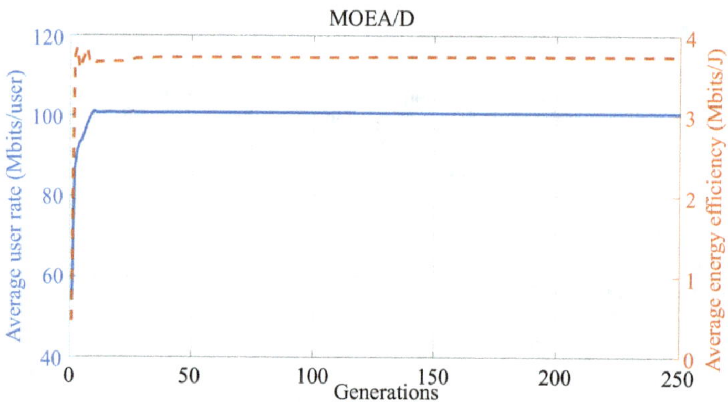

Figure 7.17 Average convergence rate by MOEAD

Table 7.2 Performance indicators of the Pareto fronts

Performance indicator	NSGA-II	SMPSO	NSGA-III	MOEAD
QI_{R2}	**4.08e−03**	5.23E−03	4.21E−02	1.57E−02
$QI_{\varepsilon}1$	7.34E−02	**6.90e−02**	5.13E−01	1.54E−01
$QI_{\bar{H}}$	**2.98e−02**	3.45E−02	5.21E−02	8.29E−02

Bold font indicates the lower values.

7.2 Joint power allocation and user association in non-orthogonal multiple access networks

Non-orthogonal multiple access (NOMA) techniques are going to have a critical role in 5G mobile networks [12]. If the network resources are assigned to users with poor channel conditions then the current orthogonal multiple access (OMA) networks suffer from low spectrum efficiency However, the above condition is entirely different in the case of NOMA systems in the power domain. The NOMA scheme allows the users to use the same frequency, time, and code simultaneously, while it allocates different levels of power [13] (Figure 7.18).

The principal concept of a NOMA scheme is that users' receivers with good channel conditions use successive interference cancellation (SIC) techniques for removing the interferences of users with poor channel conditions. Therefore, SIC cancels the intra-cell or intra-cluster interference on a users' receiver [14]. The user association problem in NOMA systems is very challenging. This is due to the fact that some unique features exist in this case such as cochannel interferences to existing networks. In [15] the user association problem in NOMA networks is modeled mathematically by grouping the users into orthogonal clusters and by associating them different resource blocks using a game theoretic approach. However, game theoretic approaches, which are commonly used in user association problems, have limitations and work under certain assumptions. Power control is a parameter that introduces additional complexity to the problem. Quite often, as in [15] the power coefficients are considered fixed for all network. In [16,17] the authors include in the optimization the suitable power coefficients as unknowns for every NOMA user. In [16] the problem is solved using the salp swarm algorithm [18]. EAs inspired by nature are suitable techniques for

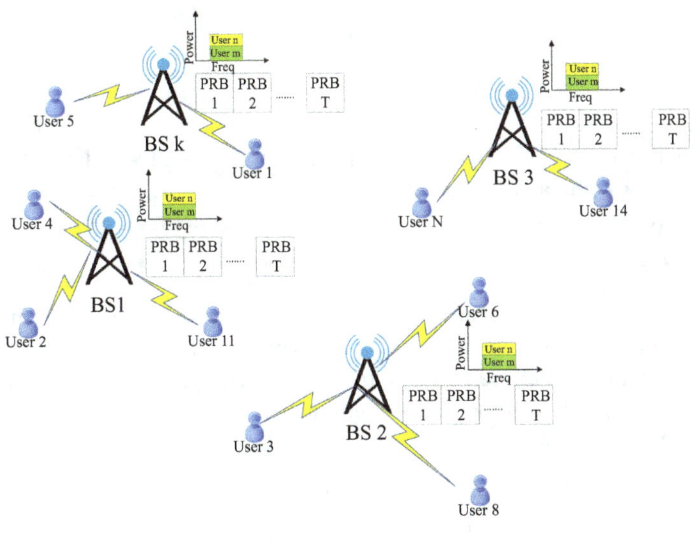

Figure 7.18 A downlink NOMA network

solving this problem. In Section 7.2.3, we apply the grey wolf optimizer (GWO) [19], and the popular PSO [20]. Thus, we apply two swarm intelligence algorithms. In the next subsection, we provide the optimization problem formulation using the network sum rate utility function.

7.2.1 System model

We make the assumption that there is a cellular network that uses NOMA scheme and consists of several BSs and several users. Additionally, the BSs transmit data by physical resource blocks (PRBs).

Thus, we assume a downlink NOMA network with a set of users $\mathcal{N} = \{1, 2, \ldots, N_u\}$, and $|\mathcal{N}| = N_u$ is the set cardinality or the number of users. Additionally, we denote the set of PRBs $\mathcal{T} = \{1, 2, \ldots, T_{RB}\}$ with cardinality $|\mathcal{T}| = T_{RB}$. Thus, there are T_{RB} orthogonal clusters. We denote the set of users associated with PRB t as $\mathcal{E}_t$ with cardinality $|\mathcal{E}_t| = E_t$. In networks that use an OMA scheme there is a simple rule, only one PRB can be assigned to one user. However, it is quite different in a NOMA scheme, and the basic idea is that more users share the same PRB with different power levels. In this event, the receiver of the user equipment uses SIC to cancel the intra-cluster interference. Another assumption we make is that a NOMA scheme is employed by all users in individual clusters. Then, we may formulate the received signal at user m in any cluster E_{tk} as [15]

$$\begin{aligned} Y_{km}^t &= \underbrace{g_{km}^t \sqrt{p_{km}^t} s_{km}}_{\text{Desired signal}} + n_m + \underbrace{\sum_{i=1, j \neq k}^{K_{BS}} \sum_{j=1}^{|E_{ti}|} g_{im}^t s_{ij}}_{\text{Inter-cluster interference}} \\ &\quad + \underbrace{g_{km}^t \sum_{i=1, i \neq m}^{|E_{tk}|} \sqrt{p_{ki}^t} s_{ki}}_{\text{Intra-cluster interference}} \end{aligned} \tag{7.10}$$

where $|E_{tk}|$ denotes the size of E_{tk}, g_{km}^t represents the channel between user m and PRB t which is allocated by BS k, s_{km} is the transmitted signal, p_{km}^t denotes the power allocation coefficient, and n_m denotes the noise.

Moreover, we model mathematically the channel power gain as

$$|g_{mk}^t|^2 = |\hat{g}_{mk}^t|^2 G_{\text{PL}}(d_{mk}) \tag{7.11}$$

where $\hat{g}_{mk}^t \sim \mathcal{CN}(0, 1)$ is the circular-symmetric complex Gaussian zero mean noise from BS k and PRB t to user m, $G_{\text{PL}}(d_{mk})$ is the propagation path loss. We model the propagation path loss between user m and the BS with path gain (loss) $GPL(d)$. In this section, we use the outdoor macro cell line-of-sight model defined in [21]. This is given by

$$G_{\text{PL}}(d_{mk}) = -103.4 - 24.2 \log_{10}(d_{mk}) \text{ (dB)} \tag{7.12}$$

where d_{mk} is the distance in kilometers between BS k and user m.

In any cluster E_{tk}, the power allocation coefficients $\sum_{i=1}^{|E_{tk}|} p_{ki}^t \leq 1$.

We can make the assumption that in each PRB there can be allocated at most M NOMA users, where the power allocation coefficients satisfy $p^t_{k,1} \geq p^t_{k,2} \geq, \ldots, p^t_{km}, \ldots, \geq p^t_{kM}$. Without loss of assumption, we assume that the well-served user in each cluster is the Mth user. The receiver at the mth user in E_{tk} will consider the ith user's signal as noise ($m > i$) and decode its own signal at the signal-to-interference-plus-noise ratio (SINR):

$$c^t_{km} = \frac{|g^t_{km}|^2 p^t_{km}}{|g^t_{km}|^2 \sum_{i=m+1}^{|E_{tk}|} p^t_{ki} + \sum_{i=1,j\neq k}^{K_{BS}} \sum_{j=1}^{|E_{ti}|} |g^t_{im}|^2 p^t_{ij} + \frac{1}{\rho}} \tag{7.13}$$

where $\rho = P_t/\sigma^2$ is the transmit signal-to-noise ratio, P_t is the transmit power, and σ^2 denotes the variance of the additive white Gaussian noise. Furthermore, the receiver at the last Mth user will apply SIC to remove the intra-cluster interference, and it will decode its own signal with SINR:

$$c^t_{kM} = \frac{|g^t_{km}|^2 p^t_{kM}}{\sum_{i=1,j\neq k}^{K_{BS}} \sum_{j=1}^{|E_{ti}|} |g^t_{iM}|^2 p^t_{ij} + \frac{1}{\rho}} \tag{7.14}$$

From the previously mentioned ones, we may notice that in order for the m-user to decode its signal, the decoding and the removal of the intra-cluster interference due to the previous user is needed. We make the assumption of perfect SIC process employed at the nth user. Therefore, the required condition for perfect SIC is that the rate of the nth user for decoding the mth user's signal is not less than the rate of the mth user for decoding its own signal, i.e., $R^t_{k,n\rightarrow m} \geq R^t_{k,m\rightarrow m}$ for $n > m$. This condition can be written as

$$\begin{aligned} &\frac{|g^t_{kn}|^2 p^t_{km}}{|g^t_{kn}|^2 \sum_{i=m+1}^{|E_{tk}|} p^t_{ki} + \sum_{i=1,j\neq k}^{K_{BS}} \sum_{j=1}^{|E_{ti}|} |g^t_{in}|^2 p^t_{ij} + \frac{1}{\rho}} \\ &\geq \frac{|g^t_{km}|^2 p^t_{km}}{|g^t_{km}|^2 \sum_{i=m+1}^{|E_{tk}|} p^t_{ki} + \sum_{i=1,j\neq k}^{K_{BS}} \sum_{j=1}^{|E_{ti}|} |g^t_{im}|^2 p^t_{ij} + \frac{1}{\rho}} \end{aligned} \tag{7.15}$$

Thus, if the following condition is true, then the nth user is able to remove the mth user signal using SIC:

$$\begin{aligned} Q(E_{tk}) \triangleq\ & \frac{|g^t_{kn}|^2}{\sum_{i=1,j\neq k}^{K_{BS}} \sum_{j=1}^{|E_{ti}|} |g^t_{in}|^2 p^t_{ij} + \frac{1}{\rho}} \\ & - \frac{|g^t_{km}|^2}{\sum_{i=1,j\neq k}^{K_{BS}} \sum_{j=1}^{|E_{ti}|} |g^t_{im}|^2 p^t_{ij} + \frac{1}{\rho}} \geq 0, \\ & \forall n \in \{2, \ldots, M\}, \quad \forall m \in \{1, \ldots, M-1\} \end{aligned} \tag{7.16}$$

Therefore, the data rate of any user m connected with BS k and with allocation of PRB t is expressed as

$$R^t_{km} = \log(1 + c^t_{km}). \tag{7.17}$$

7.2.2 Problem formulation

We describe the association of the kth BS with the ith user with the binary variable, b_{ki} given by

$$b_{ki} = \begin{cases} 1, & \text{if user } i \text{ is associated with the BS } k \\ 0, & \text{otherwise} \end{cases} \tag{7.18}$$

Additionally, the association of the ith user with tth PRB can be described with another binary variable, y_{ti} as

$$y_{ti} = \begin{cases} 1, & \text{if user } i \text{ is associated with the PRB } t \\ 0, & \text{otherwise} \end{cases} \tag{7.19}$$

As in [15,16] we use an additional weight factor based on the distance between users and BSs. This weight factor in cluster E_{tk} is expressed as $w_{kn} = (d_{kn}/\sum_{i=1}^{2} d_{ki})^{1/\alpha}$, i.e., the users with bad channel conditions and interference are affected very little, while the users with good channel conditions are affected more and their data rate is reduced. Given the previous ones, the joint power allocation and user association problem can be expressed as

$$\begin{aligned} U^* = \max_{\{b,y,p\}} \quad & \sum_{i \in \mathcal{N}} U_{kt}(R_i) \\ \text{s.t.} \quad C_1: \quad & Q(E_{tk}) \geq 0 \\ C_2: \quad & b_{kn} \in \{0,1\}, \forall n \in \mathcal{N}, \forall k \in \mathcal{K}, \\ C_3: \quad & y_{tn} \in \{0,1\}, \forall n \in \mathcal{N}, \forall t \in \mathcal{T}, \\ C_4: \quad & \sum\nolimits_{i=1}^{K_{BS}} b_{in} = 1, \forall n \in \mathcal{N}, \\ C_5: \quad & \sum\nolimits_{i=1}^{T_{RB}} y_{in} = 1, \forall n \in \mathcal{N}, \\ C_6: \quad & \sum\nolimits_{n=1}^{N} y_{tn} \leq M, \forall t \in \mathcal{T}, \\ C_7: \quad & \sum\nolimits_{i=1}^{|E_{tk}|} p_{ki}^{t} \leq 1 \forall t \in \mathcal{T}, \forall k \in \mathcal{K} \end{aligned} \tag{7.20}$$

where $U_{kt}(R_i) = w_{k,i} b_{k,i} y_{t,i} R_{k,i}^{t}$ denotes the utility function, and b and y represent the set of all indicators b and y, respectively. The perfect SIC condition is imposed by constraint C_1 that orders the users in each cluster. The association or not between users n and BS k is denoted by constraint C_2. Correspondingly, the association or not between user n and PRBs t is represented described by the constraint C_3. Constraints C_4 and C_5 represent the unique association between one user n with BS k and PRB t at the same time. Furthermore, constraint C_6 expresses the fact that in any PRB at the most M users may be served. Moreover, constraint C_7 defines that the total power allocation coefficients in each cluster should be less or equal to 1. This problem is non-convex and complex to solve. Thus, EAs can be used.

7.2.3 Numerical results

In order to solve the user association problem described earlier using EAs, several simulations were run. In every run, we create a random topology of BSs and users. The users are placed in a uniform random way within a circle of 450-m radius. We obtain results for a different user number that ranges from 12 to 24 with step 3. We make the assumption that the randomly deployed users are served by 2 BSs that have 6 PRBs each. Furthermore, we limit the maximum number of NOMA users in one PRB to $M = 2$. We consider lognormal shadowing with standard deviation 8 dB. We use 180 kHz for the PRB bandwidth as it is used in 4G/LTE.

We run the simulations with both GWO and PSO. For both algorithms we use 200 for the population size and 24 for the maximum number of iterations. For each problem case algorithms run for 500 independent simulations each one with different random topology. Thus, we obtain the average results that come from 500 different random topologies.

Figure 7.19 depicts the algorithms results for both NOMA and OMA cases for $N = 12$ users. One may observe that the sum rate distribution of the NOMA schemes is larger than the OMA ones as it can be expected. Moreover, we notice that PSO produces slightly better results than GWO. The interquartile range in both algorithms is similar; however, PSO obtains a slightly higher median value for the NOMA case. The corresponding convergence rate graphs are plotted in Figures 7.20 and 7.21 for $N = 12$ users. Both algorithms converge at similar speed. PSO seems to converge at a slightly higher value in the NOMA case, while for the OMA case GWO converges at a slightly higher value.

Similar results are obtained for $N = 15$ users. The boxplot of the results is illustrated in Figure 7.22. We notice that for the NOMA case, PSO produces better results than GWO. The median value obtained by PSO is clearly higher. The results for both algorithms seem very similar for the OMA case. We may also notice that the sum

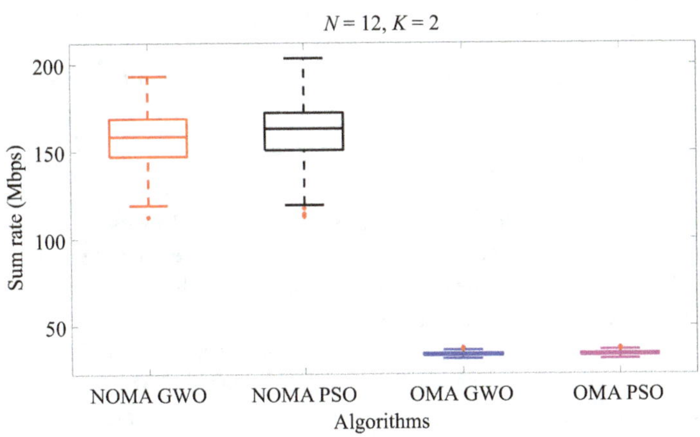

Figure 7.19 $N = 12$ users. Boxplot of all algorithms results

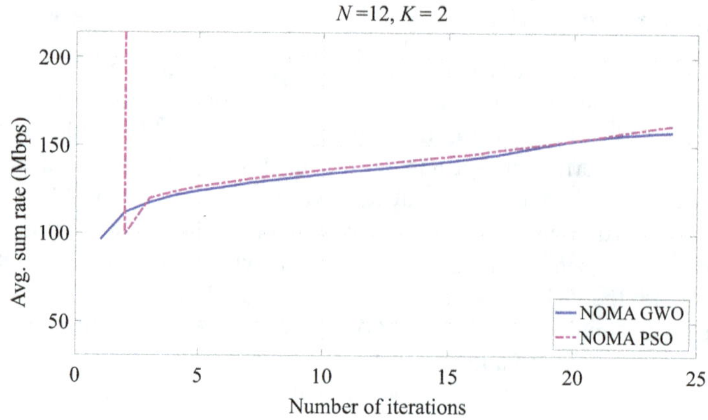

Figure 7.20 $N = 12$ *users NOMA. Convergence rate graph*

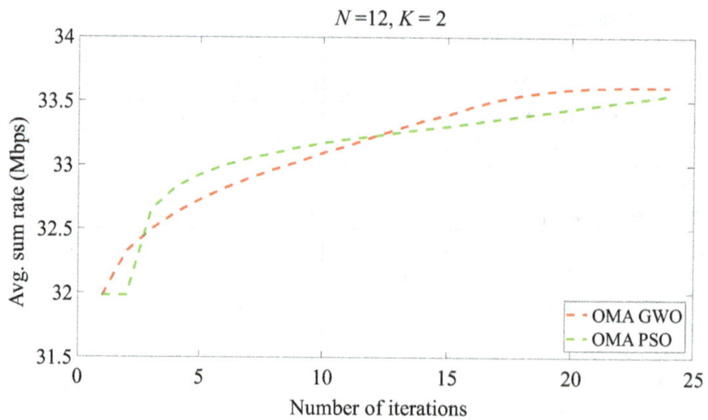

Figure 7.21 $N = 12$ *users OMA. Convergence rate graph*

rate values are in general lower than the ones for $N = 12$ users. Figures 7.23 and 7.24 show the convergence rate graphs for this case. They seem quite similar with the ones from $N = 12$ users. For the NOMA case, both algorithms converge equally fast; however, PSO converges at a higher final value. For the OMA case, the convergence rate graph is very similar to that for $N = 12$ users. We notice that for the OMA the total sum rate remains about the same for both user cases; thus, it is independent of the user number. However, for the NOMA case the total sum rate reduces when we increase the user number. This can be expected as for more NOMA users, the problem becomes more difficult and the users may connect at lower data rates.

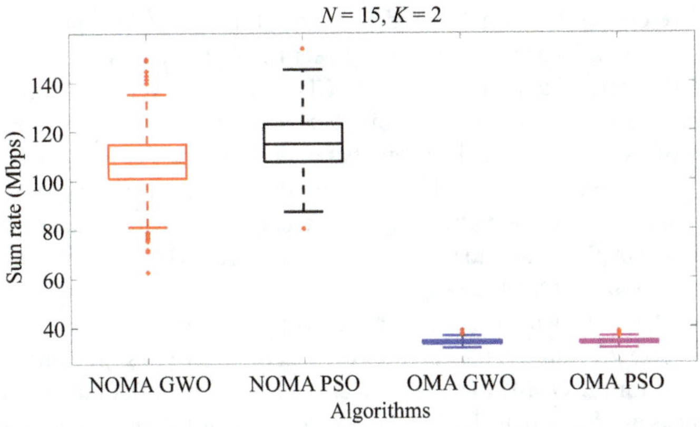

Figure 7.22 *N = 15 users. Boxplot of all algorithms results*

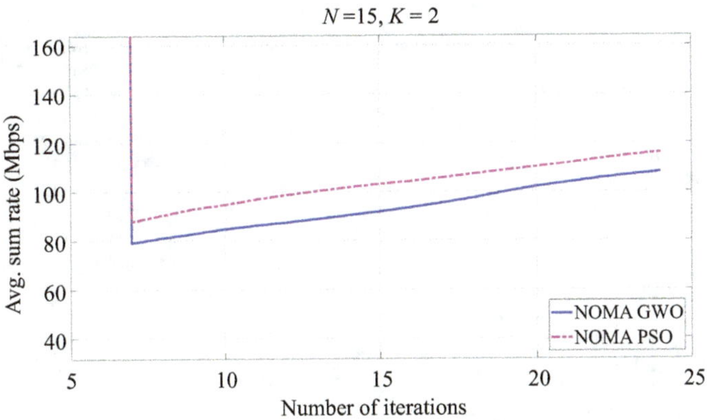

Figure 7.23 *N = 15 users NOMA. Convergence rate graph*

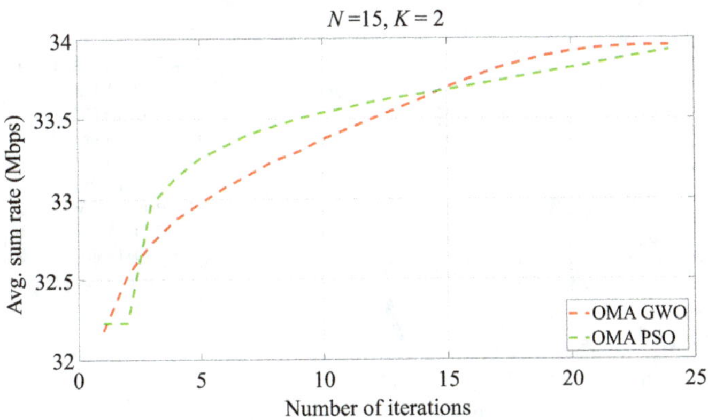

Figure 7.24 *N = 15 users OMA. Convergence rate graph*

Next, we obtain the cumulative distribution function (CDF) plots for both the OMA and the NOMA cases for increasing user number. Figures 7.25 and 7.26 show these plots. We notice that for the OMA, the CDF curves are very close one to another. The best CDF is for the smaller user number, while the worst is for the higher. As the number of users increases, the users that connect to the BS connect at lower data rates. The CDF curves for the NOMA case are distinct and away one from the other. Again, the conclusion is the network sum rate is higher for lower user number. The results obtained by PSO seem to be better than those from GWO, while the algorithms seem equivalent in the OMA case.

The effect of increasing the user number is presented more clearly in Figures 7.27 and 7.28. Figure 7.27 shows the total network sum rate versus the number of users. We notice that for the OMA, the obtained curve is an almost flat line, indicating that the differences are few small. This is due to the fact that for OMA only one user can connect to each PRB. However, this is not the case for NOMA where the total sum

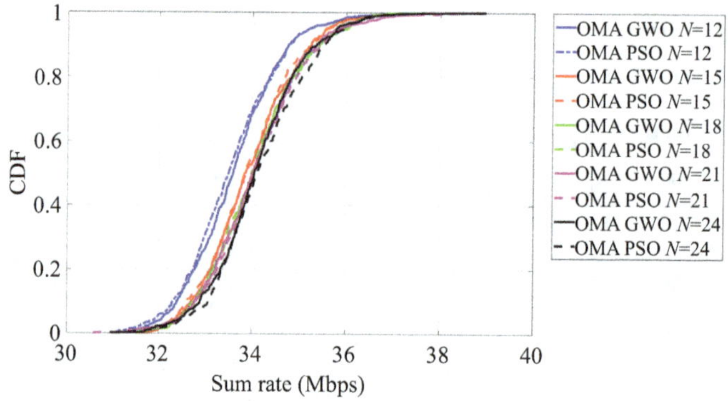

Figure 7.25 CDF of utility function results for user cases for OMA technique

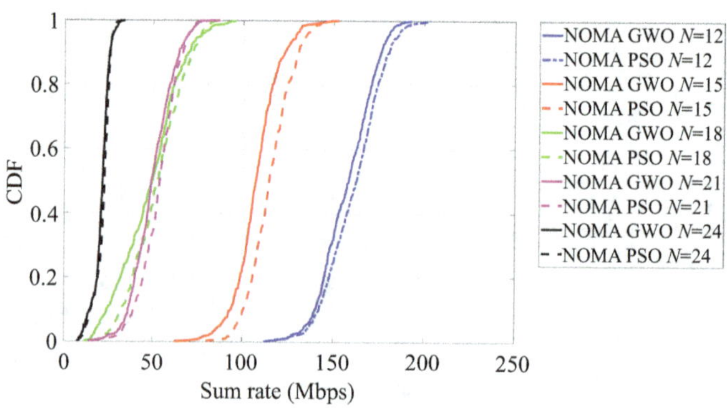

Figure 7.26 CDF of utility function results for user cases for NOMA technique

rate is reducing as the number of users grows. PSO seems to obtain higher sum rate values than GWO. This means that as the number of users increases, more network resources are required and the problem becomes more difficult to solve. Figure 7.28 depicts the user coverage percentage versus the number of users. The conclusion is similar to the previous figures. We can observe that for the OMA case, the coverage percentage is 100 for $N = 12$ and drops in all other cases. For the NOMA case, the coverage percentage remains 100 up to 18 users and then drops in all other cases. However, it remains higher than the OMA case. These figures actually show the clear advantage obtained by NOMA compared with an OMA scheme.

Overall, both GWO and PSO obtained satisfactory results. PSO outperforms GWO in all NOMA cases.

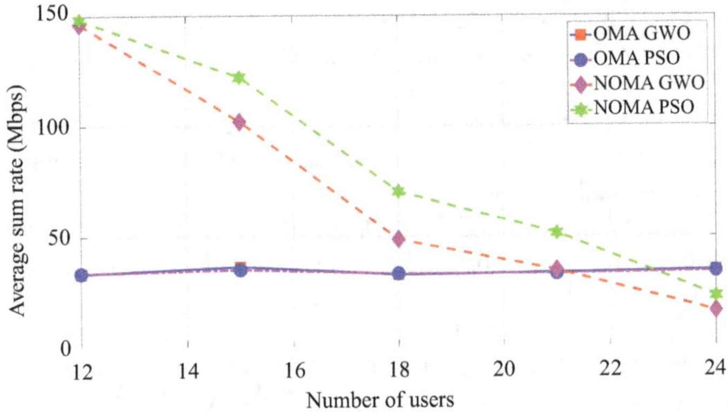

Figure 7.27 Utility function results for increasing number of users

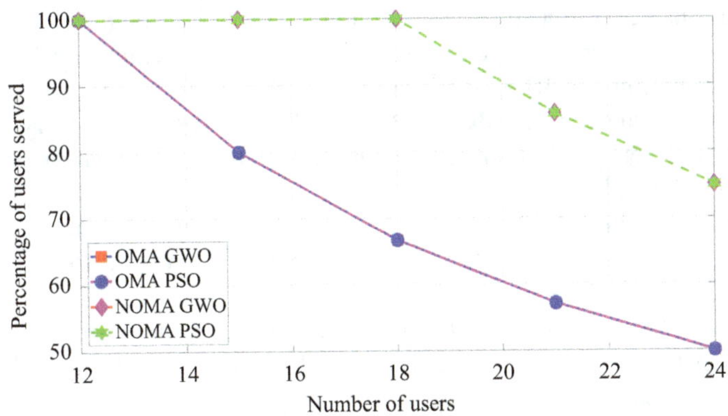

Figure 7.28 Percentage of users served for increasing number of users

References

[1] Björnson E, Larsson EG, and Marzetta TL. Massive MIMO: Ten myths and one critical question. IEEE Communications Magazine. 2016;54(2): 114–123.

[2] Zheng G, Krikidis I, Masouros C, *et al.* Rethinking the role of interference in wireless networks. IEEE Communications Magazine. 2014;52(11): 152–158.

[3] Jiang C, Zhang H, Ren Y, *et al.* Machine learning paradigms for next-generation wireless networks. IEEE Wireless Communications. 2017;24(2):98–105.

[4] Hao Y, Song Z, Hou S, *et al.* Energy- and spectral-efficiency tradeoff in massive MIMO systems with inter-user interference. In: Proc. IEEE 26th Annual International Symposium on Personal, Indoor, and Mobile Radio Communications (PIMRC); 2015. p. 553–557.

[5] Björnson E, Jorswieck EA, Debbah M, *et al.* Multiobjective signal processing optimization: The way to balance conflicting metrics in 5G systems. IEEE Signal Processing Magazine. 2014;31(6):14–23.

[6] Goudos SK, Diamantoulakis PD, and Karagiannidis GK. Multi-objective optimization in 5G wireless networks with massive MIMO. IEEE Communications Letters. 2018;22(11):2346–2349.

[7] Deb K, Pratap A, Agarwal S, *et al.* A fast and elitist multiobjective genetic algorithm: NSGA-II. IEEE Transactions on Evolutionary Computation. 2002;6(2):182–197.

[8] Nebro AJ, Durillo JJ, Garcia-Nieto J, *et al.* SMPSO: A new PSO-based metaheuristic for multi-objective optimization. In: IEEE Symposium on Computational Intelligence in Multi-Criteria Decision-Making (MCDM); 2009. p. 66–73.

[9] Marler RT and Arora JS. Survey of multi-objective optimization methods for engineering. Structural and Multidisciplinary Optimization. 2004;26(6):369–395.

[10] Deb K and Jain H. An evolutionary many-objective optimization algorithm using reference-point-based nondominated sorting approach, Part I: Solving problems with box constraints. IEEE Transactions on Evolutionary Computation. 2014;18(4):577–601.

[11] Zhang Q and Li H. MOEA/D: A multiobjective evolutionary algorithm based on decomposition. IEEE Transactions on Evolutionary Computation. 2007;11(6):712–731.

[12] Ding Z, Liu Y, Choi J, *et al.* Application of non-orthogonal multiple access in LTE and 5G networks. IEEE Communications Magazine. 2017;55(2):185–191.

[13] Ding Z, Yang Z, Fan P, *et al.* On the performance of non-orthogonal multiple access in 5G systems with randomly deployed users. IEEE Signal Processing Letters. 2014;21(12):1501–1505.

[14] Saito Y, Kishiyama Y, Benjebbour A, *et al.* Non-orthogonal multiple access (NOMA) for cellular future radio access. In: 2013 IEEE 77th Vehicular

Technology Conference (VTC), Spring 2013; Dresden, Germany; 2–5 June 2013.

[15] Wang K, Liu Y, Ding Z, *et al.* User association in non-orthogonal multiple access networks. In: 2018 IEEE International Conference on Communications (ICC); 2018. p. 1–6.

[16] Goudos SK. Joint power allocation and user association in non-orthogonal multiple access networks: An evolutionary approach. Physical Communication. 2019;37:100841.

[17] Goudos SK, Diamantoulakis PD, Boursianis AD, *et al.* Joint user association and power allocation using swarm intelligence algorithms in non-orthogonal multiple access networks. In: 2020 9th International Conference on Modern Circuits and Systems Technologies (MOCAST); 2020. p. 1–4.

[18] Mirjalili S, Gandomi AH, Mirjalili SZ, *et al.* Salp swarm algorithm: A bio-inspired optimizer for engineering design problems. Advances in Engineering Software. 2017;114:163–191.

[19] Mirjalili S, Mirjalili SM, and Lewis A. Grey wolf optimizer. Advances in Engineering Software. 2014;69:46–61.

[20] Kennedy J and Eberhart R. Particle swarm optimization. Proceedings of ICNN'95 – International Conference on Neural Networks, Perth, WA, Australia, 1995;4:1942–1948.

[21] 3GPP. TR 36.814, Further Advancements for E-UTRA Physical Layer Aspects, V9.0.0; Mar 2010.

Index

www.ingramcontent.com/pod-product-compliance
Lightning Source LLC
LaVergne TN
LVHW011844200926
848484LV00002B/398